Strategy and Communication for Innovation

Nicole Pfeffermann · Julie Gould
Editors

Strategy and Communication for Innovation

Integrative Perspectives on Innovation in the Digital Economy

Third Edition

Springer

Editors
Dr.-Ing. Nicole Pfeffermann
Pfeffermann Strategy and Research
Im Medienhafen
D-40221 Düsseldorf
Germany

Dr. Julie Gould
Princeton, NY
USA

ISBN 978-3-319-84184-7 ISBN 978-3-319-49542-2 (eBook)
DOI 10.1007/978-3-319-49542-2

Softcover reprint of the hardcover 3rd edition 2017

Printed on acid-free paper

This Springer imprint is published by Springer Nature
The registered company is Springer International Publishing AG
The registered company address is: Gewerbestrasse 11, 6330 Cham, Switzerland

Foreword

Ideas are never more valued and debated than when their realization is insufficient. So it is right now with growth and innovation. Growth drives the vast majority of the long-term value generated by corporations. Economic growth is driven by labor, capital and productivity growth or innovation. Innovation has become the critical input at a time of depressed growth prospects globally, since growth now needs to be actively created through innovation rather than mere participation. But ironically, productivity growth has stalled and some have attributed this to stalled innovation, in spite of the much vaunted technology revolution. Reviving innovation is increasingly predicated on effective communication in many important respects.

- It is necessary that organizations communicate with their increasingly dynamic environments so that they understand what it is that they need to adapt to, overcoming natural tendencies for corporations to become inward looking as they grow and age. That is, dynamic and unpredictable environments require ambidextrously balancing more inward-looking exploitation with more outward-looking exploratory activities.
- It is also necessary that the experiments on the periphery of the organization which are the life blood of increasingly popular adaptive approaches to strategy and innovation are codified, diffused and adopted. Adaptive advantage is an organizational capability predicated on the creation, selection and amplification of emerging knowledge from newly explored opportunities.
- As corporations diversify their approaches to strategy and innovation more generally, supplementing classical planning based thinking with a greater emphasis on experimentation, collaboration and environmental shaping [Reeves et al. 2015, Your Strategy Needs a Strategy: How to Choose and Execute the Right Approach], communication is also essential to establish the right cultural context, to mobilize the organization by posing the right questions and to educate employees and diffuse new dynamic capabilities and approaches.

- An increasing reliance on open source, platform or ecosystem based innovation in particular also requires new types of long range leadership and communication: the common interest and operating rules have to be communicated, and the innovation ecosystem needs to stay informed through feedback in order to be able to learn and evolve.
- Finally, the increasing role of artificial intelligence in business will inevitably impinge on strategy and innovation. It is important therefore that we do not restrict our considerations to human communication but also explore the role of human–machine and machine–machine communications in enhancing innovation.

The third edition of this volume further develops a relatively new domain springing from the convergence between innovation, strategy and communication and will be a very welcome and useful contribution for academics and practitioners alike.

It is divided into four parts dealing with strategy and innovation, communication and innovation, integrated perspectives on innovation and best practice examples. The first section on strategy and innovation contains papers dealing with open innovation strategy, creativity, open innovation business models, failure modes in innovation and cognitive diversity. The second section on communication and innovation contains papers dealing with dyadic communication, transformation in the creative industry, the role of communication innovation, strategic networking and social capital, managing emotions and communicators in innovation clusters. The third section on integrated perspectives on innovation contains papers dealing with implantation, innovation leadership and communication intelligence, collaborative innovation in the digital economy, customer centricity at the top, innovation communication as a dynamic capability and audience-centered planning. The fourth section on best practices and case examples has papers addressing impediments at scale, user centricity and agile innovation, innovation communication in the digital age, co-innovation and communication, harnessing insights from brand communities and integrating art and design.

The editors and authors are to be applauded for their rich contributions to this important emerging inter-disciplinary field and for effectively blending theoretical and practical knowledge to bridge the ever-widening gap between academics and practitioners, thereby catalyzing a flow of observations and insights between the two. And if this serves to generate usable lessons for business to drive innovation more effectively, that would serve a very noble purpose in these challenging economic times.

Martin Reeves
Director, BCG Henderson Institute, New York

Acknowledgments

The collection of works, published in this edition, aims to make a valuable contribution to the area of innovation management and communication management, covering recent and future developments in open innovation, business model innovation and innovation communication in the digital economy.

A number of people have contributed to making the editions of this book possible. First of all, we would like to thank Prof. Dr. Hülsmann, Jacobs University Bremen, and Thomas Lehnert, Senior Editor Engineering at Springer-Verlag, who saw the potential of this project and gave their early commitment to the concept of this book as well as the support that brought the book's first edition to fruition. Special thanks also go to Jan-Philip Schmidt, current Editor, Physical Sciences and Engineering, and his Book Coordination Manager Mrs. Petra Jantzen, for their support and advice in preparing the book manuscript and market launch.

Writing book chapters is especially challenging when submission deadlines compete directly with other academic, professional, and social tasks—especially in the digitalized information age. We would like to express our strong gratitude to all authors for taking the time to contribute a chapter.

Finally, our thanks go to the readers and reviewers of the first and second edition, who have supported us preparing this volume. Last but not least, we sincerely hope that researchers, students, colleagues, business managers, and innovators will enjoy reading this book and be inspired by multiple perspectives and theoretical and managerial implications provided by the thoughtful book chapters.

Cologne, Germany
June 2016

Nicole Pfeffermann
Julie Gould

Contents

Editors and Contributors

About the Editors

Nicole Pfeffermann, Ph.D. is an experienced management consultant, author, and coach, specialized in innovation communication. Since 1999 she has been working as a consultant in strategic marketing and innovation projects for companies and institutions. She is senior researcher in innovation communication, open innovation and culture (multiart) and coaches, teaches, and reviews students, PhDs and startups in digital business and methodology. Nicole was visiting scholar at UCLA Anderson and associate researcher at ESCP Europe Paris, Chair Entrepreneurship. She holds a Doctor of Engineering from the University of Bremen, scholarship from Deutsche Post, DHL, and a Diploma of Business Economics in market and consumer behaviour from the University of Wuppertal.

Dr. Julie Gould is a researcher and consultant, specialized in Supply Chain Management. As a consultant, Julie has extensive experience delivering decision support through data analysis, optimization and business process strategy and re-design, and project management. She has researched in the area of supply chain security, sustainable supply chain, logistics innovation, and intermodal logistics. She brings exceptional knowledge in modeling and optimization, as well as systems thinking and strong analytical skills. Julie holds a Doctor of Social Sciences from the University of Bremen Graduate School for Dynamics in Logistics. Julie also holds a Master of Arts in Innovative Brand Management from the University of Bremen and a Bachelor of Commerce in Decision and Information Systems from the University of Alberta School of Business.

Contributors

Dr. Grit Mareike Ahlers is Managing Director of Prof. Bruhn & Partner AG, a strategy and marketing consulting firm with its registered office in Zurich, as well as being Assistant Professor at the University of Basel and a lecturer at other higher education establishments. She studied business administration in Bayreuth and

Madrid, before joining the Department of Marketing and Corporate Management at the University of Basel in 2001 as a research assistant. Her Ph.D. thesis deals with the optimization of organizational structures in marketing with regard to consistent external and internal communications. At the marketing department, besides her scientific work she was also responsible for various consulting projects and worked as lecturer for numerous EMBA-programs. From 2006 until 2009, Mareike Ahlers worked for HUGO BOSS AG as Team Leader Business Development for the brand lines BOSS Orange & Green, responsible for brand strategies, cross-functional process management and the relaunch of the brand BOSS Orange.

Dr. Alan Belasen is Professor at the MBA program, SUNY Empire State College. He published extensively on topics such as self-managed work teams; corporate communication; motivation; women's leadership; innovation, and executive education. Belasen has provided consulting and leadership development to government, not-for-profit, and business organizations in the US and abroad. His books include *Women in Management: A Framework for Sustainable Work-life Integration* (Routledge 2017); *Confronting Corruption in Business: Trusted Leadership, Civic Engagement* (Routledge 2016); *Mastering Leadership: A Vital Resource for Healthcare Organizations* (Jones & Bartlett Learning 2015); *Developing Women Leaders in Corporate America: Balancing Competing Demands, Transcending Traditional Boundaries* (Praeger 2012); *The Theory and Practice of Corporate Communication* (Sage 2008); and, *Leading the Learning Organization* (SUNY Press 2000).

Volker Bilgram is Team Lead at HYVE, an Open Innovation company based in Munich, and Associated Researcher at the TIM Group of RWTH Aachen University. Currently, he is doing research on the use of Co-Creation as an empowerment and branding tool. Volker graduated in International Business Law at the University of Erlangen-Nuremberg with an academic focus on New Product Development and Open Innovation. Volker blogs about open innovation and publishes in journals such as the International Journal of Innovation Management, Marketing Review St. Gallen and Harvard Business Manager.

Bettina Sophie Blasini is a Ph.D. student at the University of Cambridge under the supervision of Tim Minshall. She holds a Master of Arts in Mass Communication and Journalism from Ludwig-Maximilians University Munich and an M.Phil. in Innovation, Strategy and Organisation of Cambridge Judge Business School. Besides her studies, Bettina Sophie worked at the University of Munich and as a freelance journalist.

Andrew Breen is Associate Adjunct Professor at the Stern School of Business at New York University. In addition, Mr. Breen is a Vice President of Product at American Express in New York City. He has spent over 20 years in industry building technology solutions and teams inside growth stage and large enterprises covering media, finance and communications. He is a frequent speaker at technology industry conferences on product development, innovation and organizational design.

Prof. Dr. Manfred Bruhn is Full Professor of Business Economics, in particular for Marketing and Corporate Management at the Center for Economic Sciences of the University of Basel, as well as holding an honorary professorship at the Technische Universität München. He conducts research on strategic marketing, services management, communications management, brand management, and nonprofit marketing. Many of his publications belong to the standard repertoire in marketing science and practice. He is adviser to various companies and member of numerous organizations and juries. Manfred Bruhn is editor of the Swiss magazine "Die Unternehmung" and director of the Master's program "Marketing Management and Business Administration" in Germany and Switzerland. He studied business administration at the University of Münster, qualified as a professor in 1985 and was a full professor of Marketing at the European Business School until 1995.

Noel Capon is the R.C. Kopf Professor of International Marketing and Chair of the Managerial Marketing Division, Columbia University, Graduate School of Business, New York.

Henry Chesbrough is Faculty Director of the Garwood Center for Corporate Innovation at the Haas School of Business at UC Berkeley (United States) and visiting professor at ESADE (Spain). He holds a Ph.D. in Business Administration from the University of California-Berkeley, an MBA from Stanford University, and a BA from Yale University. His research focuses on managing technology and innovation. His book, Open Innovation (2003), articulates a new paradigm for organizing and managing R&D. His second book, Open Business Models (2006), extends his analysis of innovation to business models, intellectual property management, and markets for innovation. His third book, Open Services Innovation (2011), explores open innovation in services businesses. Furthermore, Henry has co-edited Open Innovation: Researching a New Paradigm (2005) and New Frontiers in Open Innovation (2014) with Wim Vanhaverbeke and Joel West. In 2014 Henry was awarded Innovation Luminary by the European Commission and Intel Labs for achievements in open innovation.

Philip Cordes is a research associate and Ph.D. student of Prof. Dr. Michael Hülsmann in the department »Systems Management« in the School of Engineering and Science at the Jacobs University Bremen. He completed his diploma in business sciences at the University of Bremen in 2008. Since may 2008 Philip Cordes is working for the collaborative research centre 637 »Autonomous Cooperating Logistics Processes—A Paradigm Shift and its Limitations« funded by the German research foundation (DFG). Philip Cordes' exploratory focuses are autonomous co-operating logistic systems, strategic management in the music industry, and complexity science-based organization theory. He deals with complexity-based theories of complex adaptive systems, dynamic capabilities as well as dominant management Logics and path dependencies.

Jean-Philippe Deschamps is Emeritus Professor at IMD in Lausanne with 45 years of international experience as innovation practitioner/consultant and teacher. He focuses his research on the management and governance of innovation and the profile and focus of innovation leaders. He has (co-)authored three best-selling books: *Product Juggernauts: How Companies Mobilize to Generate Streams of Market Winners—nnovation Leaders: How Senior Executives Promote, Steer and Sustain Innovation*—and *Innovation Governance: How Top Management Organizes and Mobilizes for Innovation.* He has delivered lectures throughout the world, including twice in front of heads of states and entire governments, and at the prestigious Millenium Prize in Helsinki, the "Nobel" for technology. He was twice invited to speak at the World Economic Forum in Davos.

Rani Jeanne Dang is Assistant Professor at the University of Nice Sophia-Antipolis, affiliated to CNRS, the National Centre for Scientific Research in France (GREDEG Research group) where she researches and teaches in the fields of innovation, entrepreneurship, and strategy. She is also a Broman Research Fellow at the Institute for Innovation and Entrepreneurship, University of Gothenburg, Sweden. She holds a Ph.D. in Management from University of Nice/CNRS and was also a visiting researcher at the University of Cambridge, UK and a project officer at the Pôle de Compétitivité SCS (Secured Communicating Solutions), the cluster-led innovation policy in France. Rani's main research interest is the development of small firms through collaborations in regional ecosystems of innovation.

Annika Dingler is a Post-doctoral Research Fellow in Strategy and Innovation Management at Hasselt University (Belgium) and a Guest Researcher at the Technical University (TU) Berlin (Germany). She obtained her Ph.D. in Applied Economic Sciences in March 2014 from the Technical University Berlin. Her research interests span the interrelationship among high-technology strategy, intellectual property strategy, open innovation, and organizational behavior. She was a visiting scholar at the University of Illinois (USA) and the University of Utrecht (The Netherlands).

Prof. Dr. Ellen Enkel is Professor of Innovation Management and Head of the Dr. Manfred Bischoff Institute of Innovation Management of Airbus Group at the Zeppelin University. She previously headed the Competence Centre Open Innovation at the Institute of Technology Management at the University of St. Gallen (Switzerland). Her research interests include cooperative innovation processes such as open and cross-industry innovation, business model innovation, innovation networks, as well as innovation metrics systems. She has broad industry experience working with various well-known companies across industries and has published several books and academic articles in the area of innovation and technology management.

Martin J. Eppler is Full Professor of Media and Communication Management at the University of St. Gallen (HSG), where he is also the director of the MCM Institute for Media and Communication Management and the head of the International Study MBA Program. He conducts research on knowledge management, knowledge visualization, knowledge creation (creativity), and knowledge communication and has published 15 books and 200 articles on these topics. He has been a guest professor at various universities in the US, Canada, Asia, South America, and Europe. He has been an advisor to organizations such as the United Nations or the European Central Bank, Philips, UBS, the Swiss Military, Ernst & Young, Swiss Re, GfK, and others.

Dr. Sherry Devereaux Ferguson (Ph.D., Indiana University; M.A., University of Houston; B.A., Louisiana State University) is senior Professor and former Chair and Director of Graduate Studies for the Communication Department, University of Ottawa, Canada. She has acted on the editorial boards of numerous journals and yearbooks, including the *Journal of Communication*, the *Communication Yearbook*, the *Communication Studies Journal*, the *International Journal of Strategic Communication*, and the *Communication, Culture and Critique Journal*. She chaired the Public Relations Division of the ICA and held the position of Executive Board Member-at-Large for the Americas. She also chaired the Internationalization Committee for that organization. While acting as chair of the PR division, she established the Global Task Force, a research initiative involving scholars from many countries. She is an honorary board member for the *Systems Theory and Complexity* book series (Aracne Publishing, Rome), in collaboration with the World Complexity Science Academy (WCSA). Her publications include three books on organizational communication, two on public opinion and strategic planning in communication, two on public speaking, a book on civic discourse and cultural politics in Canada, and a book on interpersonal communication. Publishers she has worked with include Oxford University Press (New York and Toronto), Sage, Greenwood, Transaction, and others. Ferguson has published 30 articles in refereed journals and made 45 conference presentations. Her clients in her extensive consulting career include the Department of Foreign Affairs (Canada), the Department of Justice, Transport Canada, Health Canada, the Canadian Space Agency, the National Research Council, Office of the Auditor General, CIDA, the Canadian Management Institute, Petro Canada, and others. She served on two major federal advisory boards, which oversaw the writing of a vision statement and defining of curriculum needs for government communicators. She trained more than a thousand government communication officers in public opinion analysis and strategic planning techniques, initially at the request of the Assistant Secretary of Communications to Cabinet. She also did professional speech writing for a former Prime Minister, government ministers, and top level bureaucrats.

Dr. Albrecht Fritzsche holds a Doctoral degree in Philosophy from Technical University of Darmstadt and another one in Business Economics from Hohenheim University, Stuttgart. He has 15 years of experience in the manufacturing industry, as a systems expert for order management and production in the automotive industry, and as a strategy consultant and project manager on different assignments in Germany, China, and many other countries. Dr. Fritzsche has also been involved in various interdisciplinary research activities which are reflected in his two dissertations on the suppression of indeterminacy in technical operations and the management of heuristic search and decision procedures. He teaches courses in the field of innovation strategy and various other topics concerning the organizational and technical conditions of industrial development at the Institute of Information Systems of Friedrich-Alexander-University of Erlangen-Nuremberg. At the same time, he manages various applied research projects on open innovation, service innovation, and collaborative problem solving.

Lawrence McGrath is a research assistant at the University of St. Gallen MCM Institute for Media and Communications Management. He holds a Master's degree in Marketing from Monash University, Australia (Dean's Honour List). He has also studied at the University of Cambridge, the University of Adelaide, and the University of Canterbury, New Zealand. He has worked as an entrepreneur, teacher, and editor, and he has experience in strategic planning, vertical market support, and innovation management. His research focuses upon the role of pairwork and visual knowledge communication for creative collaboration in management teams.

Prof. Dr. Michael Hülsmann studied Business Administration at the University of Bayreuth. Then he continued as a Research Assistant at the University of Bayreuth and at the University of Bremen. He earned a Ph.D. degree on the topic "Management im Orientierungsdilemma - Notwendigkeit eines Managements rationalitätsbezogener Widersprüche von Effizienz und Nachhaltigkeit" (Orientational Dilemma in Management—The Need for a Rationality Driven Contradiction Between Efficiency and Sustainability in Management) with "summa cum laude" and received the 2nd award for studies of the Förderverein für Mittelstandsforschung e.V. (Association for Scientific Development of Small and Medium Enterprises). In his business career Prof. Hülsmann worked for notable companies. Between 2003 and 2009 Prof. Hülsmann was in charge of the "Management nachhaltiger Systementwicklung" (Management of Sustainable Systems Development) fi eld of studies in the Economics department of the University of Bremen. Since January 2004 he has been in charge of sub-project A3 "Monitoring der Selbststeuerung" ("Monitoring of Autonomous Control") and a leading member of SFB 637 "Selbststeuerung logistischer Prozesse - ein Paradigmenwechsel und eine Grenzen" (Autonomous Logistics Processes—a Change in Paradigm and Constraints) which is already in its second research level as the kernel of the scientific cluster "Dynamics in Logistics". In 2007 Prof. Hülsmann was invited as a visiting researcher at the University of California, Los Angeles, USA. In March 2009 he took over the "System Management" field of studies at the School of Engineering and Science at Jacobs University Bremen. He focuses on strategic management of logistics systems. Professor Hülsmann teaches and does research mainly in the fields of competence and technology-based positioning, accomplishing of technological change in value-adding networks and setting up of interorganizational coordination mechanisms.

Kathrin Jepsen has been working on customer integration in innovation processes for over 10 years. Together with Fee Steinhoff she is head of the User Driven Innovation (UDI) team at the Telekom Innovation Laboratories (T-Labs) in Berlin.

Together with her team she supports individual projects and T-Labs spin-offs with both, user research and innovation market research. In her role she is responsible to ensure that T-Labs develops technical innovations that are aimed at customer needs. Prior to joining T-Labs, Kathrin worked at T-Mobile, where she developed user-centered intercompany IT systems. She holds a degree in Psychology from the Technical University of Berlin.

Madeleine Kröper is a psychologist with a passion for user-centered design. She believes that creating successful and innovative products or services requires insights from the user perspective. Madeleine studied work and organizational psychology in Potsdam and Madrid where she specialized in innovation processes and creativity methods. Furthermore she enrolled at HPI School of Design Thinking in Potsdam. As part of multidisciplinary teams, she solved real-life design challenges applying user-centered design methods such as creating personas or prototypes in order to collect feedback. As Project Manager of the Innovation Research Team at HYVE, Madeleine is in charge of co-creation and innovation research projects in the early phases of new product development and conducted projects for companies such as BMW, Beiersdorf, Danone or Vorwerk.

Annika Lorenz is a Post-doctoral Research Fellow in Strategy and Innovation Management at Hasselt University (Belgium) and a Guest Researcher at the Technical University (TU) Berlin (Germany). She obtained her Ph.D. in Applied Economic Sciences in March 2014 from the Technical University Berlin. Her research interests span the interrelationship among high-technology strategy, intellectual property strategy, open innovation, and organizational behavior. She was a visiting scholar at the University of Illinois (United States) and the University of Utrecht (the Netherlands).

Elliot B. Luber, MBA is a graduate from SUNY Empire State College's MBA in Management Program, and is now Visiting Assistant Professor of Business, Management and Leadership there. He is also Director of Marketing and Communication for SupplyLogic, a national marketing execution partner, and has previously spent 16 years in various Communications roles at IBM, supporting executives, corporate acquisitions and social transformation in the company's Software Group, Global Industrial Sector and North American Sales organizations. The company funded his graduate studies. He was previously a vice president at two Manhattan public relations firms after a career as a journalist, covering the entertainment business and launching a string of successful business magazines. He wrote the Foreword to Belasen (2008).

Peter Malone is an experienced entrepreneur currently completing doctoral studies at the University of Western Australia. His research centers on developing a model of Entrepreneurial Innovation Value (EIV). He has 10 years experience as a researcher and consultant in the fields of innovation and entrepreneurship and has successfully launched a number of start-up firms in technology industries. He has taught at the University of Western Australia, delivered presentations at Curtin University and the University of Notre Dame Australia on innovation and holds an

MBA from the University of Western Australia and a Bachelor of Architecture from Curtin University. Peter has delivered several papers on EIV internationally and his research is at the forefront of understanding commercialization value in entrepreneurial SMEs.

Carsten Mangels is a final-year Master's candidate at Zeppelin University, studying Corporate Management and Economics. During his studies in Friedrichshafen, Vancouver, Lisbon and Muenster, his research has focused on innovation communication, innovation ecosystems, and entrepreneurship. In addition to his university education, he gained first work experience as an intern for different consulting firms in the fields of digital transformation and e-commerce.

Claudia Mast is Professor of Communication Science, Journalism Studies and Public Relations at the University of Hohenheim in Stuttgart. She qualified as a university lecturer at the University of Munich with a postdoctoral thesis on the structural change of communication systems in our society. Claudia Mast was a lecturer at the universities of Munich and Eichstätt (Germany) and Zürich (Switzerland) for several years. Claudia Mast studied Communication Science, Political Science and Romance Languages at the University of Munich. She attended the German School of Journalism (Deutsche Journalistenschule) in Munich where she got her training in press, radio, and TV journalism. After some time working for newspapers and in broadcasting, she joined the international company Siemens and was responsible for management information and training. Her main areas of interest include professionalism and innovation in journalism. In this field, she is best known for her handbooks on media coverage of economic topics and techniques for handling news stories. She is the author of several other books—dealing with the principles and practice of journalism, corporate communication, especially crisis communication, and the role of values in communication strategies. Claudia Mast is a member of the advisory board of Germany's international broadcaster Deutsche Welle. e-mail: sekrkowi@uni-hohenheim.de

Tim Mazzarol is a Winthrop Professor in Entrepreneurship, Innovation, Marketing and Strategy at the University of Western Australia and an Affiliate Professor with the Burgundy School of Business, Groupe ESC Dijon, Bourgogne, France. He has around 20 years of experience of working with small entrepreneurial firms as well as large corporations and government agencies. This includes strategic management, marketing, and support to commercialization. He is the author of several books on entrepreneurship, small business management, and innovation. His research has been widely published internationally. He holds a Ph.D. in Management and an MBA with distinction from Curtin University of Technology, and a Bachelor of Arts with Honours from Murdoch University, Western Australia.

Tim Minshall is a Senior Lecturer at the University of Cambridge Centre for Technology Management. His researches, teaches, writes, and consults on the topics of open innovation, technology enterprise, the financing of innovation, and university–industry knowledge exchange. He is a non-executive director of St. John's Innovation Centre Ltd., Cambridge and a Visiting Professor at Doshisha

University Institute for Technology, Enterprise and Competitiveness in Japan. He has a BEng from Aston University, and a Ph.D. from Cambridge University Engineering Department. Prior to joining the University of Cambridge, he worked as an engineer, teacher, consultant, freelance writer, and project manager in the UK, Japan, and Australia.

Kathrin M. Möslein is Professor of Information Systems at the School of Business and Economics at the Friedrich-Alexander-University of Erlangen-Nuremberg. She is also a professor of management and member of the team of directors at the Center for Leading Innovation & Cooperation (CLIC) at HHL Leipzig Graduate School of Management. In addition, she serves as member of the advisory boards of the Peter Pribilla Foundation at the Technical University of Munich (TUM), the Cambridge Service Alliance (CSA) at the University of Cambridge and the Fraunhofer IIS-SCS, Erlangen. Amongst others she is a founding member and has been a Vice President of the European Academy of Management (2006–2012). Prof. Möslein has been researching, teaching and consulting in the field of strategic innovation and innovation systems since the early 1990s. Her current research focuses on the implementation of innovation strategies and technologies as well as leadership systems in service organizations.

Letizia Mortara is a Senior Research Associate at the Centre for Technology Management at the Institute of Manufacturing, University of Cambridge (UK). Her current interests and expertise include open innovation, additive manufacturing and technology intelligence. Letizia has a first degree in Industrial Chemistry from the University of Bologna (Italy). After working as a process/product manager for the IVM group specialised in coatings, stains and primers for wood, she moved to the UK where she gained her Ph.D. on processing and process scale-up of advanced ceramic materials at Cranfield University.

Sabine Patsch has been working in the field of communications and marketing at SAP AG in Walldorf, Germany, since 2008. She holds graduate degrees in Social Sciences from the University of Koblenz-Landau and in International Business Administration and Marketing from Baden-Wuerttemberg Cooperative State University Mosbach and Hogeschool Zeeland, Netherlands. She has gained experiences in marketing, corporate communications, consulting and operations in several organizations across different sectors and industries, such as the automotive and electrical industries. Her academic field of interest is the interface between strategic communications and innovation collaborations.

Sophie Reboud is Professor of Strategy and Management of Innovation at the Burgundy School of Business (Groupe ESC Dijon Bourgogne), France and an Honorary Research Fellow at the University of Western Australia. She has over ten years of experience as a researcher and consultant in the field of management and strategy of small firms. Originally trained as an agronomist she served as a research engineer for École Nationale Supérieure des Mines de Paris for five years, and completed her Ph.D. there. Sophie's research interests include the strategic management of innovation and technology. This includes firms in the food sector and low-tech industries with specific focus on intellectual property and strategy in small firms.

Martin Reeves is a senior partner in the New York Office of the Boston Consulting Group, focused on business strategy. He is also director of the BCG Henderson Institute, which focuses on developing and applying new ideas to help frame the CEO's next game. He is co-author of Your Strategy Needs a Strategy (Harvard Business Press 2015) and numerous papers in HBR and other publications dealing with new approaches to strategy. The main focus of his research and translation is corporate resilience and longevity, biological thinking in business, operationalizing ambidexterity, and artificial intelligence and strategy.

Nadine Roijakkers is Associate Professor of Strategy and Innovation Management at Hasselt University (Belgium). Besides a career in academia she was a senior strategy consultant at KPMG Consulting in the Netherlands for several years advising global companies on their collaborative strategies and practices. Nadine published numerous articles and book chapters on alliance management and innovation management. Outlets for her work include among others Long Range Planning, Research Policy, Harvard Business History Review, British Journal of Management, European Management Journal, Technological Forecasting and Social Change, Small Business Economics, California Management Review, and Journal of Product Innovation Management.

Prof. Dr. Rolf Sachsse holds the seat in Design History and Design Theory at the University of Fine Arts Saar (HBKsaar) in Saarbruecken where he also serves as Pro-rector for Academic Affairs. He is also Associated Professor in Theories of Formation (Gestaltung) at the State University of Art and Design (HfG) Karlsruhe; in the 1980s, at the Design Department of the Niederrhein University of Applied Sciences in Krefeld, he installed one of the earliest Computational Design labs in Germany. He has been member of the Artist Placement Group (London), is member of the board at Het Apollohuis (Eindhoven) and has curated numerous exhibitions on art, media and design around the globe. His roughly 400 publications cover subjects in photography, design, architecture, and sound art—most of them describing innovative strategies in history, presence and future.

David Schonthal is a Clinical Assistant Professor of Innovation & Entrepreneurship at the Kellogg School of Management. David teaches courses in new venture creation and creativity, and leads the school's startup accelerator—the Zell Fellows program. He is also a Director in the Business Design practice at IDEO. David serves as a member of Chicago Mayor Rahm Emanuel's technology, innovation, and entrepreneurship council, and is one of the Co-Founders of MATTER, a 25,000 square foot innovation center in downtown Chicago focused on catalyzing and supporting healthcare entrepreneurship. He has been featured as a contributor to Forbes and Inc., received a Kellogg Faculty Impact Award for excellence in teaching, and was named to Crain's Chicago Business magazine's "40 Under 40" list.

Jörn Schulz is research associate and doctoral candidate in the project field User Driven Innovation (UDI) at the Telekom Innovation Laboratories (T-Labs). At UDI, he is responsible for and specialized in explorative user research methods, ideation, and foresight as well as strategy. He studied European Ethnology as well as American and English Studies at the Humboldt University in Berlin.

Christoph Senn is Director of the Competence Center for Global Account Management and Senior Lecturer of International Management at St. Gallen University, Switzerland.

Dr. Samah Shaffakat is a Post Doctoral Research Fellow in the Department of Organisational Behaviour at INSEAD—the Business School for the World. She is a member of INSEAD's Leadership and Communication Research Group. She is also a Visiting Professor at IIM-Ahmedabad where she is teaching courses in leadership and research methods. Her research interests include: leadership communication, leadership development, psychological contracts, mindfulness, gender and media. Her research is published in international management journals.

Dr. Fee Steinhoff is together with Kathrin Jepsen head of the User Driven Innovation (UDI) team at Telekom Innovation Laboratories (T-Labs) in Berlin. Together with her team, she is responsible for the development and implementation of user research tools along the innovation process. Fee Steinhoff studied Business Administration (Technical University Berlin), earned a Master of Business & Engineering (Steinbeis University Berlin) and received her Ph.D. from the Technical University Berlin. Her research interests include innovation marketing, customer orientation and radical innovation—topics which she also teaches in executive programs.

Vera Stolarski is a doctoral student at the European Business School (EBS), Oestrich-Winkel. Her research interests include cognitive diversity, innovation, and team dynamics. Within her Ph.D. thesis she focused on the complex and dynamically interacting effects of cognitive diversity in top management teams on innovation on firm level. Vera Stolarski studied psychology with a focus on "occupational and organizational psychology" at the University of Cologne. During her studies, she did various internships in the areas of human resource consulting and a research internship at the Max Planck Institute for Human Development, Berlin.

Leigh L. Thompson is the J. Jay Gerber Distinguished Professor of Dispute Resolution and Organizations at the Kellogg School of Management. She is the director of the Kellogg Team and Groups Research Center and the *Leading High Impact Teams*, *Constructive Collaboration*, and *Negotiation Strategies* executive programs. An active scholar and researcher, she has published over 110 research articles and chapters and has authored 10 books, including *Creative Conspiracy* (HBS press), *The Truth about Negotiation* (Financial Times press), *The Mind and*

Heart of the Negotiator (6th edition), *Making the Team* (5th edition), and *Creativity in Organizations*. She created a Coursera MOOC titled, *High Performance Collaboration,* and teaches executives and managers all over the world.

Prof. Dr. Meike Tilebein born 1966, is Full Professor and Director of the Institute of Diversity Studies in Engineering at the University of Stuttgart, Germany, with an interdisciplinary background. She studied Engineering Cybernetics and earned a doctoral degree in Strategic Management with a thesis on principles of complex adaptive systems and their applications to firms and networks. She served as the DPD Endowed Assistant Professor of Innovation Management at the European Business School in Oestrich-Winkel, Germany before she joined the University of Stuttgart in 2009. She has published in the fields of strategic management, complexity science, innovation, team diversity, and service science. Her current research interests focus on new business models, cooperative strategic foresight in firms and networks, and the effects of team diversity on innovation. Her research methods include agent-based and system dynamics simulation. She has been teaching since 1995, covering topics like project management, innovation management, system dynamics, and diversity management. Meike Tilebein is member of several scientific and professional organizations in the fields of strategic management and systems sciences and member of the executive committees of the Gesellschaft für Wirtschafts- und Sozialkybernetik and the Deutsche Gesellschaft für System Dynamics/German Chapter of the International System Dynamics Society.

Wim Vanhaverbeke is Professor of Strategy and Organisation at Hasselt University (Belgium) at the Department of Business Studies and is appointed as visiting professor open innovation at ESADE (Spain) and the National University of Singapore. He published in international journals such as Organization Science and Organization Studies, Research Policy, Journal of Business Venturing, and Journal of Management Studies, and serves on the editorial board of the Journal of Engineering and Technology Management, and Strategic Organization. He was co-editor with Henry Chesbrough and Joel West of the books Open Innovation: Researching a New Paradigm (2006) and New Frontiers in Open Innovation (2014). He was recently recognized by the International Association of Management of Technology (IAMOT) as one of the top 50 authors of technology and innovation management during the period 2008–2012.

Prof. Eric Viardot is permanent Professor and Director of the Global Innovation Management Centre at EADA Business School in Barcelona. He has published various books and articles on strategic management and marketing with a strong focus on technology and innovation management. He is currently the co-editor in chief of *the International Journal of Technology Marketing* and a member of the editorial board of *Technovation*. He teaches regularly in executive programs in Europe and is a consultant for various international companies on strategy and innovation management.

Dr. Johannes von Karczewski has been the communications officer at Corporate Technology, the central research and development department of Siemens, since March 2015. In this role, he helps to actively shape innovation communication at Siemens and supports the Chief Technology Officer in all communications-related issues. Following his study of medieval and contemporary history, political science, and public law in Bonn and after earning his doctorate in 2007, Dr. von Karczewski has worked for Siemens AG in a variety of corporate communications positions.

Ramona Wehlig is Director Customer & Advisor Insights at Vorwerk International/Division Thermomix located in Switzerland. In her role she is gathering insights from customers and sales representatives in the European subsidiaries to reflect them in international product and marketing strategy. Ramona holds an M. Sc. in Business Psychology.

Prof. Ian C. Woodward is Senior Affiliate Professor in Organisational Behaviour at INSEAD, directing its flagship "Advanced Management Programme". Teaching and research focuses on exceptional leadership communication, strategic/high performance leadership, fair process communication, and personal leadership transformation using innovative development techniques related to human "drivers, blockers and values". He created the "communication intelligence" concept, and also delivers innovative arts-based learning experiences for business executives (he is Associate Conductor of Singapore's *Metropolitan Festival Orchestra).* His research is published in international journals, and he was formerly Associate Faculty Director of Columbia Business School's Senior Executive Program. Before academia his business career included Board Directorships, Chief Executive, Senior Executive and Broadcasting roles in government, financial, energy and public broadcasting sectors.

Howard Yu is Professor of Strategy and Innovation at IMD. He specializes in technological innovation, strategic transformation, and change management. His teaching and research activities focus on why and how some firms can sustain new growth while others cannot. In 2015, Yu was featured in Poets&Quants as the World's Best 40 Under 40 Business School Professors. He received his doctoral degree at Harvard Business School. Stay connected on LinkedIn and Twitter (@howardhyu).

Ansgar Zerfass has been Full Professor of Communication Management at the University of Leipzig since 2006 and is Professor of Communication and Leadership at BI Norwegian Business School, Oslo. He serves the academic community as president of the European Public Relations Research and Education Association (EUPRERA), Brussels, and as editor of the "International Journal of Strategic Communication", Routledge Publishers, Philadelphia, USA. Zerfass holds a graduate degree and doctorate in Business Administration and a postdoctoral qualification (Habilitation) in Communication Science from the University of

Erlangen-Nuremberg. He has authored and edited 50 books and research reports and more than 180 articles on corporate communications, online communications, measurement and evaluation, and innovation communication.

Sandra Maria Zistl works at Siemens AG as an editor and author for innovation communications. She holds an Editor's degree in Journalism as well as a Master's degree in Political Science with a specialization in international relations. Following her study of political science, public law, Spanish and French in Munich, Sevilla and Montevideo, she attended journalism school in Munich and worked as a print and online editor and author for *FOCUS* news magazine, *Die Welt* and *Süddeutsche Zeitung* and as a correspondent for *FOCUS* news magazine in Paris. She is co-founder of hostwriter.org and advisory board member of journalists network.

Part I
Strategic Perspectives on Innovation

Chapter 1
The Importance of Connecting Open Innovation to Strategy

Wim Vanhaverbeke, Nadine Roijakkers, Annika Lorenz and Henry Chesbrough

Abstract Companies that are experienced in open innovation integrate open innovation activities as part of their strategy. By contrast, open innovation research has not been adequately integrated into the strategy literature and vice versa. In this chapter, we discuss a number of existing strategy fields that offer inroads to connect open innovation to strategy. The need to connect open innovation to strategy is illustrated by describing how companies increasingly organize new business development projects through open innovation ecosystems where an increasing variety of partners do not only act as sources of innovative knowledge but also provide input to a joint strategizing process.

1.1 Introduction

Openness in innovation and strategy has received tremendous attention from practitioners and scholars alike. In contrast to closed settings where the company was only conducting internal R&D and strategy development was secret and the exclusive task of top management, the open innovation literature emphasizes the importance of joint R&D efforts together with external partners (in ecosystems) to successfully arrive at a desired innovative outcome. Open strategy purports the idea of joint sense-making and idea generation when crafting strategy through the inclusion of internal and external expertise. These two paradigms have evolved independently in the innovation management (open innovation) and strategy (open strategy) literature although they borrow insights from each other. Despite several calls for coupling the two paradigms in recent years, the link between strategy and external collaboration has almost exclusively been studied in the literature on alliances (Gulati 1995; Gulati and Singh 1998; Lavie 2006), networks (Gulati et al.

W. Vanhaverbeke · N. Roijakkers (✉) · A. Lorenz
Hasselt University, Martelarenlaan 42, 3500 Hasselt, Belgium
e-mail: nadine.roijakkers@uhasselt.be

H. Chesbrough
University of California, Berkeley, CA 94720, USA

N. Pfeffermann and J. Gould (eds.), *Strategy and Communication for Innovation*,
DOI 10.1007/978-3-319-49542-2_1

2000), and ecosystems (Adner 2006, 2012; Adner and Kapoor 2010; Gawer and Cusumano 2014) with only a few exceptions in open innovation (Chesbrough and Appleyard 2007; Dittrich et al. 2007; Doz and Kosonen 2008; Vanhaverbeke and Cloodt 2014; Vanhaverbeke and Roijakkers 2013; Whittington et al. 2011). Mattes (2011), Vanhaverbeke and Cloodt (2014), and Vanhaverbeke and Peeters (2005) emphasize that strategic goals may be influenced and reformulated on the basis of open innovation outcomes. Chesbrough and Appleyard (2007) put forward that existing strategic frameworks (e.g., Porter 1980, 1985) are inadequate for effectively integrating open innovation activities and that there is a clear need for 'open strategy'. Whittington et al. (2011) have, in turn, related open strategy to open innovation where they consider open innovation to be a subset of open strategy. Open strategy refers to increasing numbers of firms that disclose information on strategic goals and directions and involve a greater number of internal and external actors in strategy formation processes (Whittington et al. 2011; Matzler et al. 2014). Similarly, open innovation relies on the objectives and capabilities of an organization's outside partners. Yet, in comparison to the open strategy literature, open innovation brings in a slew of new external actors, which shape an organization's (innovation) strategy through their relationships to and engagement with these actors and which we will highlight in this chapter.

1.2 Potential Inroads to Connect

This section examines how the concept of openness has been introduced separately in both the innovation management literature and the strategy literature and how the open strategy perspective can be used within an open innovation context. While both strands of literature have been studying firms' openness in their strategy formulations (open strategy) and their innovative activities (open innovation), the explicit connection between the two has not often been made yet; it seems, however, of eminent importance to find inroads to connect as companies' (open) innovation endeavors can only be effective if they are explicitly embedded in the strategizing process.

With the introduction of the concept of open innovation in 2003, Chesbrough created an umbrella term where previous important, but segregated, strands of literature, studying firms' openness in their (innovative) activities, such as the strategic alliance literature, could be integrated into a broader perspective on innovation. Chesbrough's introduction of open innovation and his visualization of openness through the 'open innovation funnel' have constituted an important integration of previously separate streams of literature. Chesbrough has drawn up the funnel to represent a firm's internal innovation process and has pointed out that, at certain points, companies in this process open up to tap into knowledge existing outside their boundaries. This mainly happens when the company is unable to create necessary capabilities internally and therefore reaches out to partners through various types of inter-firm relations such as alliances, in-licensing, spin-ins, acquisitions, etc.

A company can also decide to open up in order to externalize and commercialize knowledge that does not fit its strategic purpose and is therefore not directly used for strengthening the internal innovation process through out-licensing, spin-outs, divestments, etc. Chesbrough has implicitly linked openness in innovation to the strategic goals of the firm through the connection between the funnel and the business model (current and new, the firm's own business model and that of others). Additionally, selective revealing describes a strategy where firms in open innovation settings strategically determine to which extent and in which areas they purposefully share knowledge with others in order to stimulate the open innovation process (Henkel 2006; Henkel et al. 2014). While the open innovation literature has studied firms' openness in a detailed manner, only few attempts have been made to connect open innovation explicitly to strategy (notable exceptions are Chesbrough and Appleyard 2007; Vanhaverbeke and Cloodt 2014; Henkel et al. 2014). A number of existing strands of strategy literatures offer inroads to connect with open innovation and we will elaborate on those in the following: the unrealized/deliberate/emergent strategy typology, the induced/autonomous strategy process, and the open strategy framework. Scholars such as Mintzberg (Mintzberg 1978; Mintzberg and Waters 1982, 1984, 1985; Mintzberg et al. 1986, Mintzberg and McHugh 1985; Brunet et al. 1986) and Burgelman (1983a, b, c, 1994, 2002) have been focusing on the process of strategy formation. Strategy has mostly been conceived in terms of what the leaders of an organization are 'planning' for the future and therefore strategy formation has been treated as an analytic process for establishing long-range goals. This approach is however seriously limited since there is a variety of ways in which strategies actually take shape.

1.2.1 The Unrealized/Deliberate/Emergent Strategy Typology

Mintzberg and Waters' (1985) strategy typology—unrealized, deliberate, and emergent strategies—reflects the relationship between leadership plans and intentions on the one hand (intended strategy) and what the organizations actually realize on the other hand (realized strategy). In a deliberate strategy the realized strategy reflects the intended strategy (realized as intended). Emergent strategies emerge in an organization but are not guided by the intended strategies from the top. Emergent strategies do not mean that management is out of control—it is just more open, flexible, and willing to learn—and they are particularly important in a complex, uncertain, and changing environment. Emergent strategies enable management to act before everything is fully understood. When firms innovate in order to cope with rapidly changing technological and market environments, intended strategies will only be partially realized leaving a lot unrealized. As new business opportunities only slowly get articulated over time emerging strategies become increasingly important. Emerging strategies may find their origin within (e.g., employees or frontline managers) or outside the company (e.g., consultants, technology partners, customers, etc.). In open innovation in connection to strategy, special attention should be paid to how innovating companies can capitalize on external sources to formulate new strategies. Even more important is how they can develop management processes to use insights from external partners as building blocks for changes in the intended strategy. The benefits of open innovation have been always framed with respect to the innovation objectives of a firm but open innovation may also help companies to develop new/emerging strategies compared to the closed innovation setting where departments responsible for the innovation performance of the company are 'isolated' from external sources that can fuel emerging strategies.

1.2.2 The Induced/Autonomous Strategy Process

Burgelman (in line with Minzberg's work) has been focusing on the processes underlying strategic renewal or corporate entrepreneurship in firms. Burgelman (1983a, 1994) argues that induced strategic processes or autonomous strategic processes can drive strategic renewal. Induced strategic processes capture formal corporate entrepreneurship, which is a top-down process whereby the company's strategy and structure provide the context for renewing the firm's businesses. Autonomous strategic processes refer to entrepreneurial behavior, which surfaces informally in the firm. Both induced and autonomous strategic behaviors are important to a firm's entrepreneurship efforts, whether they intend to create new businesses or reconfigure existing ones. In his more recent work (2002), Burgelman develops an evolutionary perspective on strategy making showing that large firms tend to pursue more focus and efficiency in the existing businesses, which may come at the expense of effective exploration and strategic renewal. "Induced strategy exploits initiatives that are within the scope of a company's current strategy and that extend it further in its current product-market environment. Autonomous strategy exploits initiatives that emerge through exploration outside of the scope of the current strategy and that provide the basis for entering into new product-market environments" (p. 327). Burgelman shows that there is a risk of coevolutionary lock-in: that is, there is a positive feedback process that increasingly links previous success of a company's strategy to that of the existing product-market environment, which reduces the firm's ability to change strategic direction when necessary. The strength of top management's beliefs in the validity of the core (existing) business strategy leads to two sources of inertia associated with coevolutionary lock-in on the autonomous strategy process. The first type is that the core business' logic is applied to initiatives that need an autonomous strategy process. Second, the narrow strategic focus of the dominating/existing business prevents top management from understanding the changing strategic context resulting in a risk of immobility. Burgelman's framework for strategic decision-making processes is quite relevant within the context of open innovation. His studies about Intel are mainly focusing on autonomous strategy processes where ideas for new strategies emerge bottom-up within the company. However, as companies open up to tap into external sources of knowledge, autonomous processes should be reconsidered to include external knowledge partners as valuable sources for the cognition of new strategic opportunities. Innovation partners are connected both to the core business of the company as well as to the new businesses under development and simply connecting to external partners is not sufficient to alleviate the threats of the two types of inertia that Burgelman has identified. In our view, firms have to develop two processes in order to use knowledge partners for balancing exploration and exploitation. First, openness to partners should reach beyond simple knowledge transfers or arms-length's transactions. Strategic partnerships in which firms co-innovate and combine their capabilities in long-term agreements allow them to understand the view of partners on the strategic context and on the future technological and market opportunities. Second, there should be mechanisms in the company to voice the strategic insights of those initiatives that are focusing on new technologies or businesses that are remote from the core technologies and businesses. Companies can give these initiatives (which usually represent a small share of the R&D investments) a weight that is equal to that of representatives of core technologies/businesses in corporate decision-making processes. This is important as initiatives that are distant from the firm's core are more radical in nature and require different types of partners (more research oriented), different time spans of establishing the business (long-term), and different ways to connect to partners (research collaborations, CVC investments, etc.) compared to initiatives that strengthen or extend the current businesses.

1.2.3 The Open Strategy Framework

A final strategic stream providing opportunities to connect open innovation to strategy is the field of open strategy, which has paid significant attention to the concept of openness (see Giraudeau 2008; Whittington et al. 2011; Whittington and Yakis-Douglas 2011). These authors observe that, in the old days, strategy formulation in large corporations was a rather traditional exercise where the CEO and the Board of Directors carried the process out in a secluded way. Once the corporate strategy was drawn up, it was translated in a top-down manner into business- and operational-level goals and communicated to lower levels within the organization. There was hardly any input from within the organization included in the strategy process, let alone from outside parties. Furthermore, the manner in which the strategy came about in many organizations was not very clear to lower ranked managers and employees. Whittington et al. (2011) point out that nowadays strategy processes are more 'inclusive' and 'transparent.' Both internal parties (e.g., employees) and external parties (e.g., consultants, suppliers, complementary partners) are involved in a firm's strategy discussions and the manner in which a firm's strategy comes into existence as well as the end product is communicated more widely, both internally and externally. An example of a more inclusive and transparent corporate strategy process is the ‘Corporate Strategy Dialogues or CSD’ initiative at the large Dutch chemical company DSM: it is an initiative that the company organizes every 5 years, consulting in a bottom-up style all businesses and corporate services for interesting strategic ideas that could benefit the company. This process can be considered as an open strategy discussion with all the corporate segments: after consulting with the businesses, the Corporate Strategy unit is deriving a new corporate strategy including and integrating the interesting ideas captured during the CSD. As such, the strategy formation process at DSM exhibits a high level of openness. According to Whittington et al. (2011), open strategy can be viewed as a continuum where companies can be more or less open (i.e., inclusive and/or transparent) in terms of their strategy process. ‘Inclusiveness’ and ‘transparency’ are two interesting dimensions to connect open innovation to open strategy. In fact, by including an increasing number of open innovation partner types in the innovation process, a firm is able to extract not solely new knowledge, viewpoints, and insights with respect to technological developments, but also regarding new areas of strategic development. Open innovation partners, such as start-ups, research labs, customers or suppliers, may share their experiences and opinions regarding interesting strategic directions (besides product-related know-how) with the focal firm and through these collaborations the company may also be able to feed its strategy process with new information in addition to its innovative process. To an ever-increasing extent, open innovation initiatives have moved from being short-term arrangements to more long-term collaborative structures (or ecosystems) where a group of companies join forces semi-permanently, develop a common set of strategic plans, and where partners even specialize in particular areas to optimize complementarity. In these open innovation initiatives, the individual firm strategies are tightly linked to the ecosystem strategy and including partners’ strategic purposes and viewpoints in the individual firm’s strategy formulations is inevitable. When it comes to ‘transparency’, by involving a wide array of (open innovation) partners in various firm activities, firms’ strategic purposes become more apparent and transparent to both the inside and outside world. Engineers involved in open innovation activities and negotiations with external partners need to be very well aware of the strategic direction of the company and the areas where they can and cannot share information; as such, the strategy process and outcome within their organization needs to be well known (i.e., transparent) to them. A good example, in this respect, is the interaction of R&D personnel at a high-tech campus (HTC), such as the HTC in Eindhoven (the Netherlands), initially founded by Philips. While these scientists and engineers employed by different high-tech firms, such as ASML, NXP, Medtronic, Intel, IBM, Canon, etc., exchange knowledge in various common areas, they have been instructed well by their employers as to what/what not to reveal. Similarly, in order to make effective use of open innovation intermediaries, such as NineSigma, or to find an individual scientist or company that can fill a specific technological need, many companies publish their technology roadmaps or technology agendas online. Here, they identify areas where they are in need of a partner with the right expertise. The publication of these agendas signals to competitors what the firm is working on; the strategy inevitably becomes more visible to outside parties. It is thus important to determine under which circumstances openness should be strived for and in which areas it is advisable to be more closed (i.e., where new barriers to information exchange should be created).

The open innovation management and open strategy literature also emphasize the downside of openness. Both strands of literature have put forward that openness can lead to a loss of uniqueness. Researchers in open strategy have pointed out that by including consultants in the strategy process, i.e., advisors that assist many companies and often generalize their solutions across their client portfolio, companies run the risk of becoming increasingly similar in terms of their strategic portfolio resulting in a foregoing of competitive advantage. Likewise, in open innovation, the use of intermediaries or knowledge brokers, technology competitions, IP licensing, etc., have made widespread knowledge sources more accessible but have also standardized the knowledge bases available to most companies.

In conclusion to this section, we can state that both the strategy field and the innovation management literature could significantly benefit from greater integration at the level of openness of firms. In the following section, we illustrate the usefulness of cross-fertilization further by examining in more detail the strategizing process with open innovation partners.

1.3 Open Innovation as a Strategizing Process with Partners

The importance of connecting open innovation to strategy builds on the notion that open innovation partners not only act as a source of innovative knowledge but can also provide input to the strategy process. As such, open innovation settings can enable strategy formulation where external actors are included as building blocks for changes in the intended strategy or as sources of inspiration for fuelling emerging strategies (Mintzberg 1978; Mintzberg and Waters 1982, 1984; Mintzberg et al. 1986, Mintzberg and McHugh 1985; Brunet et al. 1986). Similarly, open innovation activities can feed the autonomous strategy formation process (Burgelman 1983a, b, c, 1994, 2002; Whittington et al. 2011) where bottom-up strategies and innovative ideas are developed with input from a new set of external partners. As the strategic process becomes increasingly open, strategy formulation is bound to be influenced by an increasing number of partners (Jarzabkowski et al. 2007; Whittington 2006). Whittington et al. (2011) argue that internal employees' personal identities, (in)formal contacts, relationships to, and engagement with the firm shape its strategy as much as external actors such as regulators, consultants, stakeholders, and customers do. Firms applying open innovation interact by definition with external partners, which, through their collaboration, are likely to have an impact not only on firms' innovation but also on their strategy process. In a closed innovation paradigm, an innovating firm's management is responsible for the development and the execution of the innovation strategy. In an open innovation setting, partners are not only useful in the execution of the innovation strategy but they can also be instrumental in the conception and development of a firm's innovation strategy. Companies practicing open innovation should not only open up their doors for ideas from their (potential) innovation partners but they should also actively shape the external conditions to facilitate the development of a joint innovation strategy where all partners can benefit. Thus, increasing integration of open innovation partners may also lead to new insights for strategic directions and future growth plans. Collaborating with internal and external partners whether on R&D or strategy formulation projects may result in inspiration for change, improvement of current strategic routines, a reassessment of strategic goals, or even a joint strategizing process where the inclusion of partners leads to increased commitment to and support of common strategic goals (Jarzabkowski and Balogun 2009).

We illustrate this—without being exhaustive—by focusing on the role of establishing strong ties with a number of important external innovation partners that can also serve as partners in strategy formation: we will focus on the role of prominent scientists in Scientific Advisory Boards (SABs), the need to build an extended ecosystem with major research institutes, and the development of a preferred relationship with the venture capital community.

1.3.1 Scientific Advisory Boards

Many companies struggle to keep up with the rapid pace of science and technology. A failure to remain up to date with the latest breakthroughs can be a serious threat to the competitiveness of a business. What can companies do to navigate the complexities of technical domains, when those domains are changing so often? How can they cope with the need to understand an ever-increasing range of technologies to build new products?

One option is to establish a SAB for the company to get informed and strategically advised about the latest scientific and technological developments. A SAB is a small (typically 3–5 people) group of elite professors at leading scientific research institutions, who are invited to advise the company on a periodic basis (typically 2–4 times a year) about new trends in their respective fields. These SAB members are a key part of the answer for how busy executives can update their knowledge of important developments in the scientific and technology areas of their business, and take appropriate strategic action in light of those developments. Diversity of expertise is important in recruiting a SAB. Companies need to identify leading experts in the various areas underlying their business. A company like Intel, for example, must track developments in materials science, electrical engineering, computer science, user interface design, and software engineering, among other fields. A pharmaceutical company needs to track emerging research in biology, chemistry, biochemistry, synthetic biology, computational chemistry, and genetics, among other fields. No single person can provide sufficient expertise in all of these disparate areas: therefore, companies instead select a handful of experts that collectively encompass the fields of knowledge they need to follow.

Once established, the SAB provides a useful additional perspective on the medium and long-term strategic plans for the company (typically SABs meet too infrequently to be of much help on short-term issues). The SAB members have knowledge and a network of contacts that differ substantially from most executives. They can validate the claims of internal technical staff when an important strategic decision must be made, or, more occasionally, they can raise issues or alternatives that may not have been adequately considered by the internal technical staff. A particularly valuable way to employ a SAB is to ask the experts to review and contribute to long-term roadmaps of the business. Roadmaps often look out several years into the future in hopes of defining the path of future products and services to be offered by the company. Here is where experts with different perspectives from those within the company can contribute a lot of useful knowledge, in time for it to be helpful for the business. Perhaps a new breakthrough is emerging in a particular area that promises to offer a significant improvement in accuracy, speed, or cost. Knowing that such a breakthrough was imminent might allow the company to position itself to exploit that breakthrough ahead of competitors. And sometimes there can be a bonus from having a SAB: they can help with recruiting new technical talent to the company. They can enrich the technical depth of internal experts in the company, and help them keep up to date. They can even find connections among different technical groups in the company that might not yet be apparent to management.

1.3.2 Ecosystem Building with Research Organizations

Research in most industries becomes extremely expensive and risky even for large multinationals. Setting up joint research projects with universities and research labs is one way to reduce costs and risks. Yet, as universities and research centers become vital for research-based competencies of a large company, it is recommended to align long-term strategic and research agendas with each other. This is, for most firms and their research partners, uncharted territory and teaming up means they also have to initiate some institutional innovations.

A nice example is the Energy Bioscience Institute (EBI)—a major joint research program between British Petroleum and a consortium of US-based universities under the leadership of UC Berkeley, set up late 2007. This research project is planned to last for 10 years and BP invests 50 million US Dollars every year. With this initiative BP intends to make a major inroad into biotech: the company had no bio-expertise (3 biologists in a company of 100,000 employees) and there was a need for a big push for biofuel research. Internal research capability development would take too much time and traditional collaborations with research institutes are not appropriate for achieving the expected goals. BP and the research consortium developed the EBI which is new in its kind as collaboration between research institutes and a large company: some major organizational innovations concerning joint governance, IP regulations, and the dissemination of the results were required to set up and manage the collaboration effectively. By teaming up with the best academic institutions, BP is developing its own inroad into bioenergy, which could not have been done through internal R&D or through traditional R&D collaborations with universities or research labs.

In a similar vein, DSM is establishing new public–private partnerships with the Dutch government, academia, and other industry partners to create a world-leading research environment in the Netherlands in research areas that are crucial for the future growth of the company. BMM and CTMM are two examples of this strategy: BMM (the BioMedical Materials program) was originally set up by DSM to support its ambitions to become a leading player in the biomedical market but developed over time into a much larger public–private partnership with 28 partners that focuses on research and development in the field of biomedical materials and their applications in a clinical environment. Focusing on four priority disease areas, new concepts are explored in a translational approach, where ideas are tested from the lab up to the proof-of-principle, involving all partners, disciplines, and areas of expertise that are required for application in a clinical environment. Centre for Translational Molecular Medicine (CTMM) is a public–private partnership for translational research, bringing new, medical related technological inventions to the patient as soon as possible. This center develops highly innovative arrangements between research institutes, universities, and firms specialized in molecular diagnostics and molecular imaging technologies up to the hospitals where patients benefit from early diagnosis and personalized treatment. In both cases, DSM is shaping the knowledge environment to its own advantage: biomedical research is crucial for DSM but internal R&D (too slow and too expensive) or traditional research collaborations with academic institutions (insufficient guarantees for breakthrough research and inefficient transfer to business partners) are no options. BMM and CTMM are examples of novel institutional arrangements through which DSM and other major stakeholders in the Netherlands codevelop a world-class environment in disciplines that are crucial for DSM's future businesses. There is tight co-alignment of partners' strategic objectives, which, if managed correctly, leads to a mutually beneficial situation.

1.3.3 Tapping into the Venture Capital Community

Large innovative companies nowadays also work closely together with venture capital-backed start-ups. Many large companies have started a corporate venturing capital unit to invest in and collaborate with these ventures (Chesbrough 2003b; Weiblen and Chesbrough 2015). The relationship between venture capitalists and large corporate investors is not an easy one, because corporate venturing capital units usually want preferential rights. An alignment between the strategic interests of the management of the venture, the venture capitalists, and the large company is necessary if the latter is to become a preferred partner. The benefits for a large company of being perceived as a preferred partner can be huge since venture capitalists may prefer to team up with it before they contact competing companies. Moreover, when a company knows how to deal effectively with start-ups, new technologies are absorbed more quickly and chances for successful commercialization increase.

In sum, firms not only have to source innovative knowledge from partners but they also have to invest in relationships with important external partners and involve them in the strategy formation process to secure lasting technological and commercial success. Not relations, per se, but their structure and governance are important and lead to a hard-to-imitate competitive advantage. Relationships, such as those with an advisory board consisting of eminent scientists, strategic alignment with top research centers, and preferred relationships with the venture capital community are hard to establish and difficult to imitate. Relational capital is a crucial element in open strategy. Dyer and Singh (1998) already mentioned that competitive advantage results from investments in relation-specific assets and from extensive knowledge exchange resulting in joint learning and strategizing. Molding the relations with important external parties is part of the strategy of companies applying open innovation in advanced ways: they shape their external environment in a world-class innovation ecosystem through co-alignment of strategic objectives with main partners such as public authorities, academic partners, and other commercial entities.

We have illustrated how large firms can shape relations with external partners to both receive crucial input to their innovation strategies as well as to jointly formulate and reach strategic goals using SABs, joint research initiatives, and preferred relationships with the venture capital community. These are only three examples: similar initiatives can be developed with other types of partners such as strategic suppliers, complementors, technology service vendors, and consultants.

1.4 Conclusions

In this chapter, we have argued that it is of utmost importance to connect open innovation to the strategy formation process within firms, as external open innovation partners are increasingly important sources of new strategic notions as well as innovative insights. More specifically, we have put forward that open innovation partners can dramatically shape firms' strategy from development to execution.

In ever more dynamic environments, firms cannot solely rely on top-down and deliberate strategies anymore. They have to quickly adapt to market needs and establish entrepreneurial bottom-up initiatives, which will only be possible if they can flexibly adjust their strategy and use internal as well as external knowledge sources. Thus, to realize a firm's full potential, it needs to connect its open innovation endeavors with its strategy development initiatives. In an open innovation era, external innovation partners are not only valuable as a source for new technologies and innovative business models, but they are also increasingly essential in the development of firms' strategies. That is especially the case when strategies of partners become interdependent because they are offering jointly a platform solution or are working as a innovation ecosystem where strategies of the partners have to be mutually aligned: open innovation leads in these cases to strategy development processes where technology and market partners jointly define strategies for the platform or the ecosystem.

Bibliography

Adner, R.: Match your innovation strategy to your innovation ecosystem. Harvard Bus. Rev. **84** (4), 98–107 (2006)

Adner, R.: The Wide Lens: A New Strategy for Innovation. Penguin, New York (2012)

Adner, R., Kapoor, R.: Value creation in innovation ecosystems: how the structure of technological interdependence affects firm performance in new technology generations. Strateg. Manag. J. **31**(3), 306–333 (2010)

Brunet, J.P., Mintzberg, H., Waters, J.: Does planning impede strategic thinking? The strategy of Air Canada 1937–1976. In Lamb, R (ed.) Advances in Strategic Management, vol. 4. Prentice-Hall, Englewood Clifis (1986)

Burgelman, R.A.: A model of the interaction of strategic behavior, corporate context, and the concept of strategy. Acad. Manage. Rev. **8**(1), 61–70 (1983a)

Burgelman, R.A.: A process model of internal corporate venturing in the diversified major firm. Adm. Sci. Q. 223–244 (1983b)

Burgelman, R.A. (1983c). Corporate entrepreneurship and strategic management: insights from a process study. Manag. Sci. **29**(12), 1349–1364.

Burgelman, R.A.: Fading memories: a process theory of strategic business exit in dynamic environments. Adm. Sci. Q. **39**(1), 24–56 (1994)

Burgelman, R.A.: Strategy as vector and the inertia of coevolutionary lock-in. Adm. Sci. Q. **47**(2), 325 (2002)

Chesbrough, H.W.: Designing corporate ventures in the shadow of private venture capital. Calif. Manage. Rev. **42**(3), 31–49 (2003a)

Chesbrough, H.W.: Open Innovation: The New Imperative for Creating and Profiting from Technology. Harvard Business Press, Boston, Massachusetts (2003b)

Chesbrough, H.W., Appleyard, M.M.: Open innovation and strategy. Calif. Manag. Rev. **50**(1), 57–76 (2007)

Dittrich, K., Duysters, G., de Man, A.-P.: Strategic repositioning by means of alliance networks: the case of IBM. Res. Policy **36**(10), 1496–1511 (2007)

Doz, Y.L., Kosonen, M.: The dynamics of strategic agility: Nokia's rollercoaster experience. Calif. Manag. Rev. **50**(3), 95–118 (2008)

Dyer, J., Singh, H.: The relational view: cooperative strategy and sources of interorganizational competitive advantage. Acad. Manag. Rev. **23**(4), 660–679 (1998)

Gawer, A., Cusumano, M.A.: Industry platforms and ecosystem innovation. J. Prod. Innov. Manage **31**(3), 417–433 (2014)

Giraudeau, M.: The drafts of strategy: opening up plans and their uses. Long Range Plan. **41**(3), 291–308 (2008)

Gulati, R.: Does familiarity breed trust? The Implications of repeated ties for contractual choice in alliances. Acad. Manag. J. **38**(1), 85–112 (1995)

Gulati, R., Singh, H.: The architecture of cooperation: managing coordination costs and appropriation concerns in strategic alliances. Adm. Sci. Q. **43**(4), 781–814 (1998)

Gulati, R., Nohria, N., Zaheer, A.: Strategic networks. Strateg. Manag. J. **21**(3), 203–215 (2000)

Henkel, J.: Selective revealing in open innovation processes: the case of embedded Linux. Res. Policy **35**(7), 953–969 (2006)

Henkel, J., Schöberl, S., Alexy, O.: The emergence of openness: how and why firms adopt selective revealing in open innovation. Res. Policy **43**(5), 879–890 (2014)
Jarzabkowski, P., Balogun, J.: The practice and process of delivering integration through strategic planning. J. Manage. Stud. **46**(8), 1255–1288 (2009)
Jarzabkowski, P., Balogun, J., Seidl, D.: Strategizing: the challenges of a practice perspective. Hum. Relat. **60**(1), 5–27 (2007)
Lavie, D.: The competitive advantage of interconnected firms: an extension of the resource-based view. Acad. Manag. Rev. **31**(3), 638–658 (2006)
Mattes, F.: How to make open innovation work for your R&D? Applied InnovationManagement #3 (2011)
Matzler, K., Füller, J., Koch, B., Hautz, J., Hutter, K.: Open strategy—a new strategy paradigm. In: Matzler, K., Pechlaner, H., Renzl, B. (eds.), Strategie und Leadership, pp. 37–56. Springer Gabler, Wiesbaden (2014)
Mintzberg, H.: Patterns in strategy formation. Manage. Sci. 934–948 (1978)
Mintzberg, H., McHugh, A.: Strategy formation in an adocracy. Adm. Sci. Q. **30**(2), 160–197 (1985)
Mintzberg, H., Waters, J.A.: Tracking strategy in an entrepreneurial firm. Acad. Manag. J. **25**, 465–499 (1982)
Mintzberg, H., Waters, J.A.: Researching the formation of strategies: the history of Canadian Lady, 1939-1976'. In: Lamb, R. (ed.) Competitive Strategic Management. Prentice-Hall, Englewood Cliffs, NJ (1984)
Mintzberg, H., Waters, J.A.: Of strategies, deliberate and emergent. Strateg. Manag. J. **6**(3), 257–272 (1985)
Mintzberg, H., Otis, S., Shamsie, J., Waters, J.A.: Strategy of design: a study of architects in co-partnership. In: Lamb, R. (ed.) Competitive Strategic Management. Prentice Hall, Englewood Cliffs, NJ (1986)
Porter, M.E.: Competitive Strategy: Techniques for Analyzing Industries and Competitors. Free Press, New York (1980)
Porter, M.E.: Competitive Advantage: Creating and Sustaining Superior Performance. Free Press, New York (1985)
Vanhaverbeke, W., Cloodt, M.: Theories of the firm and open innovation. In: Chesbrough, H.W., Vanhaverbeke, W., West, J. (eds.) New Frontiers in Open Innovation. Oxford University Press, Oxford, U.K. (2014)
Vanhaverbeke, W., Peeters, N.: Embracing innovation as strategy: corporate venturing, competence building and corporate strategy making. Creativity Innov. Manage. **14**(3), 246–257 (2005)
Vanhaverbeke, W., Roijakkers, N.: Enriching open innovation theory and practice by strengthening the relationship with strategic thinking. In: Pfeffermann, N., Minshall, T., Mortara, L. (eds.) Strategy and Communication for Innovation, pp. 15–25. Springer, Berlin (2013)
Weiblen, T., Chesbrough, H.W.: Engaging with startups to enhance corporate innovation. Calif. Manag. Rev. **57**(2), 66–90 (2015)
Whittington, R.: Completing the practice turn in strategy research. Organ. Stud. **27**(5), 613–634 (2006)
Whittington, R., Yakis-Douglas, B.: Strategic disclosure: the communication of strategy. In: Barnett, M., Pollock, T. (eds.) The Oxford Handbook of Corporate Reputation. Oxford University Press, Oxford (2011)
Whittington, R., Cailluet, L., Yakis-Douglas, B.: Opening strategy: evolution of a precarious profession. Br. J. Manag. **22**(3), 531–544 (2011)

Chapter 2
Setting the Stage for Creativity: Upstream, Mid-stream, and Downstream

Leigh Thompson and David Schonthal

Abstract The most common question we get as researchers and practitioners is: how to instigate creativity in teams and organizations? Some managers are focused downstream, on the development of a product; others are focused upstream—such as how to start thinking big. However, managers often forget about the mid-stream process. In this chapter, we combine insights from scientific research with actual examples and case studies to identify specific stage-setting processes that leaders and managers need to take in order to keep the raft moving through the mid-stream rapids and get downstream, and finally ashore. We point out the key challenges at each point in the creative river and how to leverage the talents of the team and organization.

2.1 Introduction

The most common question we get as researchers and practitioners in the area of creativity and innovation is: *how to instigate creativity in the team and organization*? Most people have a particular point of view when they ask that question. Some are focused downstream—such as how to quickly get a product to market. This is often the key to survival for many startups. Yet, others are focused upstream—such as how to even start a conversation or think big. This is particularly true for established organizations looking to appeal to new segments and evolve business models in the face of mounting competition. However, organizations often forget about the *mid-stream* process—this is where managers and teams can stall-out, get lazy, and operate on automatic pilot. As it turns out, there are specific stage-setting processes that leaders and managers need to take to keep the raft moving from upstream through the mid-stream rapids and get downstream, and finally ashore. In this chapter, we combine insights from scientific research with actual examples and

L. Thompson (✉) · D. Schonthal
Kellogg School of Management, Northwestern University,
Evanston, IL 60208, USA
e-mail: leighthompson@kellogg.northwestern.edu

N. Pfeffermann and J. Gould (eds.), *Strategy and Communication for Innovation*,
DOI 10.1007/978-3-319-49542-2_2

case studies. We point out the key challenges at each point in the creative river and how to leverage the talents of the team and organization.

The upstream portion is usually referred to as ideation—the generation of ideas that are ultimately novel and useful. This is a two-part criterion that may not be obvious at the time. The downstream portion is usually referred to as innovation—the realization of actual ideas in the form of products and services (Thompson and Choi 2006). Often, much greater weight or emphasis is put on innovation. For example, how many times have you heard that "*in business, success is 5% strategy; 95% execution?*"

However, everything starts with an idea. For example at Ideo Design Firm, it is stated that "you start with your head in the clouds (ideation) and you end up with your feet on the ground (innovation)" (Kelley and Littman 2005). In David's experience as a Director of Business Design at IDEO, a global design and innovation consultancy headquartered in Palo Alto, teams must resist their natural temptation to tame a wild idea. It may not be the outrageous first idea that ultimately makes it to market, but these concepts often point teams in a novel direction that can lead to the commercialization of a product that is desirable, feasible, and viable. For example, one IDEO team aiming to develop a pain-free drug injection device began with the idea of inventing a "mechanical mosquito" that would quickly deliver medication when it bit the patient. Not a feasible idea unto itself, but it ultimately led to a device that looked like a mechanical mosquito. Had the mechanical mosquito idea been "tamed" too early, this device may never have been designed.

Upstream Creativity

By "upstream," we refer to the beginning of the creative process. At this point, the team is on shore, getting ready to embark. The key mistake that leaders and teams make here is that they often never get past the first idea. We refer to this as *idea myopia*. One of us observed a company brainstorming session, and observed that the group never got past the first idea. In this particular case, a pharmaceutical company was looking to develop a technology to drive greater patient adherence to a drug regimen. The initial idea offered during the group brainstorm was an app that provided the patient with reminders to take their medication at certain times. Once the group was anchored in thinking about "apps" each idea that followed as also app-based—significantly limiting the universe of opportunity to address the underlying problem (e.g., smart pill bottles, service-based innovations, time-release tablets, etc.)

There are several social-cognitive factors that conspire to produce idea myopia. In our previous research, we have referred to this as the "4 horsemen of the apocalypse" to try to make the dramatic point that if the group is only working with one or two ideas, then the chance of crafting an exciting end product is severely curtailed.

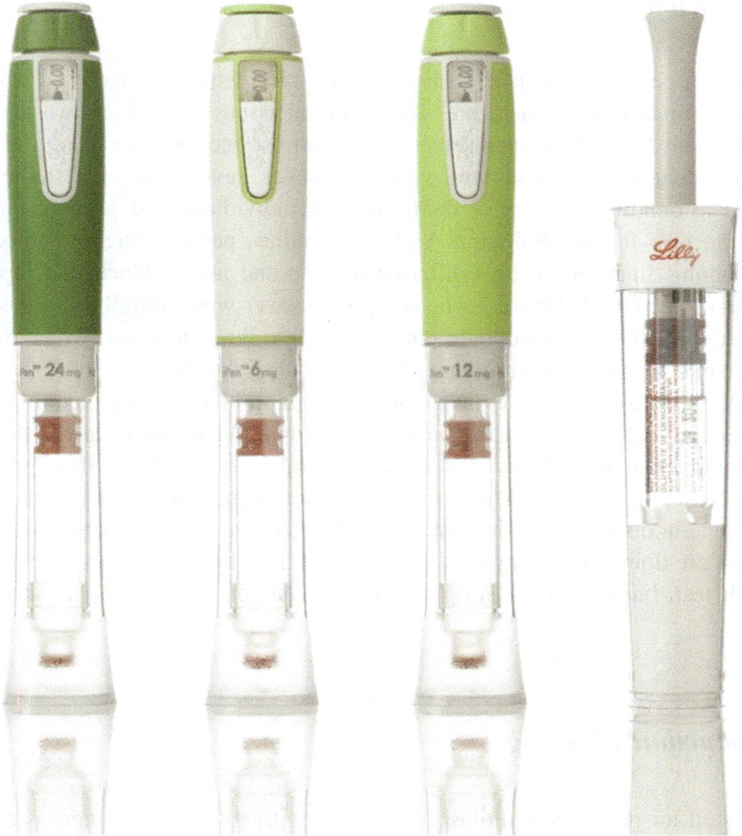

2.1.1 Social Loafing

When people loaf, they are not contributing as much as they can or working as hard as they could. Social loafing leads to the fact that in groups, the whole is often less than the sum of the parts (Karau and Williams 1993). In business situations, this is referred to the 80–20 rule, which means that 20% of the people do 80% of the work. To be sure, most of the time, people are not consciously aware that they are not working as hard (as they could); rather they have a self-serving bias to view their own contributions as more impactful (than they actually are). Psychologists and management scientists have identified some strategies to reduce social loafing and increase social striving—which is essentially the polar opposite phenomenon—people work harder in a team than they do individually because they are identified with the team (Ferguson 1989). Unfortunately, social loafing is much more common than social striving. In our research, for example, when we tell people that their individual contributions to a group effort will be monitored, we greatly increase creative output. Thus, accountability is a key factor (Bouchard 1972).

2.1.2 Conformity

This occurs when people bring their behavior in line with what they feel will win them acceptance in a group. There are some cultural issues here. Most western people do not want to describe themselves as conformist because they take great pride in their individuality as a cultural value. However, several research studies have revealed astounding degrees of conformity in individuals and groups. For example, in the now-famous Solomon Asch line studies, people were presented with three stimulus lines of (wildly) differing lengths and asked which one best matched a "test" line (Asch 1956). To be sure, the answer was painfully obvious. Anybody with even 20–60 vision could choose the right line. However, when the group participants heard other group members choosing the "wrong line," they conformed. Why? According to the theory of normative influence, people conform because they *want to be liked* and they *want to be right*. What this means is that people are smart enough to know that when they agree with others, they gain immediate social approval. And it also means that sometimes, we think we have gone, well insane in our own thinking, and so the best course of action is to simply do what others are doing. This may be a good instinct in a fire emergency or even taking a math test, but it is a catastrophic course of action to take in a creative enterprise.

2.1.3 Production Blocking

It is very difficult for people to multitask. In fact, cognitively speaking, there is no such thing as multitasking; rather there is a *rapid switching of attention* that hinders creative idea generation. In a group situation, production blocking occurs in two key ways: first, people must take turns speaking, and this conversational turn-taking means that people must engage in the politeness ritual—and essentially lose their own train of thought while listening to others (Mikula and Schwinger 1978). A second way it occurs is that sometimes, one person in the group is taking notes or recording the group's ideas—the self-appointed scribe—and this person is essentially taking dictation, which means that they cannot fully engage in the creative idea generation process. The advent of brainwriting was developed as an antidote to the production-blocking process.

2.1.4 Performance Matching

People's level of productivity in groups quickly converges toward an average. Consider the following pejorative expressions: *tyranny of the average, lowest-common denominator, regression toward the mean, and the like*. Essentially, these processes all describe the tendency for high performers to dumb themselves down in groups and for the group to descend into mediocrity. Again, this is not something that people are necessarily aware of at a conscious level, but preconsciously, it is happening.

We have described a rather depressing litany of factors that can dramatically curtail the group's best upstream effort, so the question is can the manager or leader take any proactive steps? The answer is yes. The question is how.

2.1.5 Set Norms

It is a fact that groups who are left to their own devices end up with the most dysfunctional norms. In the absence of clear norms, two things happen: First, the least socially aware people set bad norms (e.g., coming late to meetings, interrupting people, etc.) and the most socially aware people constrain themselves—in an effort to try to determine the norms of behavior. Nearly any norm or rule can enhance the performance of creative groups. For example, Paulus and his colleagues found that by introducing the following rules, there was a marked increase in creative idea generation: "stay focused," "don't tell stories," "encourage others to contribute," "keep people talking," and "remember not to criticize" (Paulus et al. 2006).

2.1.6 Enforce Rules

Though it seems contrary to the idea of free-spirited creativity, constraints and rules are important (in fact desired) by designers and innovators. Rules can be as basic as "one conversation at a time," or can guide creative constraint, such as focusing a brainstorm on generating ideas that can likely get to market in the next 12 weeks. In either case, it is up to the leader to keep sessions and conversations focused, making sure social contacts and design constraints are respected.

2.1.7 Keep the Group Small

In our research, we find that groups are too big. And they are growing in size. Unfortunately, there is a steady, linear decline in the number of per-person ideas generated as group size increases. For example, in our own research we found that teams of size three generated over 50% more ideas in a brainstorming task than did teams of size six or more.

2.1.8 Select a Diverse Team

We are mentally programmed to work with people who are, well, too much like ourselves. Cognitive diversity increases creativity. For example, Kevin Dunbar found that laboratories in US and UK that had diverse hiring practices generated more patents (Dunbar 1997). And one study found that diversity in cognitive styles (ready-fire-aim versus analysis-paralysis) make a good combination (even though it feels maddening) (Mauro et al. 2009).

At IDEO, project teams, are in large part, assembled based on diversity. The firm employs "designers" from a variety of disciplines and backgrounds (anthologists, psychologists, engineers, business people, visual artists etc.), and on each project, teams are curated to make sure that a variety of points of view are brought to bear beyond just design discipline.

2.1.9 Don't Waste Your Time Trying to Recruit "Creative" People

There is no such thing as a 'creative" person. Rather, some people are extremely talented in their chosen domain—speaking, marketing, engineering, etc. Find people who love what they do and make them work with different kinds of people.

2.1.10 Start with the Right Question

Oftentimes brainstorms and other ideation formats yield uninspiring results because they begin with asking the wrong question. This usually takes the form of something too narrow with limited degrees of creative freedom, or too broad resulting in ideators being paralyzed by possibility. At IDEO, teams begin the generative processes by starting questions that are broad enough to inspire lots of ideas, but narrow enough to keep participants focused on the problem at hand. An example of a bad question on which to brainstorm might be: "how can we create a better tasting toothpaste for kids?" The answer is: create a better tasting toothpaste! A better question on which to start ideation is "How might we inspire kids to be more excited about brushing their teeth?". This question is narrow enough to keep a team on target, but broad enough to inspire a range of potential solutions (some of which may not be toothpaste at all!).

2.1.11 *Look at the Calendar*

Don't schedule long meetings with a lot of down time in between, rather, schedule frequent, short meetings. Groups work to fill the time allotted; if you have a 2-h meeting scheduled, you will accomplish no more than if you have a 1-h meeting scheduled. Indeed one investigation found that teams are more productive when they have two, 15-min brainstorming sessions versus one 30-min session; indeed about 75% of ideas emerge in the first 15 min of a 30 min brainstorming session (Howard et al. 2010).

2.2 Mid-stream Creativity

If you have come this far, you are mid-stream in the creative process. You have pulled the small, diverse group together, set some norms and are ready to enforce the rules. Mid-stream is where leaders and managers can get lazy; they often feel that they've put in their time, recruiting the ideal team and then they want to take their seat in the stands and watch the magic show. For example, in one large company we worked with the leader of a business unit hired an outside design consultancy to create a "game changing new product offer" for her core customer. Once she had signed the contract, she assumed that she had "ticked the box" on innovation. As a result, she was absent from key moments in the innovation process (user research, synthesis, ideation) and relied on the consultants and her team to manage the process to success. What she failed to realize was that though she may have invested in the *project*, she was not invested in the *process*—and as a result never felt ownership responsibility for driving the new ideas forward within the organization. The outcome of the project was a couple of novel product solutions that sat on a shelf in prototype form waiting for someone at a senior level to feel passionate and invested enough to make them real. We find that feeling "ownership" for new ideas is something that only truly comes from full participation in the innovation process. It's very hard to develop entrepreneurial passion for a solution one has no part in creating.

Leaders need to be very proactive and hands on during the mid-stream. There are a few rules of thumb that are useful here:

Don't run the "control group" meeting
The control group meeting is just like it sounds. It is the meeting in which no steps are taken to direct or guide people. It is a placebo meeting. The data are in: Brainstorming is better than control group; and brainwriting beats brainstorming. Let's consider the research. In one investigation, the performance of brainstorming groups was compared to control groups—who were not given specific instructions. The quantity and quality of each groups' output was compared. Hands down, the groups who followed brainstorming rules were more creative (Parnes and Meadow

1959). The rules for brainstorming are fourfold: expressiveness, quantity, no judgment or evaluation, and building and leveraging.

Brainwriting

This is a technique in which group members individually scribe ideas for a fixed amount of time. During this time, there is no cross talk between group members. We like to enforce two rules during this time: *no guessing* and *no confessions*. No guessing means that group members should not attempt to judge who-said-what. Instead, it is meritocracy of ideas, not a popularity context. No confessions means that no group members should not reveal what their ideas were. After the group members individually generate ideas (for say, 5–10 min), we like to post the ideas. Group members should vote for the ideas they find the most exciting.

One process that works well for merit-based idea selection is "dot-voting". All ideas are posted on a wall, and participants are given three colored dot stickers with which to vote on their favorite ideas. Criteria for selection are clearly articulated (e.g., desirability, feasibility and viability) and then each person will use their dots to select their favorites. To add some additional weight, participants may use all three dots on one idea if they feel like it is the standout favorite, or distribute their dots more evenly across several ideas if they feel there is more parity. This type of clustering can help tease out clear first-round "winners" from the broader field far more efficiently than some other voting methods.

Draw, don't write

Writing is great for a Ph.D. thesis, but drawing quickly gets a point across. This is a golden rule at IDEO. When participants think visually they inspire other team members to build on their ideas. It's pretty awkward to rewrite someone else's sentence during a brainstorm, but it's easy to add a twist to someone else's drawing. For example, I may think a good idea for getting kids to brush their teeth an automatic brush that reminds kids when it's time to brush their teeth. Somebody else may think that a cool build on that idea is to connect the toothbrush to the internet so that parents can be notified when their kids are brushing. To expand on the idea, the team member can add a "little" Wi-Fi icon to the drawing. Imagine how much time (and space on a post-it) it would take to describe the combined idea through prose!? Pictures are truly worth a thousand words.

Focus on quantity not quality

The data are in. Groups that focus on quantity outperform groups that focus only on quality; they outperform groups that focus on quantity and quality and groups who don't have a focus (Paulus et al. 2011). And it is far better to have a specific, difficult quantity goal than an amorphous or vague quantity goal. As Nobel Prize winner Linus Pauling is famous for saying, "The way to get good ideas is to get lots of ideas and throw the bad ones away. You aren't going to have good ideas unless you have lots of ideas and some sort of principle selection" (as quoted in Barnett 2002).

Engage in short sprints, not long-distance endurance
Creative muscles, like physical muscles, can become fatigued. As a result design firms and other creative organizations use "sprints" and "charrettes" to get the most of our of creative teams before idea lethargy sets in.

Standup (vs. sit down)
They say that sitting is the new smoking. Sitting hurts creative energy as well. It is for this reason, that many managers are scheduling walking meetings, as opposed to sitting meetings. For example, evolutionary biologist Andrew Parker applied designs from his walks through the Australian outback to solve problems in engineering, materials science, and medicine. In one investigation, the iridescence in butterflies and antireflective coatings in moth eyes led to the development of brighter cellular phone screens and an anticounterfeiting technique (Mueller 2009).

Debate versus Demoralize
For years (make that decades) people have been strictly cautioned to not challenge or criticize others ideas. However, by eliminating all debate, then all deviance is eliminated too. Teams in which a single member proposes an unusual or even incorrect solution outperform teams in which no such "deviance" occurs. And, teams instructed to "debate" are more creative than teams instructed to "brainstorm" (Nemeth and Ormiston 2007)! Furthermore, these performance advantages generalize to subsequent, unrelated tasks, even when the vocal, cognitively deviant member is not present (Smith et al. 1996). A study of 71 IT project teams revealed that task conflict had curvilinear effect on creativity, with creativity highest at moderate levels of task conflict (Farh et al. 2010).

Recent research indicates that you can mobilize the creative process by engaging in debate, rather than by self-censoring all doubts. Debate is depersonalized criticism and is fact-based. In contrast, demoralizing others, also called "jeer pressure," is personalized and hinders creativity. In one study, groups who debated one another were more creative than did those who did not debate (Personnaz et al. 2004). At IDEO, a method often used to give feedback in a non-demoralizing way is to offer critique with the phrases "I like" and "I wish." This reminds us that we are evaluating the idea, not the ideator.

Encourage outlandish ideas
Deliberately encourage outlandish, politically incorrect, outrageous ideas; these can often be the seeds for workable ideas (recall the mechanical mosquito!). It is not enough to simply encourage the group to "not be timid," rather must behaviorally engage in this. One thing we do is pre-brainstorming warm up in which group members each share the most embarrassing thing that has happened to them in the last month (Wilson et al. 2016).

Downstream Creativity
An excellent sign of a creative trip down the river is the presence of a lot of ideas. Groups often feel a need to select one or two of those ideas for further development. One issue that groups face is how to select among a plethora of ideas. If groups are successful in generating a lot of ideas, how do they best select among those ideas?

As it turns out, idea selection is best done individually, not within groups. For the simple reason of groupthink, which is defined as "excessive like-mindedness." Idea selection is also another great time to introduce constraints. Asking questions such as "which of these ideas feel like the best fit with our core competence?" and "which ideas feel like they are the most inexpensive to test?" are excellent ways to focus attention and keep participants thinking about the business goal behind the creative process.

Groups may select "safe" ideas that represent the tyranny of the average; least offensive to everyone. The research is pretty clear on idea selection: groups are not as good at recognizing creative ideas as are individuals. So, a better practice is to allow individuals to choose the idea. The second mistake groups make is that they choose ideas prematurely, in the absence of evidence. A far better practice is to conduct an evidence-driven experiment to test ideas.

One of the scenes that breaks our hearts is looking at a conference or meeting room after a day-long brainstorming session and seeing the flipcharts, post-its, party horns, and all the paraphernalia and then realizing that all the ideas have stayed in the room to die a quiet death. To prevent this quiet death, we advise that teams and organizations create a boneyard (place where ideas are archived for future use). Note: Mattel Media got an idea for an application of hotwheel toys by digging through the boneyard (Thompson 2003). In addition, a 1-2-3-4 plan should be put in place: what will take place in next week (1)? In two weeks? In 3 weeks and a month out, all members need to conference with one another.

Going Ashore

It is not practical to always be "on" when it comes to the creative enterprise. Just like the athlete, who trains for an event, and then has some recovery period and some training time, leaders and managers need to do similar when in the creative performance mode. Indeed groups who take a short break (2–5 min in length) halfway through a 20- or 30-min brainstorming session increase their productivity following the break compared to teams that brainstorm continuously without a break (Mitchell 1998). Breaks also allow brainstorming groups to overcome mental blocks through the process of incubation. Breaks can stimulate a different approach to a problem. In one investigation, individuals and three person groups attempted to solve sets of rebus puzzles; following incubation (taking a break from the problem), groups improved (Smith et al. 2010).

2.3 Conclusion

We have examined creativity at the upstream, mid-stream, and downstream points. Too often well-meaning leaders and organizations become preoccupied with innovation and outcomes and they sabotage the ideation process. We have made the argument that ideation, or the generation of a high volume of ideas without concern for utility or even uniqueness, opens the door for the generation of truly creative ideas that are novel and useful. We have deliberately set aside the allure of quality and put our bet on quantity. There are several steps that managers and leaders can take to encourage idea generation in their groups and teams.

Bibliography

Asch, S.E.: Studies of independence and conformity: I. A minority of one against a unanimous majority. Psychol. Monogr.: General Appl. **70**(9), 1 (1956)

Barnett, C.: The Verdict of Peace. Pan Books, London (2002)

Bouchard, T.J.: Training, motivation, and personality as determinants of the effectiveness of brainstorming groups and individuals. J. Appl. Psychol. **56**(4), 324–331 (1972)

Dunbar, K.: How scientists think: online creativity and conceptual change in science. In: Ward, T. B., Smith, S.M., Vaid, S. (eds.) Creative Thought: An Investigation of Conceptual Structures and Processes, pp. 461–493. American Psychological Association, Washington, DC (1997)

Farh, J.L., Lee, C., Farh, I.C.: Task conflict and team creativity: a question of how much and when. J. Appl. Psychol. **95**(6), 1173–1180 (2010)

Ferguson, E.D.: Adler's motivational theory: an historical perspective on belonging and the fundamental human striving. Individ. Psychol. **45**(3), 354 (1989)

Howard, Dekonick, Culley: The use of creative stimuli at early stages of industrial product innovation **21**(4), 263–274 (2010)

Karau, S.J., Williams, K.D.: Social loafing: a meta-analytic review and theoretical integration. J. Pers. Social Psychol. **65**, 681–706 (1993)

Kelley, T., Littman, J.: The Ten Faces of Innovation: IDEO's Strategies for Beating the Devil's Advocate & Driving Creativity Throughout your Organisation. Ransom House, New York (2005)

Mauro, R., Pierro, A., Mannetti, L., Higgins, T., Kruglanski, A.: The perfect mix: regulatory complementarity and the speed-accuracy balance in group performance. Psychol. Sci. **20**(6), 681–685 (2009)

Mikula, G., Schwinger, T.: Intermember relations and reward allocation: theoretical considerations of affects. In: Brandstatter, H., Davis, H., Schuler, H. (eds.) Dynamics of Group Decisions. Sage, Beverly Hills, CA (1978)

Mitchell, C.K.: The effect of break task on performance during a second session of brainstorming. Masters thesis, University of Texas at Arlington (1998)

Mueller, T.: Biomimetrics: design by nature. National Geographic. nationalgeographic.com. Apr 2009

Nemeth, C.J., Ormiston, M.: Creative idea generation: harmony versus stimulation. Eur. J. Social Psychol. **37**(3), 524–535 (2007)

Parnes, S.J., Meadow, A.: Effect of "brainstorming" instructions on creative problem-solving by trained and untrained subjects. J. Educ. Psychol. **50**(4), 171–176 (1959)

Paulus, P.B., Nakui, T., Brown, V.R., Putman, V.L.: Effects of task instructions and brief breaks on brainstorming. Group Dyn.: Theory, Res. Pract. **10**(3), 206–219 (2006)

Paulus, P., Kohn, N., Arditti, L.: Effects of quantity and quality instructions on brainstorming. J. Creative Behav. **45**(1), 38–46 (2011)

Personnaz, M., Personnaz, B., Goncalo, J.A.: The liberating role of conflict in group creativity: a study in two countries. Eur. J. Social Psychol. **34**(4), 365–374 (2004)

Smith, C.M., Tindale, R.S., Dugoni, B.L.: Minority and majority influence in freely interacting groups: qualitative vs. quantitative differences. Br. J. Soc. Psychol. **35**(1), 137–149 (1996)

Smith, C.M., Bushouse, E., Lord, J.: Individual and group performance on insight problems: the effects of experimentally induced fixation. Group Process. Intergroup Relat. **13**(1), 91–99 (2010)

Thompson, L.L.: Improving the creativity of organizational work groups. Acad. Manag. Executive **17**(1), 96–109 (2003)

Wilson, E.R., Thompson, L., Lucas, B.: Embarrassment versus pride and creative idea generation (2016)

Thompson, L.L., Choi, H.S. (eds.): Creativity and innovation in organizational teams. Psychology Press, Routledge (2006)

Chapter 3
When Business Model Meets Open Innovation

Howard H. Yu

Abstract This chapter will first examine why open innovation is increasingly common. Rather than a mere spread of management know-how across the field, I argue that it is the increasing connectivity that results in plummeting transaction cost, which in turn, has driven the widespread adoption. I will then examine what principles an organization must adhere when building the collaboration platform for open innovation before reaping maximum payoff. The analysis leads to testable propositions for future researches.

3.1 Introduction

In August 2011, Chinese mobile phone maker Xiaomi (pronounced SHAO-mee) created a media stir when it unveiled its first smartphone. Pacing the stage in a Steve Jobs-like iconic outfit—blue jeans, dark shirt, and sneakers—Lei Jun, Xiaomi's founder and CEO, made a shock announcement. The company's first smartphone, Xiaomi Mi1, equipped with a 1.5G dual core processor, 4-inch touch screen, 900 min of talk time and an 8 megapixel camera, would sell for RMB 1999 (US$320), almost half the price of a comparable Samsung model. In a country that produces copycat versions of everything from Nike shoes to Gucci handbags to Disney theme parks, the outside world viewed Xiaomi with skepticism. The *New York Times* referred to Lei Jun as a "knockoff of one of the gods of American ingenuity: Steven P. Jobs" (Barboza 2013).

But the strategy seemed to work. By 2012, the firm had sold $2 billion worth of phones and was aiming to double sales the following year. In 2014 Xiaomi raised $1.5 billion in an investment round, pushing the company's valuation to a staggering $40 billion, exceeding the combined value of Sony and Lenovo. By 2015, it commanded the largest market share of mobile phones sold in China, larger even

H.H. Yu (✉)
IMD—International Institute for Management Development, Chemin de Bellerive 23, 1003 Lausanne, Switzerland
e-mail: howard.yu@imd.org

N. Pfeffermann and J. Gould (eds.), *Strategy and Communication for Innovation*, DOI 10.1007/978-3-319-49542-2_3

than Samsung and Apple. In less than five years, a mobile start-up had become China's largest phone maker.

At the heart of a Xiaomi phone is its preinstalled Android-based operating system (OS), MIUI, which allows consumers to customize the user interface (UI) and enables advanced users to tweak the otherwise hard-coded firmware of their handsets.

For Lei Jun, Xiaomi's success lies in its ability to rely on its "hardcore fans" to generate word-of-mouth marketing. Every week, the company releases a new version of MIUI that it co-develops with leading users, responding to their feedback via the Internet and online user forums, a feature that resonates especially with young Chinese. By outsourcing the design and features of its OS to an online fan base, Xiaomi can lower the cost of software development and keep its end users excited. "When Apple develops its iOS 7, you have no idea what they will do with it before the release. It's not like that for us. We will first ask what you want. [...] Xiaomi is not selling a product, but an opportunity to participate," explained Lei Jun (Kan 2013). "Steve Jobs told everyone Apple's OS changed once a year. Now I am telling my customers, 'It's really no big deal; we can change your OS every week.'"

Xiaomi—which means "little rice" in Chinese—does not seem all that small now. How can its sudden success be explained and to what extent does its business model represent an increasing acceptance of open innovation in industries?

3.2 What Is Open Innovation

In his pioneering work, Chesbrough (2006) defines open innovation as a paradigm that "assumes that firms can and should use external ideas as well as internal ideas, and internal and external paths to market, as the firms look to advance their technology." The basic idea is to bring in external ideas, technologies, and resources to augment internal development in order to accelerate the commercialization of products and services. In this regard, the concept is not new. The literature on strategic alliance has a long history of examining how companies can forge research relationships with scientific communities: universities, contracted research organizations, individual scientists, or end users (von Hippel 1988). What is new, however, is the context in which collaboration is facilitated and managed. Once commonly a highly exclusive collaboration between a few parties, open innovation now has a far wider scope (Majchrzak and Malhotra 2013; West and Borgers 2014). As exemplified by the case of Xiaomi, product ideas are often generated by a large number of individuals outside organizations; ideas are systemically incorporated into the product pipeline on a continuous basis. In short, organizations have changed not only the way product ideas are sourced but also the way products are commercialized and marketed, or in industry parlance, there has been a shift in the business model. What has spurred this radical shift?

This chapter sets out to examine: What enables companies to increasingly embrace open innovation with external partners in a way that is ever widening in scope? What are the benefits? What changes do they need to make in terms of resources, processes and value propositions?

3.2.1 Plummeting Transaction Costs

In the "Nature of the Firm," Ronald Coase (1937) asks why large firms exist in a modern market economy. If markets are so good at allocating resources, as free-market economists usually suggest, why are there "islands of conscious power in this ocean of unconscious co-operation like lumps of butter coagulating in a pail of buttermilk." The answer lies in transaction cost (Williamson 1981). There are tremendous benefits to internalizing market transactions into administrative coordination within the firm, including lower costs and improved efficiencies. Historian Alfred Chandler (1977) describes managerial coordination, which accounts for the rise of big corporations at the turn of the century, as the visible hand. To capitalize on the rise of affluent consumers, the arrival of mass transportation and logistics networks, and the dawn of modern telecommunications, organizations gobbled up small players, expanded their operations across every step of the production process, and became vertically integrated enterprises. The day-to-day activities of countless managers, directed by strategy and reinforced by the structures of the firm, have spurred the growth of the American economy. For the same reason, pioneers of numerous industries were once highly vertically integrated players: IBM in computing, Motorola in mobile phones, Marriott in hospitality, and inevitably, the Yellow Cab Company in taxis.

But the world has changed. Ubiquitous connectivity is commonplace today, resulting in the plummeting of transaction costs (Rifkin 2014).

Ever since the advent of semiconductors, computers have become ever more powerful, with computing power doubling approximately every 18 months—a phenomenon known as "Moore's Law" after Intel co-founder Gordon Moore. The total computing power of all the electronic equipment and calculating devices that NASA used in the spacecraft for the Apollo moon mission in 1969 is packed into a single iPhone. There has also been a giant leap in data storage technology and connectivity bandwidth, leading to increased capacity, higher performance, more portable phones, and lower costs. These developments have occurred so rapidly that our society has barely come to grips with the breathtaking progression: ubiquitous connectivity, connectivity between humans, connectivity between humans and things, and connectivity between things—or as some industry experts say, "the internet of everything."

As the economy continues its march toward zero margin costs, we see new models that have relied heavily on open innovation, facilitated by coordination across organizations and individuals. Xiaomi's ability to outsource its software development to millions of hard-core fans is enabled by the seamless communication of its online communities where members constantly discuss, explore and act upon promising ideas. Airbnb is able to mobilize millions of homeowners to turn their apartments and houses into vacation rentals by identifying vacancies in any neighborhood and providing relevant information to potential visitors, such as online reviews and photos. Uber is able to turn any car owner into a temporary cab driver thanks to the high accuracy of the real-time GPS system and the geolocation function that finds available vehicles and links potential customers to specific drivers at the right time and place.

To formalize this observation, here is proposition #1.

The lower the market transaction cost, the higher the tendency of companies to adopt the practice of open innovation.

However, while the lowering of transaction costs in the marketplace is necessary for a firm to benefit from open innovation, it is not enough on its own. A company cannot naïvely post an open call on the web and then expect unsolicited solutions to emerge automatically from a diverse range of individuals. Organizational problems are often domain-specific and can only be resolved by managers with a wealth of requisite experience. To leverage external resources and to make collaboration feasible, narrowly defined organizational objectives, which are often internally oriented, must be transformed into a more general abstract problem statement. In other words, problems need to be clearly articulated and framed in a way that can be understood by external talents with different backgrounds (King and Lakhani 2013).

3.2.2 Enabling Tool: Modularity and Decontextualization

In October 2012, DARPA, the US Defense Advanced Research Project Agency, invited all American innovators to design a new amphibious infantry fighting vehicle—the Fast, Adaptable, Next-Generation Ground Vehicle (FANG). It called upon innovators of all stripes with expertise in designing and engineering drivetrain and mobility systems. The objective was to crowdsource the design of the next-generation fighting vehicle: a tank capable of traveling in water and seamlessly continuing ashore, making it more secure for Marines to transition from sea to land.

The risk involved in an exposed beachhead operation is so great that the Marines had not stormed a hostile beach since the Korean War. So the idea of developing an amphibious fighting vehicle had been top of mind among military experts for decades. The goal for DARPA was to build a vehicle capable of self-deploying from a home ship and delivering 17 Marines from a launch distance of at least 12 miles away from shore.

At the height of the debt-ceiling crisis in 2011, former defense secretary Robert Gates, under congressional pressure, promised to rein in the budget by saving $100 billion over the next five years and was forced to shut down the amphibious vehicle project, which had already cost the department $13 billion over 20 years. "The only entities able to contribute to the design of these important systems are corporations able to build multimillion dollar prototypes. That limits us to a handful of companies and maybe a few hundred brains," said Army Lt. Col. Nathan Wiedenman, DARPA program manager, pointing out the downside of over-relying on just a few defense contractors, such as Lockheed Martin and Boeing. "In a nation of 300 million people, we can do better."

To run the competition, DARPA with the help of Vanderbilt University set up an online portal that served as a collaborative work system. The portal consisted of a range of collaborative tools that sound like an all-you-can-eat candy store for engineering aficionados: qualitative reasoning, static constraint analysis, time-varying models using ordinary differential equations, detailed CAD, computational fluid dynamics, structural analysis using the finite element method, blast analysis using nonlinear FEA, manufacturability analysis, probabilistic analysis for uncertainty quantification, among others. "It's not about connecting just the people," observed Mark Gersh, a program manager at Lockheed Martin Advanced Technology Center in Palo Alto, California. "It's also about integrating the analysis capabilities with the tools, models, and simulations that integrated project teams use." Users were able to download the tools and explore the component model libraries to start putting their designs together, simulating the performance of thousands of potential mobility and drivetrain subsystems. As Wiedenman put it, "FANG is applying a radical approach to the design and manufacture of a military ground vehicle while seeking to engage innovators outside of the traditional defense industry."

Since the beginning of the first FANG challenge in January 2013, more than 1000 participants within 200 teams participated. The 15 finalist teams spent an average of 1200 h in total. Assuming a hypothetical hourly rate of $200, including administrative overheads, the collective design effort would have cost about $3.6 million, an absolute bargain compared to the cash prize of $1 million pre-tax payout.

The winner turned out to be a team of three people dispersed across the US, known as Ground Systems. While the design relied almost entirely on existing components in the library, spotting the right one still represented a significant amount of work. Ground Systems eventually settled on a design that traded automotive performance for manufacturing lead time by selecting components, including the main engine, that were available from the supply chain much faster—a decisive factor for its victory—speeding up manufacturing by a factor of two compared with the propositions of the second- and third-ranked teams.

"We really want to open the aperture to non-traditional design entities; folks who have the skills, but don't traditionally have a mechanism by which they can participate in the development of military vehicles," explained Wiedenman. "I can be a transmission engineer for a civilian firm who really knows my stuff, [and] I can log on, join and participate in a meaningful way."

The struggle to identify and locate the right talent to solve specific problems has long plagued all big science endeavors (Guinan et al. 2013). Like DARPA, NASA —National Aeronautics and Space Administration—has struggled for years to develop an algorithm to predict a solar particle event, or what is commonly known as the solar flare. Solar flares are powerful bursts of radiation that can affect crew members on space missions, and disturb GPS, satellite, and other radio systems. A significant solar flare-up can also trigger a life-threatening geomagnetic storm. Decades of research efforts have resulted in predictive models that give reliable warning only up to two hours in advance. What scientists need is an algorithm that can accurately predict an event at least 4–24 h in advance, with 50% reliability.

In 2009, when NASA partnered with InnoCentive, an online clearinghouse for scientific problems and problem solvers, over 2800 solvers from over 80 countries participated in the open challenge. The winning solution was a predictive model that provided an accurate reading up to eight hours in advance, with 85% accuracy rate, well beyond NASA's expectation by orders of magnitude. And to everyone's surprise, the solution came from a semi-retired radio frequency engineer living in New Hampshire, who was subsequently paid a modest $30,000 for his ground-breaking insight. For NASA, the realization of the wisdom of the crowd is tantamount to "a deep philosophy change." "Your main responsibility is [now] to seek for solutions and they may come from the lab, from open innovation, or from collaboration, you should not care! You are the solution seeker!" concluded a NASA engineer. More than cost saving, crowd sourcing also allows organizations to get help from creative geniuses whom big companies would otherwise miss (Malhotra and Majchrzak 2014). For example, through the platform Connect and Develop, Procter and Gamble tapped into external expertise from university to industry researchers, from supplier networks to customers (Huston and Sakkab 2006).

To formalize this observation, here is proposition #2.

> The more diverse the background of the crowd, holding the number of participants constant, the better the quality of the final result.

To harness the wisdom of the crowd effectively, however, participants cannot simply be told about the technical problem and then sent home to work it out. The famed Gates Foundation has advanced science on the health challenges of the poor by establishing "Grand Challenges." These challenges target previously intractable problems, encouraging scientists to collaborate in a single place where "half-baked" solutions and partial advances are accessible across the entire scientific community (McGahan 2013).

DARPA, as we saw, went much further. In the FANG challenge, the competing teams were given constant feedback on their designs which they could then resubmit; the winner was chosen based on the final score against the requirements for system performance and manufacturability. Critical to making crowdsourcing work was the creation of an online platform that spanned the entire design process, thereby allowing contestants access to DARPA's component model library, and virtually simulating and validating a specific design. It was this online platform that decontextualized a military-related question from a generic engineering problem. Participants did not need to know every detail of a military deployment. The requirements of such an operation were already embedded in the online simulation, built into the system as objective criteria. All contestants needed to focus upon was making a creative tradeoff, attaining the kind of optimization that no one had previously thought about. As one contestant commented, "[Just like] … all multidisciplinary design spaces, this was an awesome experience and has my team reconsidering our **tool sets** [*emphasis added*] for our day jobs. I think it was a great experience and I really enjoyed working the FANG project."

Naturally, providing toolkits for customers to develop new product ideas often necessities sharing proprietary intellectual property (IP). Nvidia, a leading computer chip maker, creates a few "open" modules or developer zones that anyone can access, supporting customization and user modification (Henkel et al. 2013). It shows that companies with a strong IP policy do not have to cripple open innovation (Alexy et al. 2009). Enabling tools such as application programming interfaces (APIs)[1] and software development kits (SDKs) helps encourage external developers to increase a product's functionality (Kohler 2015). And more importantly, as seen with DARPA, with the right tools in hand and basic building blocks available, a non-military engineer can suddenly rival a system expert at Lockheed Martin.

Xiaomi CEO Lei Jun was, in many ways, practicing just that. Xiaomi's competitive advantage is neither manufacturing—since it has always outsourced everything from components to product assembly—nor software development—for it has no more than a fraction of Apple's programmers. The problem is modularized through the fragmentation of the value chain. All Xiaomi has done is to commercialize the MIUI user interface, which allows non-software experts to tweak the product's features without having to understand how the company works internally.

By completely decontextualizing the engineering problem from an organization-specific problem, Xiaomi fulfills a popular saying in China: Three smelly cobblers (with their wits combined) equal one Zhuge Liang—a chancellor in 200 AD widely recognized as the greatest and most accomplished strategist of his time. Xiaomi's greatest achievement thus lies in the cultivation of an online following who relentlessly contribute to new phone features in creative ways, making the company's mobile phone a continuous delight for its fan base; and thereby catapulting the Beijing start-up into the world's third largest mobile phone maker in less than four years.

To formalize this observation, here is proposition #3.

> The more an organization-specific problem is decontextualized, the better quality the final result.

---

[1]API are interfaces that allow other computer programs to make requests for services or exchange data.

3.2.3 *What Motivates Participation?*

At some level, we all know that salary alone will not sufficiently motivate people. This is particularly true for individuals who decide to respond to a challenge that has been posted online (Boudreau and Lakhani 2009). Some research by economists suggests that monetary incentives can sometimes backfire. Offering incentives for blood donation, for example, can offend people who want to give just for the sake of giving. Wikipedia has been more successful than Encyclopedia Britannica not because it compensates its writers and editors better. It pays its contributors nothing. Yet shortly after Wikipedia was founded, tens of thousands of people joined in, many contributing 20–30 h per week without remuneration. Anyone can access what is available online and make further comment. It was estimated in 2008 that 100 million hours were spent on the project. At the time of writing, some 31,000 "Wikipedians" have been identified as active editors, spending on average of one hour per day editing, seven days a week. The top 20% spend more than three hours per day in addition to their day jobs. Factors, other than monetary reward, are clearly at play—what are they?

In a fascinating study, Mark Muraven, a psychology PhD at Case Western Reserve University at the time, came up with an experiment to understand how willpower operates: Why some people exhibit stamina when facing a tedious assignment and persist, while others quickly become distracted and give up? Is "grit" an inner capacity or something strongly influenced by the environment?

Muraven recruited 77 undergraduates and tempted them with cookies. They were asked to skip lunch. (Cookies and undergraduates remain the constant staple of psychology experiments.) As the students arrived in the windowless laboratory with empty stomachs, they were shown two bowls of food. One was filled with warm, freshly baked cookies, soft to the touch and sweet to smell. The other was heaped with cold, stale radishes in a pile. Without stating the true intention of the experiment, a researcher stepped in and told half of the participants to eat the cookies and ignore the radishes. For the cookie eaters, concentrating on the cookies did not feel like a task at all. They savored every moment of it. Then the second group was told to ignore the cookies and concentrate on eating the radishes.

After five minutes, the researcher returned and asked the two groups of participants to solve a puzzle that appeared easy but was actually impossible to solve. It involved tracing a geometric pattern in one go without lifting the pen off the page or repeating a line twice. Muraven wanted to find out whether willpower behaves more like a reservoir of resource or a personal disposition. Would the radish munchers, who had used a lot of willpower to resist the temptation of the sweet cookies, be so depleted of their willpower that they would no longer have the stamina to stay focused on solving the impossible puzzle?

The relaxed cookie eaters, some humming songs, attempted the puzzle repeatedly. They spent an average of 19 min before giving up. The radish eaters, however, looked agitated, showed palpable frustration, shifted restlessly in their seats and complained about the setup of the experiment. They quit after just 8 min, contributing 60% less time to the experiment than the cookie eaters did. One even bombarded the researcher with insults.

As it turns out, willpower can be depleted like any other resource—once it is exhausted, we can no longer focus on difficult tasks. We lose our patience and succumb to temptation of all kinds. After a long day at work, closing the accounting books and filing endless complicated expense reports, most of us end up slouching in front of the TV eating a tub of ice-cream. Going to the gym is for tomorrow. Let's relax tonight.

When Muraven became a professor at the University of Albany, he ran the same experiment with a new twist. After a new group of unsuspecting undergraduates had been asked to skip lunch, Muraven assigned the cold, stale radishes to the first group and asked them to ignore the warm, sweet cookies. This time, however, a kind-looking researcher came into discuss the purpose of the experiment and described how the research team was trying to understand the human ability to resist temptation. She asked the students nicely, expressing gratitude for their time and effort in advancing modern psychology and told them that they would have the opportunity to provide feedback on the setup of the experiment to the research team.

After much radish munching by participants, the researcher came back and asked the students to sit in front of a computer screen. Random numbers were flashed for 500 ms each. Every time a four followed a six, participants were instructed to hit the spacebar on the keyboard. It was a classic mundane task designed to measure people's ability to stay focused. The computer program lasted for 12 min. To everyone's surprise, this group of students was able to concentrate the entire time, despite their supposedly depleted reservoir of willpower.

The second group of students was subjected to the same treatment with one exception—they were not told purpose of the experiment. The researcher looked rushed and unconcerned, flipping through papers and instructing them in an abrupt manner not to touch the cookies. When the students were seated in front of the computer screen, they performed terribly during the test. They missed hitting the spacebar, despite explicit instructions, they complained of feeling tired and that the numbers were flashing too quickly. They were exhausted.

As Muraven later explained, "If people feel like it's a choice or something they enjoy because it helps someone else—it's much less taxing. If they feel like they are just following orders, their willpower muscle gets tired much faster." (Duhigg 2012).[2]

This conviction of helping others may well explain the Wikipedia phenomenon. Like the spreading of open-source software, altruism looms large (Zeitlyn 2003). What follows is the very personal remark of one Wikipedian, a young woman from Iraq named Ravan Jaafar Altaie:

---

[2]For more details, please refer to Duhigg (2012) for a succinct explanation of the series of fascinating experiments conducted by Professor Mark Muraven at the University of Albany.

At the beginning of my joining Wikipedia in 2008, I started many articles. One of them, I think it was about a lady called Mariam Nour. [...] I then later forget about it. Maybe two or three years later, I passed by this article. I was shocked... More than 100,000 people read this article. They used it, so they got their information from this article. They passed [on] this article, so you feel like you affected and influenced more than 100,000 people.[3]

And so the purpose of Wikipedia of "bringing free educational content to the world" has been fulfilled. This is in stark contrast to the popular belief that "gamification"—using entertaining methods—can suffice to engage employees and customers in creative ways (Robson et al. 2016). What Wikipedia has shown is that humans are capable of performing great feats without remuneration, even when they do not find the task particularly entertaining, as long as they feel a deep sense of purpose.

Of course, not everyone can be sufficiently motivated by a sense of purpose alone. For some people, the act of altruism and sensing that they are contributing to and cataloging the growing body of human knowledge is enough to justify spending countless hours on Wikipedia. For others, it is because Wikipedia gives them a social bragging right about what they have written online. We all want to look good in front of others and feel good about ourselves. There is nothing more powerful than disclosing a piece of interesting information unknown to others. This is why the water cooler is the place where secrets are exchanged and rumors sported. Somehow, we are hard-wired to dispense insider information in exchange for no more than a vague sense of admiration from our peers. We do it to gain social recognition and status (Boudreau and Lakhani 2013). Wikipedia essentially transforms the mundane task of writing and editing factual articles into a golden opportunity for anyone to tell anything to the entire world. In addition, top contributors receive recognition for the sheer effort they put in. They see how many times their articles have been viewed and their relative position in terms of the amount of entries that they made, all published in a league table for public examination. For a hard-core Wikipedian, this is a big deal.

To formalize this observation, here are propositions #4a and 4b.

Monetary rewards have an inverted U shape influence in attracting individuals to respond to a company challenge.

The more social recognition one receives when participating in an open challenge, the higher the response rate a company will encounter.

[3]http://www.youtube.com/watch?v=4E5T94ThevE.

3.3 Conclusions: Beyond the "Not Invented Here" Syndrome

"Without political savvy," Kanter (2004) writes, "no one can get a project beyond the proposal stage." The implementation of open innovation is often thwarted by organizational inertia that favors internal effort, characterized by a "not-invented-here" tendency (Lichtenthaler et al. 2011). Tackling this problem has been the subject of many studies (Whelan et al. 2011; Heimans and Timms 2014). This article, however, focuses on the external interface, namely how to enable a diverse pool of nonexperts to contribute to an organizational problem and motivate them to do so.

The decline in transaction costs as a result of ubiquitous connectivity has removed the traditional imperative for being vertically integrated. Problems-solving can be leveraged and outsourced to external parties more effectively. But to achieve such a lofty ambition, organizations must invest in toolkits and resources that effectively decontextualize a domain-specific problem from a more general and abstract issue. In addition, open innovation also challenges our economic model which traditionally regards monetary reward as the primary incentive. From DARPA to Xiaomi, from Wikipedia to NASA, such simplistic assumptions cannot be more wrong.

But perhaps most intriguing is the prospect of integrating open innovation into the core aspect of a business model that might allow a new firm to rewrite the rules and boundaries of a traditional industry. Xiaomi, in spite of being a start-up, has almost single-handedly rewritten the way mobile phones can be brought to the marketplace. It owns no factory for manufacturing; possesses few software programmers; and does little, if any, above the line advertising. Yet with millions of advanced users who endlessly tweak its product features, in less than four years, the company has become the world's third largest mobile phone maker—and the biggest in China. As Lei Jun explained, "Every user becomes your R&D, every user becomes your sales, every user becomes your friend, that's the company we want to make" (Kan 2013).

Bibliography

Alexy, O., Criscuolo, P., Salter, A.: Does IP strategy have to cripple open innovation. MIT Sloan Manage. Rev. **51**(1) (2009)

Barboza, D.: In China, an empire built by Aping Apple. New York Times (2013)

Boudreau, K., Lakhani, K.: How to manage outside innovation. MIT Sloan Manage. Rev. (Magazine: Summer 2009 Research Feature) (2009)

Boudreau, K., Lakhani, K.: Using the crowd as an innovation partner. Harvard Bus. Rev. **91**(4) (2013)

Chesbrough, H.: Open innovation: a new paradigm for understanding industrial innovation. In: Chesbrough, H., Vanhaverbeke, W., West, J. (eds.), Open Innovation: Researching a New Paradigm, pp. 1–14. Oxford University Press, Oxford (2006)

Chandler, A.: The visible hand Cambridge. Massachusetts (1977)
Coase, R.H.: The nature of the firm. Economica **4**(16), 386–405 (1937)
Duhigg, C.: The Power of Habit: Why We Do What We Do in Life and Business. Random House, New York (2012)
Guinan, E., Boudreau, K., Lakhani, K.: Experiments in open innovation at Harvard medical school. MIT Sloan Manage. Rev. **54**(3), 45–52 (2013)
Henkel, J., Balwin, C., Shih, W.: IP modularity: profiting from innovation by aligning prodcut architecture with intellectual property. Calif. Manage. Rev. **55**(4) (2013)
Heimans, K., Timms, H.: Understanding new power. Harvard Bus. Rev. (2014)
Huston, L., Sakkab, N.: Connect and develop: inside procter & gamble's new model for innovation. Harvard Bus. Rev. **84**(3), 68–76 (2006)
Kan, M.: China's Xiaomi takes crowdsourced phone development model abroad. Netw. World (2013)
Kanter, M.: The middle manager as innovator. Harvard Bus. Rev. **82**(7–8), 150 (2004)
King, A., Lakhani, K.: Using open innovation to identify the best ideas. MIT Sloan Manage. Rev. **44**(1) (2013)
Kohler, T.: Crowdsourcing-based business models: how to create and capture value, Calif. Manage. Rev. **57**(4) (2015)
Lakhani, K.R., Lifshitz-Assaf, H., Tushman, M.: Open innovation and organizational boundaries: task decomposition, knowledge distribution and the locus of innovation (Chap. 19). In: Grandori, A. (ed.) Handbook of Economic Organization: Integrating Economic and Organization Theory, pp. 355–382. Edward Elgar Publishing, Northampton (2013)
Lei Jun Challenges Idol Steve Jobs and Sells Mobiles at Forty Years Old. Chinatimes (2011)
Lichtenthaler, U., Hoegl, M., Muethel, M.: Is your company ready for open innovation? MIT Sloan Manage. Rev. **53**(1) (2011)
Majchrzak, A., Malhotra, A.: Towards an information systems perspective and research agenda on crowdsourcing for innovation. J. Strateg. Inf. Syst. **22**(4), 257–268 (2013)
Majchrzak, A., Malhotra, A.: Managing crowds in innovation challenges. Calif. Manage. Rev. **56** (4) (2014)
McGahan, A.: Unlocking the big promise of big data. Rotman Manage. Mag. (2013)
Rifkin, J.: The zero marginal cost society: the internet of things, the collaborative commons, and the eclipse of capitalism. Macmillan (2014)
Robson, K., Plangger, K., Kietzmann, J.H., McCarthy, I., Pitt, L.: Game on: engaging customers and employees through gamification. Bus. Horiz. **59**(1), 29–36 (2016)
von Hippel, E.: The Sources of Innovation. Oxford University Press, New York (1988)
West, J., Borgers, M.: Leveraging external sources of innovation: a review of research on open innovation. J. Prod. Innov. Manage. **31**(4), 814–831 (2014)
Williamson, O.E.: The economics of organization: the transaction cost approach. A. J. sociol. 548–577 (1981)
Whelan E., Parise S., de Valk J., Aalbers, R.: Creating employee networks that deliver open innovation. MIT Sloan Manage. Rev. **53** (2011)
Zeitlyn, D.: Gift economies in the development of open source software. Res. Policy **32**(7), 1287–1292 (2003)

Chapter 4
Classic Root Causes of Innovation Failures—Things We All Know but Sometimes Forget

Jean-Philippe Deschamps

Abstract Innovation failures happen when management neglects to follow a number of well-known innovation rules. This chapter reviews some of these rules and it identifies frequently found—root causes behind their non-observance. We will first focus on the non-respect of a critical strategy rule—the imperative need to anticipate and react to external changes—and a basic execution rule—the necessity to build integrated innovation process roadmaps. Failures also come from not abiding by the rules that determine technology leadership, product leadership and access to market, three combined conditions for innovation success. In most cases, the root causes are managerial. They often point to a disconnected approach between the technical and business sides of the organization resulting from a deficient innovation governance.

4.1 Introduction

There is so much hype about innovation, both in companies and in the business media, that it is wise to remind ourselves that a large proportion of innovative new products or services actually fail to either build their own market space or are quickly displaced by competitors. Many articles have listed the most common reasons why new products fail, but they seldom go deeply enough in identifying the root causes behind these failures. So, at the risk of adding yet another explanation for innovation failure rates, this chapter will revisit some of the most basic innovation rules that lead to success when applied, or to failures when neglected. Success and failure are indeed the two sides of the same set of innovation rules.

Innovation rules cover both the front end and back end of the process, i.e. innovation strategy formulation *and* execution. Strategy failures tend to be irreparable, hence more costly and sometimes more lethal than execution failures. Let us review a couple of these neglected rules.

J.-P. Deschamps (✉)
1 Chemin Des Chênes, CH-1135 Denens, Switzerland
e-mail: deschamps@imd.ch

N. Pfeffermann and J. Gould (eds.), *Strategy and Communication for Innovation*,
DOI 10.1007/978-3-319-49542-2_4

4.2 Neglected Strategy Rule: Anticipating and Reacting Fast to External Changes

One of the most dangerous traps companies face is to stay blind or to react too late to major changes in the company's industry and market environment, thus to assume that current favorable conditions will prevail. Such management blindness can relate to a variety of changes;

- Changes in industry structure and regulations;
- Convergence of segments or creation of new segments or even entirely new industries;
- Emerging trends in customer behavior;
- New pressures on the viability of current business models;
- Emerging technologies and competitors.

All these changes can happen at the same time, thus endangering the future of whole industries. The big Switzerland-wide debate about the future of connected watches and their implications for the Swiss watch industry is a good illustration of that blindness danger. Was the Swiss industry too complacent about the future prospects of its worldwide markets to see the connected watch phenomenon coming? Are Swiss watchmakers downplaying Apple's power to create yet another watch category? Do they appreciate the fact that connected watches could redefine entirely the boundaries of their industry, leaving the Swiss brands in small "nostalgia" or luxury niche segments? The most critical Swiss media do not hesitate to talk about arrogance and complacency on the part of most Swiss watchmakers.

Ravi Arora, vice-president of innovation at the Tata Business Excellence Group introduces two important concepts related to this blindness danger in his new book *Making Innovations Happen* (Arora 2015). He talks about "prescience errors" and "obstinacy errors". Companies make *prescience errors* when they neglect or downplay the importance and potential impact of an emerging phenomenon, a new disruptive technology or business model. They make an *obstinacy error* when, having seen the writing on the wall, they persist with their strategy without sufficient correction or adaptation.

4.2.1 Root Cause N°1: Management Inward-Looking Attitude

Prescience and obstinacy errors happen when a management team lacks both the curiosity to monitor early change signals and the humility to continuously challenge its own vision of the business environment. Such managers may be devoid of what Intel's charismatic former chairman, Andy Grove, called "paranoia" (Grove 1996). Paranoid managers, according to Grove, look continuously for insights and, most importantly, build foresight. They observe, question all hypotheses and challenge their staff regularly to ensure they read correctly and interpret early warning signals and are prepared to change tack accordingly. Paranoid leaders ask different types of questions

Not just *what?* but *what if?* and *what else?*
Not just *why?* but *why not?*
Not just *who?* but *who else?*
Not just *how?* but *how much?* and *how else?*

Leaders who lack this ability or willingness to challenge their own assumptions and those of their staff will typically be unable to "unlearn." They will tend to stick to their strategy—an obstinacy error—and face serious consequences.

4.2.2 Root Cause N°2: Overconfident Leaders, in Love with Their Technology and Concepts

Being proud of the company's technology and design is a good thing and a motivating factor. But beware of the risk of hubris, which is often prevalent, particularly with industry leaders who believe that they fully deserve their leadership position and feel safe about it! Hubris is apparently a risk for the leading Swiss high-end and luxury watchmakers. Many probably said: "If the world has appreciated our beautiful mechanical time pieces for decades, why should consumers buy an Apple watch?" The same can be said about Nokia mobile phones: "Since we are the world's largest cell phone company and biggest camera producer, why should we feel threatened by Apple's iPhone when Apple didn't know much about mobile phones anyway?"

Overconfident leaders tend to downplay their emerging competitors' choices. They may be unwilling, therefore, to scrutinize the future validity of their own past strategies and success formulae, with the consequences we have all recently observed in the transformation of the mobile phone industry into the Smartphone industry. Both Nokia Mobile Phones and Research-in-Motion (RIM)/Blackberry dominated their segment prior to the arrival of Apple's iPhone and the emergence of Samsung. Convinced of the superiority of their technology and the market acceptance of their concepts, they were also both slow to realize that these new competitors were about to redefine their industry. They both tried to resist the change but they made a fundamental prescience error by not seeing (1) how mobile phones were turning into platforms for multiple applications; and (2) the merging of consumer and professional segments around broad-based smartphones. They worsened their situation by making an obstinacy error, i.e. by being reluctant to relinquish their proprietary operating software and their superiority in mini keyboards in order to adopt tactile screens. The combination of these errors ultimately led to their demise as major industry players.

4.3 Neglected Execution Rule: Mapping and Integrating the Process

Whereas disregarding the key innovation strategy rule described above can be a "company killer," neglecting innovation execution rules and proceeding with a botched process is repairable. It is more of a "retarder"' than a "killer" per se.

Best practice advocates insist on the benefits of a carefully mapped innovation process and that the resulting map should be easily understood and respected by all intervening functions. Innovation process mapping is a delicate exercise because it needs to combine different parts of the process with different dynamics and requirements. A good process map recognizes the peculiarities of the creative "fuzzy front-end," with its unavoidable uncertainties, iterations and loops, and those of the more predictable "speedy back-end," which requires a lot of discipline. A good process map should provide elements of structure and solidity while maintaining adaptive flexibility. A comprehensive high-level process map is shown in Fig. 4.1. It has the merit of showing how all the pieces in the process are interrelated. But companies need to detail each of these sub-processes into individual activity chains and tasks.

Experience shows that many companies are still deficient in their innovation process mapping discipline. Management leaves many questions unanswered, such as

- What should be the overall process architecture?
- How will uncertainty be reduced over time?
- How will investment/funding evolve?

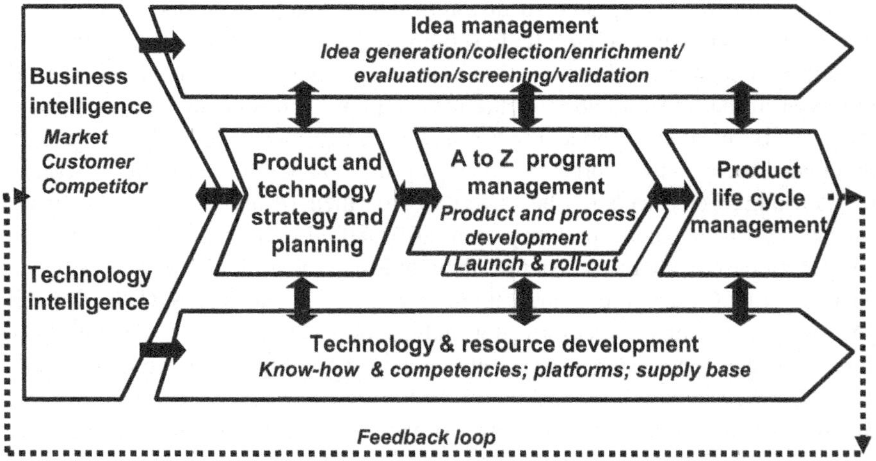

Fig. 4.1 High-level process innovation model. *Source* Innovation Leaders—How Senior Executives Stimulate, Steer and Sustain Innovation by Jean-Philippe Deschamps, Wiley/Jossey-Bass pp. 103

- How will continuity be maintained in project management?

This may result in an ad hoc process.
There are typically two root causes behind this situation.

4.3.1 Root Cause N°1: Leaders Unaware of the Importance of Integrated Roadmapping

Roadmapping—a sequential listing of expected activities or outcomes over time—has become a key buzzword in the field of innovation management. Unfortunately, many companies only practice a limited kind of roadmapping. R&D managers and technologists often convey the result of their technology intelligence exercise in the form of technology roadmaps and/or competence roadmaps. Many hi-tech companies discuss their anticipated product roadmaps with their suppliers to enable them to tool up ahead of orders. But although they are useful for providing a glimpse of the future, these roadmaps generally remain quite limited and are rarely fully integrated in that they are not compatible with each other. Rare, indeed, are management teams that demand from their staff an integrated set of roadmaps as illustrated in Fig. 4.2.

Although this figure might seem complex, it is actually quite straightforward and logical. Starting from the left of the picture, the tangible outcome of the *market intelligence process* consists of a hierarchy of customer (or consumer) needs and preferences over time. This type of roadmap seldom exists, often because marketers dislike sticking their neck out by predicting customer expectations regarding product attributes. In parallel, the *strategy process* typically results in a prioritized list of market segments targeted by the business. These two roadmaps serve as primary inputs to the *product strategy and planning process* since together they convey both

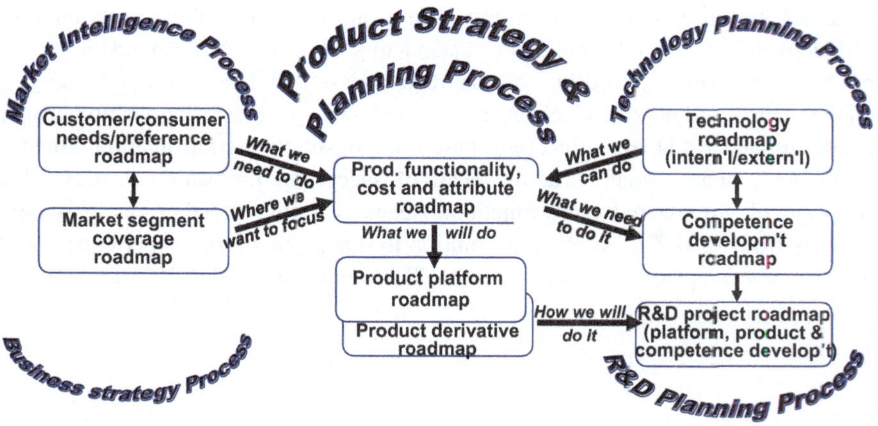

Fig. 4.2 Integrated process roadmapping. *Source* IMD class materials from Jean-Philippe Deschamps

what the company needs in order to compete effectively and which segments it wants to focus on. In parallel, on the right side of the picture, the technology planning process will generally come up with technology roadmaps that reflect both the output of the company's internal technology development efforts and the evolution of external supplier or partner capabilities over time. By indicating what the company can do on the technology side, this roadmap becomes another input into the critical *product strategy and planning process.*

Based on these two types of input, the *product strategy and planning process* starts with roadmapping the evolution of desirable product functionalities, as well as their costs and other attributes. For example, Apple is likely to be seriously working on a roadmap for enhancing the battery performance of its Apple Watch, a critical element of its offering. This type of roadmap is the cornerstone of the product plan. It can then lead to specific product platform and derivatives roadmaps that will indicate how the business offering will evolve over time. In turn, these functionality, platform and product roadmaps will be translated into competence development roadmaps and actual R&D projects as part of the *R&D planning process.*

Building an integrated set of roadmaps, like the one indicated in Fig. 2, is becoming a key success factor in the development of complex innovative products. Many botched projects simply reflect a lack of functional coordination and ad-hoc planning.

4.3.2 Root Cause N°2: Poor Functional Interface in the Specification Phase

A lack of integrated roadmapping is often caused by an ineffective marketing/R&D collaboration in the most critical part of any innovative project: the specification phase. When the specification process evolves through a series of hard-ball negotiations between entrenched functions, each group trying to safeguard its interests and demands, the results are rarely satisfactory. The process can result in dull product compromises, drifting or muddled specs, and late changes leading to market delays and poor product success. This is why a healthy relationship between marketing and R&D or engineering, based on constructive confrontation aimed at achieving a shared goal, is at the heart of successful innovation projects. The adoption of empowered cross-functional teams working together on innovative projects from A to Z is an obvious solution to this problem, but experience shows that it is not happening in all companies. Traditional functional silos die hard!

4.4 Other Neglected Rule: Not Recognizing What It Takes in Combination to Succeed

If success and failure are two sides of the same set of innovation rules, then it is critical for management to build an overall mental model of things that determine success, and check whether all these conditions, without exception, are being met. Ad Huijser, the charismatic former CTO and Research President of Philips, the Dutch technology-based company, developed his own formula to illustrate the conditions that need to be met to guarantee success in the market, i.e. the combination of technology leadership, product leadership and access to market (Fig. 4.3).

The probability of achieving *technology leadership*, according to Huijser, is conditioned by at least three factors:

- The overall technological position of the company;
- The proportion of proprietary technologies in its portfolio; and
- The patentability of key product technologies.

The probability of reaching *product leadership*, he claims, results from three additional factors:

- Uniqueness of the product concept;
- Manufacturing scale and cost advantages;
- The patentability of key process technologies.

Finally, the probability of gaining *market access* depends on three conditions:

- Internal familiarity of the company with the market;
- Access to adequate distribution; and
- Ease of transfer of the corporate brand reputation to the new product.

By presenting his success formula as a multiplication of three probabilities, each one expressed as a number between 0 and 1, Huijser purposely showed to his staff that the overall probability of market success would always be minimal unless the company worked on all three elements of the equation. R&D leaders, he felt, did not question the need to achieve technology leadership, but they needed to realize the extent to which other factors also conditioned success. Applying his formula to Philips' innovative projects had a sobering effect on the attitude of his staff and drove their determination to ensure all other elements were addressed. Ultimately,

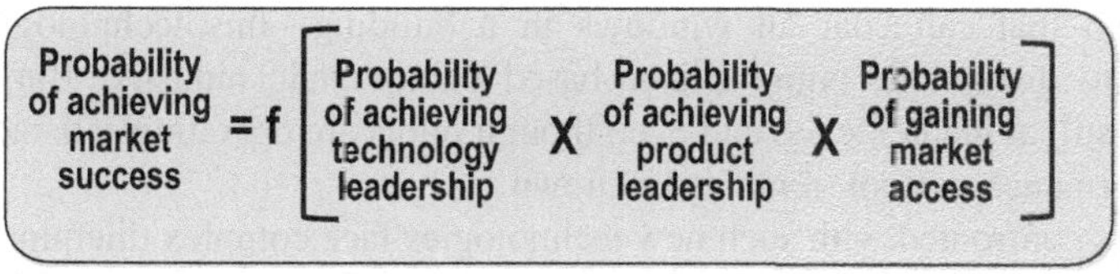

Fig. 4.3 The Huijser innovation success formula. *Source* IMD class materials from Ad Huijser, former President of Philips Research

the Huijser formula became a kind of reference model in Philips to judge new business projects. He explained: "Leaders used it for that purpose, not to say 'yes or no' to a particular project but to guide the activities and to put money and attention to where it is most needed. Focus indeed shifts over time from left to right in the formula".

The rest of this chapter will review a few of the classical issues in achieving technology leadership, product leadership and market access and their root causes.

4.5 Addressing Issues in Achieving Technology Leadership

There are a dozen reasons why companies can fail to build a technology leadership position, the base on which most innovation efforts are rooted. The most obvious ones are the lack of competences and the dearth of resources to build or acquire them. Others relate to poor technology choices, often reflecting an ineffective management of R&D. Given the focus of this chapter on strategic decisions that contribute to innovation failures, two cases are worth stressing: (1) being unclear about the impact and timing of disruptive technologies, and (2) ignoring or downplaying customers' perception of the value of a particular technology. Once again, several root causes are at play in each of these two cases.

4.5.1 Being Unclear Regarding the Timing and Impact of Disruptive Technologies

For most companies, the rapid development of technologies in all kinds of disciplines is hard to follow. Some of the new technologies that are announced by scientific media are years away from concrete economic applications. Some may actually never make it. Others, the truly disruptive ones, will revolutionize industries. Take the case of a producer of traditional photovoltaic panels looking at the outlook for Grätzel cells, a new dye-sensitized solar cell technology being developed at the Federal Polytechnic Institute of Lausanne by Prof. Grätzel, laureate of the Millenium Prize, Finland's equivalent of a Nobel Prize for technology. Given its specific characteristics—the possibility of producing electricity through a transparent film that can coat all windows in a building—this technology has the potential to disrupt the entire silicon-based photovoltaic industry. But the technology is still at the lab level, and even though pilots are operating satisfactorily, its cost/performance outlook remains unknown.

Leaders confronted with such new technologies face complex dilemmas: Should we invest or not? And if we do, within what time frame and at what level? The danger comes from investing either too soon, too late or too conservatively. Failures

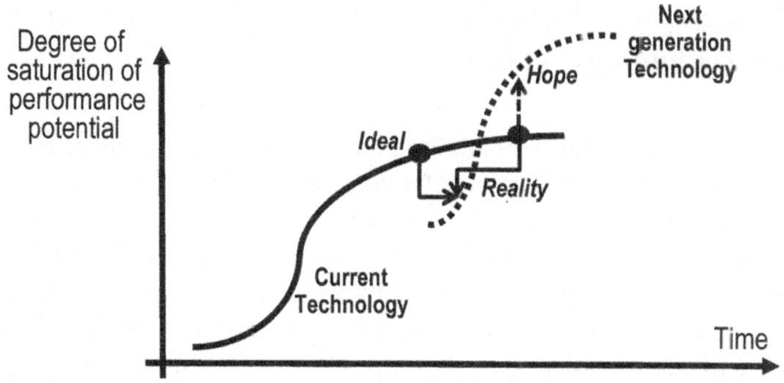

Fig. 4.4 The "jumping the curve" dilemma. *Source* IMD class materials from Jean-Philippe Deschamps

will typically result from leaders being unprepared to jump the curve, or if late, from being unrealistic regarding their own learning curve, as illustrated in Fig. 4.4.

Assuming the next technology—often known as the next S-curve—shown on this illustrative graph is perceived as promising in terms of its cost/performance outlook, farsighted leaders will typically jump the curve at an early stage (albeit not too early, at least not alone). They will invest in the new technology to shape it for the desired application. Slower or more conservative companies will wait for the validation of the potential of the new technology before jumping the curve. Their hope may be to catch up without wasting time, which often proves unrealistic. In practice, these companies cannot bypass the learning curve that their competitors followed and they lose precious time. In some instances they may even be unable to catch up.

The main root cause behind this situation is a lack of technology intelligence, sometimes combined with an unwillingness to give up investments in the old technology, either for emotional or, more often, economic motives. The well-known story of Kodak's reluctance to embrace digital photography is an excellent example of a company failure caused by its management's unwillingness to jump the digital curve in a big way and its choice to stick with its huge investments in its highly profitable silver-halide photo technology. The sad side of the story is that clear-sighted leaders within Kodak had pioneered digital photography at its inception but were later left hanging without resources. The Kodak failure story is yet another illustration of prescience error—i.e. not seeing that digital photography was bound to replace almost entirely the traditional photo technology—and obstinacy error—i.e., refusing to write off its huge network of chemical plants and labs in order to invest heavily in digital applications.

4.5.2 Ignoring or Downplaying Customers' Perception of Technology Value

For most scientists and engineers, all the new technologies that their lab develops or implements are exciting. If a new technology works cost-effectively and provides tangible benefits to their products or systems, they will not ask further questions. They may not be aware of—nor will they pay particular attention to—how their customers and end-users, often influenced by the media, will perceive that particular technology. This attitude may explain why certain technologies—consider for example genetically modified agricultural products—face strong resistance in some markets. Customers may well see the benefits being targeted by the scientists, but they imagine or perceive other drawbacks in that technology and thus abstain from adopting it. In other cases—consider probiotic yogurts—the dilemma for manufacturers is to explain to consumers the nature of the benefits provided by their technology. Some fresh-dairy managers at Nestlé recognize that the success of their competitor—Danone—is largely due to its smart marketing communications, which enabled consumers to visualize the benefits of probiotic yogurts. Nestlé had a similarly effective technology but was less able to communicate its benefits to consumers, hence the lesser success of its LC1 brand.

A marketing specialist from a leading European car manufacturer explained this phenomenon by suggesting that the adoption of a new car technology is conditioned by two parallel elements: (1) the perceived value provided by the technology, measured according to the customer's own value system; and (2) the perceived complexity, cost, risk or other negative aspect of that technology, rated against the customer's subjective mental model. Figure 4.5 illustrates how these two elements apply to various automotive technologies.

This car marketer advocates that this kind of table explains why certain ancient technologies, like airbags, were easily accepted by the market. Consumers quickly perceived the benefits of the technology because they could visualize a balloon inflating instantly to protect their face in case of frontal shock without realizing the

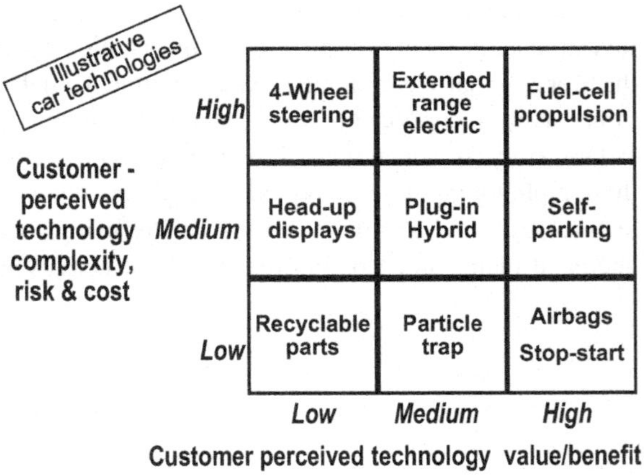

Fig. 4.5 Customer value map of new technology. *Source* IMD class materials from Jean-Philippe Deschamps

technical complexity of implementing that technology nor envisaging its other negatives. The same can be said about the stop-start technology. It is wrongly perceived as being simple but its benefits on fuel consumption can be easily grasped. At the opposite end of the spectrum, 4-wheel steering, which provides tangible benefits when cornering, was not appreciated by consumers, who saw it as a costly and unsafe solution and could not imagine driving with a technology that they perceived as overly complex for a wobbly result. The Japanese car manufacturers who developed it at great expense had to shelf the technology, and its use was long limited to a few niche sports applications. So, customers' perception of the value of a given technology matters, irrespective of whether or not that perception is objectively founded. Companies that do not pay attention to this factor may be in for a surprise when their product hits the market.

The main root cause behind product failures linked to customers' reluctance to buy or trust a new technology is a lack of research on technology adoption patterns, which often reflects excessive confidence in the power of technology irrespective of its perceived usability. Technology leaders need to understand the environment of their customers and users and their typical attitudes and expectations regarding technology. The challenge is, however, to avoid trusting blindly their customers' expressed views and declared purchase intent for a new and unknown technology. When developing a new technology, leaders should ask the following questions: (1) Will our technology address a priority need that is either unmet or ill met? (2) Will it have a strong impact on that need? And most importantly, (3) will it be seen as valuable and low risk? In other words, will the tangible benefits of our technology be fully perceived by customers?

4.6 Addressing Issues in Achieving Product Leadership

Even when the technologies adopted by a company are promising, there are many ways it can fail to achieve product leadership. Leaving aside quality or economic factors like an untenable cost position vis-à-vis competitors, the majority of these failures have a managerial cause. In my experience, two of these cases stand out: (1) accepting an ill-defined product concept and/or value proposition; and (2) investing in the wrong product design attributes or features.

4.6.1 Accepting an Ill-defined Product Concept and/or Value Proposition

Logitech is a remarkably innovative company. Yet despite—or because of—its creativity, it has experienced a few product failures. One of them, years ago, was its io Personal Digital Pen™, a technological marvel that was supposed to revolutionize the way people take, record and file notes, and how they share them across distances

(Pahwa and Deschamps 2003). The big dilemma that Logitech innovators faced was identifying priorities in terms of users and applications. Their io Pen could be used by students in class, lawyers in their cabinet, engineers working and sharing their work in labs across the world, doctors writing prescriptions, sales people writing their reports, and many other users for endless other applications. The beauty of it was its unlimited application potential, but this is also what made the innovation almost unsalable! How can a company promote such a product when each potential user has a different application pattern? And how can it convince these different users to move on from the ubiquitous pencil, or the felt or roller pen? After trying for several years to market its product, the company had to drop it.

Besides a few product issues that constrained its sales potential, Logitech's problem was essentially its product concept and value proposition. Confronted with an apparently unlimited market potential—replacing the pen in all its applications—the company's strategy was to adopt the classical response: *raise the flag and see if anyone salutes!* In other words: let us see which market segment buys it first and for what usage, and then we shall adapt our marketing accordingly. In doing so and by bypassing thorough customer and applications research, Logitech was unable to fully define its product concept. It could answer two of the classical questions that help define a product concept:

What is the product? (i.e., Product description/Objective product identity/ Product configuration/Components/Underlying technology);

- *What does the product do?* (i.e., Application/Functionality/Key attributes/ Performance (objectively defined);

But by failing on customer research and segmentation, the company was unable—or unwilling—to define the last two product concept elements:

- *What customer groups is the product targeting in priority?* (i.e., Target markets in terms of nature, geography, channels/Most valued customer segments);
- *What will the product mean to those targeted customers?* (i.e., Perceived image/Identity/Personality/Perceived performance and benefits).

Similarly, Logitech did not fully clarify its value proposition. It only managed to clearly define one element of its value proposition by answering the first question:

- *What is unique in the concept?* It managed to do this in terms of the product itself and the customer experience.

However, the absence of a thorough customer usability study prevented it from being able to answer the second question:

- *What is attractive in the concept?* Notably for which customer segment and against what alternative solution?

This, in turn, made it difficult for Logitech to answer the last question:

- *For what price positioning?* i.e., How will customers perceive the real usage value of the product and, hence, how much will they be prepared to pay for it?

The main root cause behind product failures linked to poor product concept definition or a shaky value proposition is a deficiency in the early stages of the product decision process. This reflects either a lack of constructive challenging on the part of senior product managers, or shaky compromises. This is often the case when one function—which could be marketing, R&D or engineering—dominates the product decision body.

4.6.2 Investing in the Wrong Product Design, Attribute or Feature

Innovation success often results from introducing products or services that delight customers, particularly in areas that they value most. Apple's iPhone was adopted rapidly, largely because of the appeal of its design. It redefined the standards that had been proposed for years by mobile phone leaders such as Nokia, Motorola, RIM/Blackberry and others. Of course, many other aspects or features played their role, but Apple's unique design gave it a differentiating advantage. In most industries the basis of competition is becoming increasingly broad. Products have to meet the expectations of their customers on a growing number of attributes. The challenge is double: On the one hand, companies have to reach a balance by covering all meaningful attributes reasonably well without incurring a major deficiency. On the other hand, and to avoid a bland positioning, they have to choose one or a few differentiating attributes on which they can offer a superior performance or quality. The difficulty is of course to privilege those attributes or features that will be most appealing to customers. The choice of these privileged attributes is therefore critical, and companies that choose the wrong ones can expect disappointing results.

To help make this tricky choice, management can follow the value innovation methodology advocated by Kim and Mauborgne (1997) and create a new value curve by answering the following questions:

Which of the factors taken for granted by our industry should be eliminated?
Which factors should be reduced well below the industry's standard?
Which factors should be raised well above the industry's standard?
Which factors not previously offered by the industry should be created?

As an alternative—because the two approaches share the same philosophy—they can use the Kano model that categorizes attributes on the basis of their impact on customer satisfaction.[1] As shown in Fig. 4.6, attributes can be categorized in three ways

[1]The Kano model is a theory of product development and customer satisfaction developed in the 1980s by Professor Noriaki Kano, which classifies customer preferences in various categories.

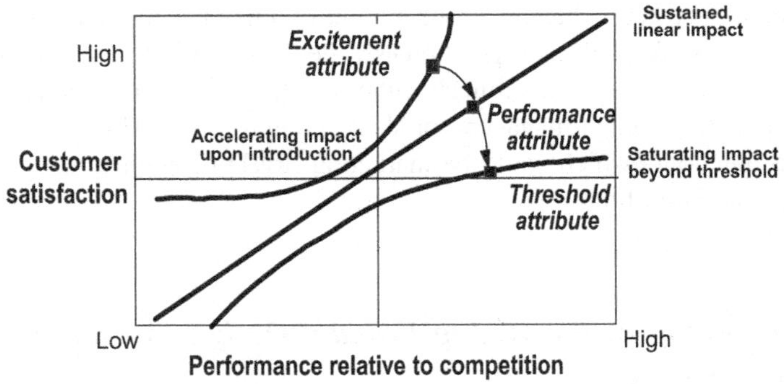

Fig. 4.6 Kano diagram—evolution pattern of various attributes over time. *Source* IMD class materials from Jean-Philippe Deschamps, based on Kano diagram concepts

Performance attributes. They have a sustained linear impact, i.e. the more a company outperforms its competitors on these attributes the higher will be its customer satisfaction.

Threshold attributes. They have a saturating impact beyond a certain threshold, i.e. offering more performance beyond a certain level does not lead to a proportional increase in customer satisfaction.

Excitement attributes. They have an accelerating impact on customer satisfaction upon introduction and can trigger customer delights, possibly because these attributes were neither expected nor available before.

By using this approach, marketers will naturally focus on identifying possible new excitement attributes—like design for Apple. They will also be careful about over-investing in threshold attributes that will only marginally increase customer satisfaction, and they will become aware that most attributes, even excitement attributes, become threshold attributes at one point in time.

Logitech provides a good illustration of a successful selection of attractive product attributes for its small new "cordless presenter" (Barnett Berg and Deschamps 2003). Unlike what happened with its io Pen, which resulted from a typical technology-push effort, the head of engineering personally conducted customer research by holding a number of insightful one-to-one interviews with people who made frequent presentations with the objective of understanding their experience as speakers. The purpose of these interviews was to identify the most important outcomes expected from a cordless presenter—or "clicker" as these products are sometimes called. This research led to a short list of critical attributes most interviewees expected, in that case extreme simplicity of use and the possibility of controlling the speaker's timing. These attributes happened to be quite different from those originally expected by Logitech's developers in engineering, and they clearly contributed to the product's ultimate success in the market.

In conclusion, the main root cause for failing to choose the most relevant attributes for customers is a reliance on a traditional—and sometimes simplistic—approach to customer research, basically asking them what they want. As shown in Fig. 4.7, in a customer-centric approach, management goes far beyond customer

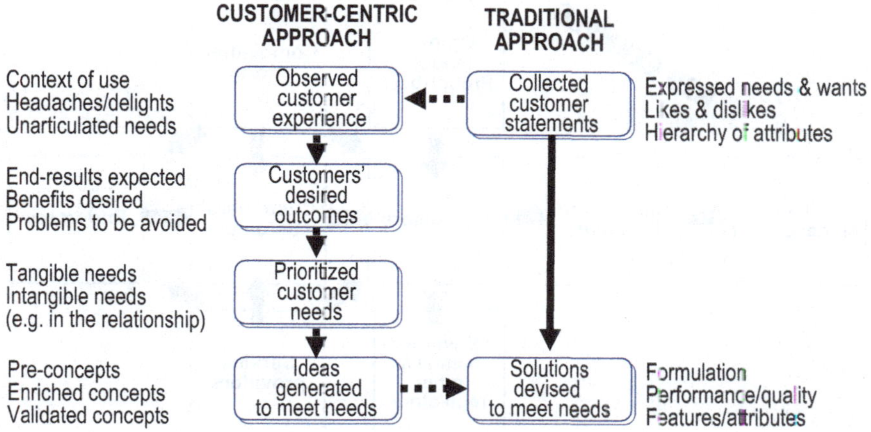

Fig. 4.7 Customer-centric approach to design new products. *Source* IMD class materials from Jean-Philippe Deschamps

statements about their wishes or dislikes. They encourage their staff to immerse themselves in the customers' world to understand their experience and probe for unarticulated needs. The approach then focuses on clarifying the customers' desired outcomes and implicit prioritized needs.

Because customer needs do evolve, failures also stem from a lack of dynamic modeling of these customer preferences. In most markets, and due to changes in the customer environment and the impact of social networks, past customer data tends to age rapidly and relying on them leads to poor product positioning and failures.

4.7 Addressing Issues in Getting Access to the Market

Once again, even with a great technology and product, there are many obstacles on the way to the market. Putting aside obvious problems linked to poor marketing communications or inadequate salesmanship, experience shows that there are two strategic marketing issues that can ruin a smooth market launch and lead to failures: (1) neglecting to seek the acceptance of all parties in the value chain; and (2) seeing acceptance by early adopters as a market green light. Let us explore these issues and their frequent root causes.

4.7.1 Neglecting to Seek Acceptance by All Parties in the Value Chain

The successful launch of an innovative new product or service is always conditioned by the need for that product or service to receive a succession of "YESes," or

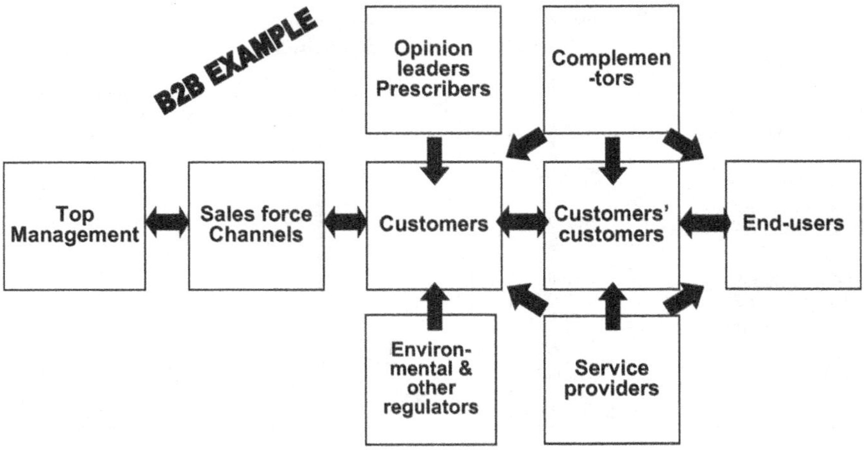

Fig. 4.8 The product adoption chain. *Source* IMD class materials from Jean-Philippe Deschamps

at least favorable comments, from the many parties who have to adopt it, both within the company and in the market. A single "NO" will typically break the new product or service adoption chain. Figure 4.8 shows a typical adoption chain for a new product in a B2B environment.

On the horizontal axis are all the parties that have to accept the innovation in sequence. It starts internally with the top management team because they have to fully back the launch and accept the risks involved in the process, e.g. launch costs; possible damages to the company brand or to its relationships; product/service liability, etc. Then the sales force also has to adopt the new product or service, which is not always the case as their compensation package often induces them to push the easy-to-sell, profitable products rather than unknown ones with uncertain potential. Distribution channels—dealers, importers, agents, etc.—have to accept to take the risk of introducing the new product in their catalogues and make the effort and spend the money to convince their customers to accept it. Of course, customers have to be approached and be willing to give it a try despite the risks, and they may have to also convince their own customers down the value chain to adopt it as well. Finally, the end-users will have to adopt the new product as well.

But there are other parties who may also have their word to say on the launch. First, opinion leaders and prescribers, who often play a key role in promoting technical products. Then, environmental and other regulators for certain products. Finally, complementors—think of suppliers of accessories—and service providers, who also have to accept the product and gear up to support it. If any of these parties refuse to accept the product or service, or to back it up, its future success is greatly compromised.

Mastering the new product adoption chain can be tricky when the interests of the various parties in the chain are in conflict, i.e. when the demands of one element in the chain go against the wishes or preferences of another. The challenge for management is therefore to identify potential differences of opinion across the adoption chain and try to mitigate them. For example, if the sales force is reluctant to spend much time promoting the new product, management can probably review its measurement and incentive system. It can also decide to privilege one element in the chain, sometimes against the interests of another, as Tetra Pak did on the issue of package openings. Originally, the company's milk or juice package, Tetra Brik™, had no opening. The package was perfectly rectangular and therefore easily stackable on pallets. The savings in materials costs and logistics were of great interest to Tetra Pak's customers—the dairies—and their customers—large distribution chains. But the lack of opening in the package was strongly resented by the end-consumers. Tetra Pak was fully aware of these conflicting interests, but stuck to its no-opening strategy until the pressure from consumers became more intense and competitors started offering openings on their packages. This change of strategy had to be effectively explained and sold to its customers and their distributors.

Neglecting to seek the acceptance of all parties in the value chain usually stems from a lack of understanding of the position of these various players, notably their reasons for accepting or rejecting an innovation. It generally reflects a predominant focus by the company on one element in the chain, typically the direct customers because they are the ones who pay for the product or service. The attitudes of other elements in the chain may be ignored or not taken into consideration despite their potential impact on the success in the market. It is therefore highly advisable to carefully investigate the attitudes of all elements in the chain on the basis of three considerations and questions:

- All parties in the value chain have a preferred solution to meet their needs. What is their current solution?
- The company's new product or service would probably require them to change their current approach. Why should they reject their satisfactory status quo?
- Switching solution may be costly, risky and troublesome for them. What would induce them to switch and adopt the company's new product or service?

4.7.2 Seeing Acceptance by Early Adopters as a Market Green Light

This second issue in gaining market access explains why many new products or services are not successful in the market despite positive initial reviews. There are always groups of customers ready to give any new product or service a chance, but these early adopters are by far not representative of the entire market. In his book *Diffusion of Innovations*, Everett Rogers proposed a categorization of adopters on the basis of innovativeness (Rogers 1995). Rogers categorization has been widely circulated, and it was at the origin of another book, *Crossing the Chasm*, by Moore (2009) (Fig. 4.9).

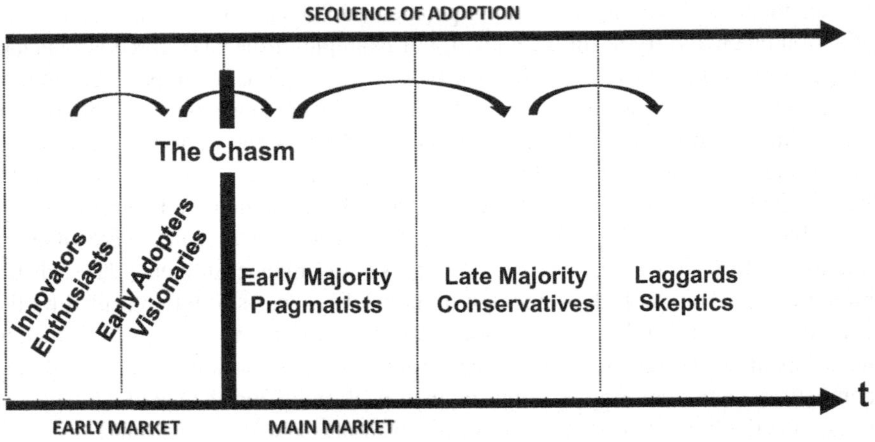

Fig. 4.9 Adopter categorization based on Rogers and Moore. *Source* IMD class materials from Jean-Philippe Deschamps, based on concepts from Everett Rogers and Geoffrey Moore

Both authors propose five identical adopter categories with a slight difference in terminology: Rogers talks about *Innovators, Early Adopters, Early Majority, Late Majority,* and *Laggards,* while Moore calls them *Enthusiasts, Visionaries, Pragmatists, Conservatives,* and *Skeptics*. Moore's major contribution is to stress the "chasm" that exists between the two first categories of adopters and the rest of the market. Crossing this chasm, he claims, is a condition for "bringing cutting-edge products to increasingly larger markets."

Many companies struggle to cross this strong market barrier that separates early adopters from the rest of the potential market and innovation success. This usually reflects a lack of management understanding of the different attitudes and perceptions of each adopter category. A product that will be strongly backed by early adopters, for example owing to its technical sophistication, may actually be rejected by more conservative customers for the opposite reason, i.e. its technical complexity. If the company is very technology-intensive, its developers will naturally monitor those they see as their "fans," in most cases early adopters. They may therefore decide to launch a product without testing it on more "average" customers. So the root cause behind failures of that type may be the lack of a convincing argument for each type of adopter. Companies do not invest enough in characterizing the needs and reactions of each category of adopter.

There are different approaches to anticipate the probability of a new product being adopted by a majority of customers. Rogers, for example, has proposed five criteria that have a strong influence on a new product's adoption rate. These criteria are particularly useful to determine customers' reactions in a B2B environment:

- *Relative advantage*. How much better is the new product versus the product that customers currently use?
- *Compatibility*. Can the customer use the product without changing its organization and infrastructure?
- *Complexity*. How easily and fast can the customer deploy the new product without too much training?
- *Triability*. Can the new product be tried easily and at no or low cost?
- *Observability*. Where can one see the new product being used?

There is another list of five criteria for B2C products that some marketing leaders call the “5-A” checklist:

- *Awareness*. Are customers aware of the existence of the new product?
- *Attractiveness*. How will customers appreciate the benefits of the new product?
- *Affordability*. Will customers be able to afford the product without too much sacrifice?
- *Availability*. Where will the new product be available?
- *Accessibility*. How easily accessible will the new product be for each type of adopter?

The two approaches are obviously complementary. Applying these various criteria to each adopter category provides insightful results on the chances of success of a new product. For this reason, successful innovators tend to raise these questions at the product conception stage, i.e. before the product is actually developed.

In other words, it is essential to reflect on adoption criteria very early in the new product development process. This is the best way to avoid bad surprises at the commercialization stage.

4.8 Conclusion

Innovation failures are often caused by a functional disconnect, primarily between the company’s R&D and business sides. This disconnect is usually more profound on long-term issues. Organizational silos typically accentuate this situation. They favor a “functional isolationism” whereby each function does “their things” without sufficient collaboration. They also prevent people from building on each other’s ideas and can even lead to infighting on ideas and budgets. The problem is often aggravated by domineering attitudes that prevail when a function or a particular side of the business dictates its conditions to the rest of the organization. Such situations reflect a lack of leadership on the part of the top management team and deficiencies in the innovation governance system put in place by the CEO.

Many of these failures can be overcome by encouraging a true partnership between the R&D, engineering functions and marketing as well as more generally with the business sides of the organization. This type of partnership is particularly necessary for collecting market intelligence and defining customer problems and concerns worth addressing. Management should also encourage humility and learning, notably through thorough and honest "post-mortem" project debriefings, particularly in the case of failures. As Oscar Wild reportedly stated, "Why make the same mistakes twice when there are so many mistakes to choose from!" Geoffrey Moore added, "If you are going to fail, at least have the courtesy to do so in a new and interesting manner!"

Bibliography

Arora, R.: Making Innovations Happen. Fostering Innovations by Inducing Foresight. Retrieved at makinginnovationshappen.com/blog/1 (2015)

Barnett Berg, M., Deschamps, J.P.: Logitech—Learning from Customers to Design a New Product, IMD case No. IMD-5-0679 (2003)

Grove, A.: Only the Paranoid Survive. Currency/Random House, New York (1996)

Kim, W.C., Mauborgne, R. Value Innovation, the Strategic Logic of High Growth. Harvard Bus. Rev. 103–112 (1997)

Moore, Geoffrey A.: Crossing the Chasm. HarperCollins Publishers, New York (2009)

Pahwa, A., Deschamps, J.P.: Logitech—Getting the io Digital Pen to Market, IMD case No. IMD-5-0622 (2003)

Rogers, E.M. Diffusion of Innovations, p. 281, Free Press/Simon & Schuster, New York (1995)

Chapter 5
The Evolution of Strategic Options, Actors, Tools and Tensions in Open Innovation

Kathrin M. Moeslein and Albrecht Fritzsche

Abstract Open innovation describes innovation processes that span across the boundaries of organizations or research and development (R&D) departments. It integrates different types of innovators—regardless of their institutional affiliations—to generate creative ideas, innovation concepts and novel solutions. The purpose of this chapter is to describe underlying mechanisms and strategic options for realizing open innovation. It presents three types of innovators and their functions in innovation endeavors and introduces five classes of tools that facilitate open innovation. Finally, challenges and core tensions are discussed as a basis for the successful management of strategic open innovation initiatives.

5.1 Towards Open Innovation

Open innovation is a subsuming concept for a large variety of different activities which involve interactions across organizational boundaries. It can be understood as an antithesis to internal research and development, performed exclusively within certain company departments (Chesbrough 2012). Most of the activities addressed as open innovation already have a long tradition (Trott and Hartmann 2009). However, they remained without a common denominator until the rise of the internet in the late twentieth century and the increasing awareness for the new forms of collaboration it made possible (Huizingh 2011). Open innovation can, in this sense, be considered as a strategic concept for the design and management of inter-organizational innovation activities. During the last years, industry and

This chapter draws on previously published content by Moeslein and Bansemir 2011, Moeslein et al. 2012, Moeslein 2013 and Fritzsche 2017.

K.M. Moeslein · A. Fritzsche (✉)
Friedrich-Alexander Universität Erlangen Nürnberg, WI1 – Innovation
Und Wertschöpfung, Lange Gasse 20, 90403 Nuremberg, Germany
e-mail: albrecht.fritzsche@fau.de

N. Pfeffermann and J. Gould (eds.), *Strategy and Communication for Innovation*,
DOI 10.1007/978-3-319-49542-2_5

academia have gained a lot of experience with the application of this concept in practice, the actors that are involved, and the tools that are used.

The numerous managerial approaches to open innovation which have been developed over the past decade are quite diverse (West et al. 2014). They have many different starting points, follow different paradigms and use different toolsets and methods. Generally speaking, two fundamentally different perspectives can be distinguished (Huff et al. 2013). First, inspired by the open source world of software development, open innovation is seen as an emerging phenomenon of largely self-organized and self-motivated, internet-based, dispersed but collaborative actors. Research and practice focus on independent participants and emphasize their open interactive collaboration.

The broadcasting of problems and potential solutions to potentially unknown participants is a defining element of open innovation from this perspective (Lakhani et al. 2007). Some even ignore or reject the role of organizations to coordinate joint activities. Eric von Hippel (MIT in Cambridge) postulates in his visionary deliberations that organizations as institutions and the importance of management for implementation of innovations could become obsolete (von Hippel 2005). This vision of dissolving organizations is inspiring and breaks with established thinking. Put into perspective, the concept of "democratizing innovation" outlines one possible extreme: the perspective of *"emergent open innovation"*.

An alternative view, described by Henry Chesbrough, sees open innovation as an antithesis of often hermetically separated R&D departments, so called "closed innovation". Traditionally, organizations bundle innovation activities in separate units, employ highly educated experts, entrust them with sophisticated innovation tasks, protect their intellectual property and rarely look outside for other solutions (Chesbrough 2003). Going in hand with the professionalization of functional organizational departments in the last decades, innovation from this perspective reflects a general preference for internal expertise and solutions, and a distinctive "not-invented-here" mentality towards external stimuli. Too often these characteristics become amplified into a "vicious circle". Open innovation then becomes a way of challenging the status quo to search for, find, develop and exchange ideas, concepts and innovations in all areas of concern. The possibilities for corporate innovation are enhanced by Web 2.0 and social software that facilitate collectively evaluating and selecting, and even implementing and merchandizing new goods and services around the world.

Both perspectives have become a reality in global innovation activities: *emergent open innovation* as a largely self-organized and self-motivated, internet-based, dispersed and collaborative innovation effort as well as *corporate open innovation* as a strategic process of opening up firm-sponsored innovation activities across organizational boundaries (Moeslein et al. 2012).

> Reichwald & Piller combine both perspectives and define open innovation as "[…] a multi-level and open search and solve process, which spans across organizational boundaries to include multiple innovators." (Reichwald and Piller 2006; translation by the authors).

Following this paradigm, organizations tend to integrate external experts, purposefully allow internal innovators to participate in external innovation activities, communicate internal innovation projects at early stage, or try to initiate common innovation projects with customers, suppliers, or even competitors. This is not only true for product, process or organizational innovations, but to a similar extent for innovations in services, systems and even strategies (Table 5.1).

On the one hand, Apple employed Tony Fadell as project manager and thus bought in his expertise; on the other hand, Apple relied on components largely developed by external innovation partners to launch the iPod. As the development of Apples iPod illustrates, there is already a tendency in industry to open up innovation processes towards services, systems and even strategies.

Table 5.1 The Apple iPod story: System innovation by integration of outside and core inside innovators

"(…) The iPod originated around a business idea dreamed up by Tony Fadell, an independent contractor and hardware expert who helped to develop handheld devices at General Magic and Philips. 'Tony's idea was to take an MP3 player, build a Napster music sale service to complement it, and build a company around it' Knauss [Editor's note: Knauss has been a close companion to Fadell] said (…) 'Fadell left Philips and set himself up as an independent contractor to shop the idea around.'
Apple hired Fadell in early 2001 and assigned him a team of about 30 people. (…) Fadell said, 'This is the project that's going to remold Apple and 10 years from now, it's going to be a music business, not a computer business.' Tony had an idea for a business process and Apple is transforming itself on his whim and an idea he had a few years ago.
Fadell was familiar with PortalPlayer's [Editor's note: At the time PortalPlayer was a cooperation partner of Apple] reference designs for a couple of MP3 players, including one about the size of a cigarette packet. And though the design was unfinished, several prototypes had been built. 'It was fairly ugly,' he said. 'It looked like an FM radio with a bunch of buttons.' The interface, Knauss said, 'was typical of an interface done by hardware guys.'
'(PortalPlayer) was attractive to Apple because we had an operating system,' said Knauss. 'That was a real selling point for Apple. We had the software and the hardware already done, and Apple was on a tight schedule.'
Knauss said the reference design was about 80 percent complete when Apple came calling. For example, the prototype would not support playlists longer than 10 songs. 'Most of the time building the iPod was spent finishing our product' Knauss said. At the time, PortalPlayer had 12 customers designing MP3 players based on the company's reference design. Most were Asian hardware manufacturers, Knauss said, but also included Teac and IBM.
Big Blue planned a small, black MP3 player, based on the company's own mini hard drives, which featured a unique circular screen and wireless Bluetooth headphones. 'The design for IBM was a lot sexier,' Knauss said. (Kahney 2004)

5.2 Actors in Open Innovation

Organizations that open up their innovation processes beyond organizational R&D departments build on the involvement of three types of innovators (Neyer et al. 2009): those members of an organization who have "innovation", "research" or "development" as an official task in their job description (*"core inside innovators"*), those external actors who get involved in the innovation processes of an organization often via open innovation platforms, while not being part of the organization itself (*"outside innovators"*) and—last, but not least—those employees across all functions, levels, and units of an organization who innovate without being "official innovators" by job description (*"peripheral inside innovators"*). These are employees within the organization who are not directly involved in the innovation process of their organizations by formal role, but nevertheless have enough information about needs and solutions to act as innovators. They engage in innovation mainly due to curiosity, proactivity and interest in the well-being of the organization.

The story of the bubblegum in 1928 shows that peripheral inside innovators ever since participated in innovation activities. Walter Diemer did not actually work as an innovator, but as an ordinary accountant. In his spare time he experimented with new chewing gum recipes. In 1928, bubblegum was launched and grew to global success. The bubblegum story shows that not only externals provide valuable input as innovators because of their enhanced knowledge in regard to needs and user information. In addition peripheral inside innovators show extraordinary engagement and motivation for innovating. To conclude, integrating this type of innovators seems especially fruitful, as they know the organization's products, processes, services and strategies, feel obliged to the company and have an undisguised perspective (Table 5.2).

In open innovation initiatives, these employees can take a key role in bridging between the inside and the outside of the organization, in balancing between different interests of openness and closeness, and in making outside-in, inside-out and coupled innovation processes work seamlessly for improving innovation performance. Table 5.3 summarizes the three essential types of innovators in open innovation.

All three types of innovators play a crucial role in innovation processes: *Core inside innovators* are by definition entrusted with developing new products, processes, services and strategies. We find that the vast majority of business offerings stem from inventing, discovering and developing capacity of this type of innovators. Designers, researchers, developers and innovators of the marketing, business operations or corporate strategy department usually belong to this group. Whilst organizational R&D departments, widely organized as closed departments used to be the "holy heart" of organizations, they open up ever more. *Outside innovators* largely comprise customers, suppliers and value creation partners, as well as

Table 5.2 The bubblegum story: Innovation by inclusion of peripheral inside innovators

"In 1928, bubblegum was invented by a man named Walter E. Diemer. Here's what Walter Diemer, the inventor himself, said about it just a year or two before he died: 'It was an accident.' 'I was doing something else,' Mr. Diemer explained, 'and ended up with something with bubbles.' And history took one giant pop forward. What Mr. Diemer was supposed to be doing, back in 1928, was working as an accountant for the Fleer Chewing Gum Company in Philadelphia; what he wound up doing in his spare time was playing around with new gum recipes. But this latest brew of Walter Diemer's was—unexpectedly, crucially—different. It was less sticky than regular chewing gum. It also stretched more easily. Walter Diemer, 23 years old, saw the bubbles. He saw the possibilities. One day he carried a five-pound glop of the stuff to a grocery store; it sold out in a single afternoon
Before long, the folks at Fleer were marketing Diemer's creation and Diemer himself was teaching cheeky salesmen to blow bubbles, to demonstrate exactly what made this gum different from all other gums. The only food coloring in the factory was pink. Walter used it. That is why most bubblegum today is pink
Gilbert Mustin, President of Fleer named the gum Dubble Bubble and it controlled the bubblegum market unchallenged for years, at least until Bazooka came along to share the wealth. Walter Diemer stayed with Fleer for decades, eventually becoming a senior vice president
He never received royalties for his invention, his wife told the newspapers, but he did not seem to mind; knowing what he had created was reward enough. Sometimes he would invite a bunch of kids to the house and tell them the story of his wonderful, accidental invention. Then he would hold bubble-blowing contests for them" (http://www.ideafinder.com)

Table 5.3 Three types of innovators in open innovation

Type of innovator	And where to find them
Core inside innovators	Employees of e.g. the R&D department or the strategic innovation unit for whom "innovation" is part of the job description
Peripheral inside innovators	Employees across all functions, levels and units of an organization for whom "innovation" is NOT part of the job description
Outside innovators	Customers, suppliers, value creation partners, universities, institutional research departments who reside outside the boundaries of the organization

university members, researchers from research institutes, or even competitors. Within the strategy of open innovation, organizations implement manifold methods and tools to integrate outside innovators. E.g. innovation toolkits, innovation contests or open innovation communities are frequently used. Often, the greater public is inspired, motivated or even explicitly invited to participate. *Peripheral inside innovators* innovated ever since based on self-motivation, engagement and confidence, but are often not or only scarcely supported by tools. Idea management systems still prevail in companies as a main mechanism to integrate peripheral inside innovators. Although modern updates that include elements of social software and Web 2.0 are available, they are usually not yet well integrated in organizational innovation strategies (Berger et al. 2005).

The fact that even today all three types of innovators have usually not yet been incorporated into an integrated innovation strategy and are not supported by integrated tools and platforms poses essential constraints for the innovation capacity of organizations. As early as in 1984, Robert Rosenfeld, founder of the office of innovation at Kodak, is cited with the following words: "The failure of large organizations in America to innovate is primarily the result of a communication gap, not a decline in ingenuity" (Rosenfeld and Servo 1984). Despite a dramatically grown variety of communication technologies (mainly enabled by Web 2.0 and social software) and innovation tools, this problem rather intensifies than declines because of increasing specialization. In the following, we introduce a set of five classes of tools to strategically implement open innovation.

5.3 Tools for Open Innovation

Historically, the creation of innovations across the boundaries of organizations has always been possible. Before the rise of corporate R&D departments this co-creation of innovations in networks of innovative craftsmen, inventors, entrepreneurs, hobbyists or discoverers has long been the dominant pattern of innovation. Standardized mass production in large corporations has turned the attention to expert engineers and business analysts as the main actors in the field. The spread of modern information technology and its shift from mere industrial application to private households put a different dynamic in operation (Fritzsche 2016). With the rise of the interactive web, now virtually everyone has the chance to get involved in highly specialized innovation activities with large numbers of collaborators on a global scale. In addition, it has lately been possible to pursue these activities practically everywhere, due to the spread of mobile digital devices with harldy any restriction on the actual location of its user. This way, it becomes possible to capture and record inspirations, ideas and solution procedures at the very moment they surface, and thus tap into a much richter set of knowledge than ever before. Digitalized innovation activities are stimulated and supported by a range of open innovation tools. In the following we present a morphology of five classes of tools that primarily support and promote open innovation. These are:

(1) innovation contests,
(2) innovation markets,
(3) innovation communities,
(4) innovation toolkits and
(5) innovation technologies.

The development, diffusion and implementation of these kinds of innovation tools are mainly driven by the attractiveness, usability and inclusiveness of Web 2.0 and social software features. This is evident when looking at each of the tools, as well as their features and functions in open innovation processes as illustrated by short examples in the following. Well established, "traditional" innovation tools which specifically refer to innovation activities *within* corporate R&D functions (e.g. computer aided design (CAD) used by engineers etc.) are not in the focus of this chapter (Fig. 5.1).

5.3.1 Innovation Contests

Innovation contests in their basic structure have a long-standing tradition and have influenced industries or even societies for a long time. For example, in 1869, Emperor Louis Napoleon III of France offered a prize to anyone who could make a "satisfactory substitute for butter, suitable for use by the armed forces and the lower classes." Still, neither Michel-Eugene Chevreul nor Hippolyte Mege-Mouris (historians are uneven about the inventor) were paid when they came up with margarine, since Napoleon died before. In the 19th century, innovation contests leave the realm of political organizers as they are increasingly adopted by industrialists as a powerful means of problem solving. Famous examples of this period include the "Rainhill trials" (1829) which were used by the directors of the Liverpool and Manchester Railway Company to decide whether hauling trains should be powered by stationary engines or locomotives. During the next century, realizations of innovation contests started to be held with increasing frequency in general business contexts. In the beginning, they were often driven by marketing considerations and focused on challenges with a comparably small level of technical difficulty. Over the years, however, they became more demanding. An example form the year 1997

Fig. 5.1 Five classes of tools for open innovation

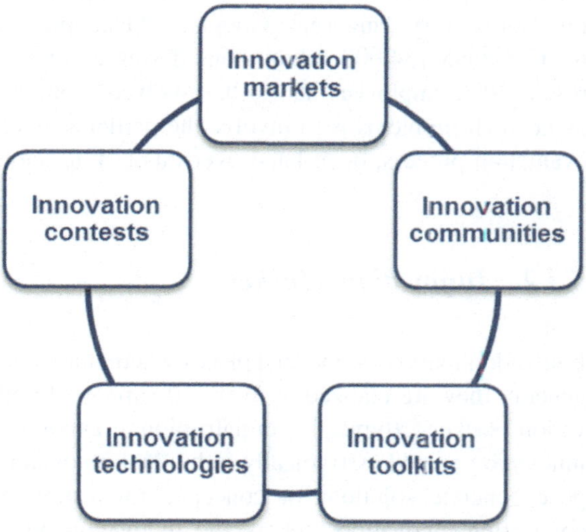

is the "Fredkin Prize for Computer Chess" granted USD 100,000 for building the first computer to beat world chess champion Garry Kasparov (Haller et al. 2011).

So, what is actually new and what makes innovation contests an important tool for open innovation? Social software features and Web 2.0 enable a multitude of actors to announce contests for exciting innovation challenges, with global reach at minimal costs. We define an innovation contest as a web-based competition of innovators who use their skills, experiences and creativity to provide a solution for a particular contest challenge defined by an organizer (Bullinger and Moeslein 2010).

Innovation contests are implemented in great variety. The continuum of different contests starts with simple idea and design contests, but also includes challenges that target radical innovations or marketable solutions. Many contests go far beyond targeting product or process innovations, but address innovative services, solutions or whole business model innovations. Formulating the innovation challenge for the contest is crucial. The way a challenge is framed and phrased finally decides about the range of innovators attracted, their expertise and motivation. At the same time, the framing and phrasing does not reveal competence deficits or strategic knowledge elements of the organizer. One main question that arises is, how does an organization formulate its innovation problem in such a way that the description motivates competent innovators to participate, does not reveal own competence deficits or strategic information, but still is concrete enough to deliver a relevant contribution for own innovation activities? To answer this question, the organizers of the contests need knowledge of the subject matter and some insight into the customs and values of the target groups. Such knowledge is best acquired by experience.

Another aspect that has lately proven to be increasingly important to run successful innovation contests is the evaluation process. Underestimating the response rate can be very dangerous. Google, for example, received an unexpected number of more than 154,000 submissions during a contest called "10 to the 100". As a result, 3000 employees had to be involved in the evaluation process. One way to avoid such problems is to involve the participants of the contest themselves in the evaluation process, or to hand over this task to another crowd (Haller 2013).

5.3.2 Innovation Markets

Innovation markets are virtual places where innovation supply and demand meet. In general, they are realized as Web 2.0 supported online platforms, on which innovation seekers (typically organizations) announce innovation challenges and innovation providers (typically individual innovators or teams of innovators) propose concrete solutions or concepts. Innovation markets act as intermediaries, connecting innovation seekers and innovation providers (often called "solvers"). A growing number of innovation markets are available online for "seekers" to announce innovation challenges and for "solvers" to contribute.

One renowned and well established innovation market is Innocentive. It was founded in 2001 by Eli Lilly and declares itself as "the world's first open innovation marketplace" (www.innocentive.com). It originally focused on innovations in the chemical industry. As of today, Innocentive reports more than 375,000 registered "solvers" from nearly 200 countries, more than 2000 challenges posted to the public, and numerous others being performed within closer industrial circles, and most of the challenges allowing the setup of team project rooms for collaborative innovation in groups, more than 59,000 solutions submitted, 866 awards, ranging from about 5000 to over 1 million USD based on the complexity of the problem and a total of more than 48 million award money spent over the past years.

Other examples of innovation markets are NineSigma (www.ninesigma.com) and Battle of Concepts (www.battleofconcepts.nl). Whilst these markets focus on organizations publishing innovation problems, e.g. Planet Eureka (www.planeteureka.com) establishes a provider-driven innovation market. Solution providers get the possibility to look for both a suitable problem and a possible purchaser for their innovative solution. Organizations directly search for promising innovative concepts; an approach especially attracting small and medium-sized organizations.

During the last years, various initiatives have been started to transfer the idea of innovation markets from the virtual environment of the internet to the physical environment of meeting spaces in city centers, science parks, airports and other highly frequented locations—or to combine virtual and real in the operation of innovation markets. Examples include many living labs (www.openlivinglabs.eu), maker faires (www.makerfaire.com) or the service-manufactory JOSEPHS® (www.josephs-service-manufaktur.de/en).

5.3.3 Innovation Communities

Innovation communities enable innovators to collectively share and develop ideas, discuss concepts and promote innovations. Web 2.0 and social software based innovation communities normally bundle interested and specialized innovators for particular issues and thus support collective development and enhancement of innovation concepts. They originate from grouping together voluntarily and independently to create innovative solutions in a joint effort, embracing a family-like spirit. Open source communities are typical examples. Great success of these communities led to a growing number of issue-related communities in various industry sectors.

One interesting example of an innovative, self-organized and internet-based innovation community could be seen in the project *OScar* which was described by the community members in the following way (see Moeslein 2013): "The idea behind the OScar project is simple: A community of people virtually plans and develops a new car. The idea is about the goal to develop a simple and innovative car, but also about the way how this goal is achieved. We would like to convey the idea of open source to 'hardware' and we want *OScar* to be the precursor for many different projects in this field." The project was launched in 1999 on the basis of a published manifesto: "To built a car [...] without any factory, any CEO, any funds, any boundaries, [...] but instead with the support of lots of creative people in the web [...] with a global spirit of optimism, [...] representing absolute empowerment. Altogether, this meant to face huge challenges and to use 'the tool' internet in its essential sense." Today, community-based car development is implemented in a highly successful way by Jay Rogers and his team at LocalMotors. And this is how LocalMotors describes itself as a key player in the world of vehicle innovation: "Local Motors leads next-generation, crowd-powered automotive design, manufacturing and technology to enable the creation of game-changing vehicles. Through open-source principles, Local Motors helps solve local problems, locally; makes transportation more sustainable, globally; and delivers, through distributed manufacturing, innovative co-created vehicles and components with its virtual community of more than 30,000 designers, fabricators, engineers and enthusiasts from around the world" (www.localmotors.com).

UnserAller, a facebook-based online innovation community founded in 2010, on the other hand, aims for simpler products, e.g. mustard, salad dressing, snacks, shower gel, and the like, and does it with impressive success. After just one year more than 10,000 members have joined this community as innovation developers at the unserAller community, run by the Munich based start-up "innosabi". Today, companies widely discover innovation communities as a tool for strategically creating innovations. While in the early days emergent and self-organized innovation communities were the typical pattern for innovation communities, we see today a growing trend in strategic, firm-sponsored innovation communities which are often connected directly to the kind of products, services and technologies that these firms offer. Furthermore, many companies have discovered innovation communities as a new means to engage with their own employees and their direct business partners (Mahr and Lievens 2012).

5.3.4 Innovations Toolkits

Innovation toolkits provide an environment in which users develop solutions step-by-step. Through the application of online toolkits for the design of individualized T-shirts (e.g. www.spreadshirt.com) or the customization of your new car or its equipment (e.g. www.mini.com) basic knowledge about this open innovation tool is already available in literally every household. While the application of toolkits is quite widespread for the configuration of predefined solutions, the mass-customization of predesigned products or services or the selection of variants from a broad range of offerings, there is still a lot to be learned about the wider potential of toolkits to facilitate open and shared innovative activities. On the one hand, toolkits can expand the solution space that users can explore; on the other hand, they can organize the solution space in a suitable way to simplify the task of finding good results, in particular when different users are involved at the same time (Naik et al. 2016).

In order to implement toolkits for open innovation on the seeker's as well as on the solver's side efficiently, toolkits have to fulfill five basic requirements (see Reichwald and Piller 2006):

Full line of trial and error: Toolkit users tend to be more satisfied with their developed solution, when they can go through the entire cycle of problem solving. This requires that users receive simulated feedback on each step of the development process. Simulations enable users to evaluate the current solution, to improve on it in an iterative process. In this way, cognitive and affective learning processes are activated (learning by doing), which improve the quality of the solution.

Defined solution space: A toolkit's solution space defines all variations and combinations of feasible solutions. Basically, the solution space only permits innovation solutions, which take specific technical restrictions into account and are "feasible" from a technical point of view. Depending upon the type of toolkit, these constraints are more or less strict (as we will discuss below).

User friendliness: User friendliness describes how users perceive the quality of interaction with the toolkit. The challenge is to find the right degree of complexity (between over-complexity and over-simplicity), openness and interactivity as to perfectly stimulate the innovators creativity and motivation. If expectations or capabilities are heterogeneously distributed among the potential innovators, the seeker may want to make different types of toolkits available to the potential target groups of innovators.

Modules and components: Modules and components are the basic building blocks of a toolkit, (f.i., programming languages, visualization, help menus, drawing software, text boxes, libraries), which make up its operational functions and are made available to potential innovators for supporting innovation activities. Modules and components make up the toolkit's solution space and determine its user friendliness.

Solution transfer: After innovators have developed a toolkit-based innovation, their concept or solution is transferred to a seeker or manufacturer. This toolkit based transfer has to allow for a perfect translation of the solver's or user's solution in the "language" of the seeker or manufacturer.

Toolkits currently available in the market can be broadly classified in the three categories which differ in their strategic goals, design principles and users to be targeted. *Toolkits for user innovation* support the generation of innovation ideas based on a 'chemistry kit'-like interface to enable complete trail-and-error cycles, featuring a broad solution space with high costs of usage. *Toolkits for idea transfer* foster application of existing ideas in a new context based on a 'black boards'-like interface with unlimited solution space and low costs of usage. *Toolkits for user co-design*, by contrast, facilitate product configurations based on a 'Lego kit'-like interface, using a restricted solution space and standardized modules and are mostly used as a sales tool.

Regardless of specific configurations, toolkit-based innovation essentially needs feedback (either by the system or by users), simulates possible solutions (regarding design, performance and costs) and fosters the quest for solutions. The core challenge in configuring innovation toolkits is to define the degrees of freedom in relation to the complexity of the solution process and the capabilities and experiences of the users.

5.3.5 *Innovation Technologies*

Innovation technologies enable to progress from concept level of a (product) innovation to prototyping or even production. Innovation technologies comprise all kinds of additive manufacturing technologies (e.g. 3D-printers), but also 3D-scanners or laser cutters. All these technologies are associated with the prospect of an ongoing democratization of innovation activities and with the often proclaimed trend towards 'personal fabrication'.

Whilst in programming of open source software programmers globally cooperate on their software code, innovation technologies enable us to collectively and globally develop intangible specifications for real products, services, and solutions. Two-dimensional software controlled cutting with CNC-cutters allows creating fascinating designs for furniture, fitments and accessories (e.g. www.ronen-kadushin.com or www.movisi.com). 3D-printers (i.e. printers that create three-dimensional objects out of plastic powder under CNC-control) allow a fast and immediate "printing" of prototypes during development processes and thus enable the so-called 'rapid prototyping'. They also enable individuals or organizations to design and produce customized products instantly. Innovators envision first to design and specify three-dimensional models of innovations and second to order these directly from their household computers. Prospective visionaries like Neil Gershenfeld, Director of the MIT's Center for Bits and Atoms, anticipated this trend of a future of 'personal manufacturing' or 'personal fabrication' quite a while ago. His book "FAB: The Coming Revolution on Your Desktop - From Personal Computers to Personal Fabrication" vividly describes a future where anyone of us could become a producer or manufacturer, even at home (Gershenfeld 2005). While many did not believe in the proclaimed revolution of 'personal computing" at a time when computers were mainframes and only available for companies or governments, today we still struggle to imagine a future of 'personal manufacturing' or 'personal fabrication' where every household can become a production unit in a future world of value creation.

From a technical perspective this future is less hard to imagine: additive manufacturing (as opposed to our traditional subtractive manufacturing) is mainly digital manufacturing as 3D-computer models are translated in 3D-printouts. 3D-printers are already available more and more widely, prices are falling and application areas growing. It is obvious, that CAD software or 3D-scanners in combination with laser cutters or 3D-printing open up new possibilities of potential personal production. The analogy between personal computing and personal fabrication, in this context, is a helpful one to understand the anticipated future developments. Similar to conventional printing, simple printers with limited quality can be bought for personal use, while better ones are accessible is special shops and facilities like fab labs, which are nowadays already established in most larger urban areas. In addition, all kinds of internet companies start to offer manufacturing services for the products and innovations created by simply anyone, including as well the abovementioned car producer Local Motors, which has recently introduced a 3D-printed automobile.

Creating complete products, however, is by far not the only way how innovation technology like 3D-printing has proven to be useful. Quite in the contrary, it also allows innovators to make better use of standard parts and combine them to create new problem solutions which are easy to afford and to reproduce. This provides an enormous potential for innovation in developing countries, or in market niches where low demand does not justify huge investments in research and development. Among other examples, 3D-printing can add the necessary interfaces to create human prostheses from simple parts that can be bought in every hardware store. This is extremely helpful for children who lost arms and legs in wars and grow too fast to spend lots of money on sophisticated artificial limbs.

5.4 Tensions of Open Innovation

All presented tools for open innovation include four common and novel effects. They allow (1) for large numbers of innovators to contribute, (2) empower these innovators to collaborate in widely distributed settings, (3) foster high-speed interaction that radically accelerates innovation processes and (4) provide a global memory for innovators to build on. These four effects clearly facilitate collaboration across organizational boundaries and symbolize new opportunities to create innovations. Additionally, the five innovation tools open up spaces for novel strategies, which allow integrating all three types of innovators described earlier (Fig. 5.2).

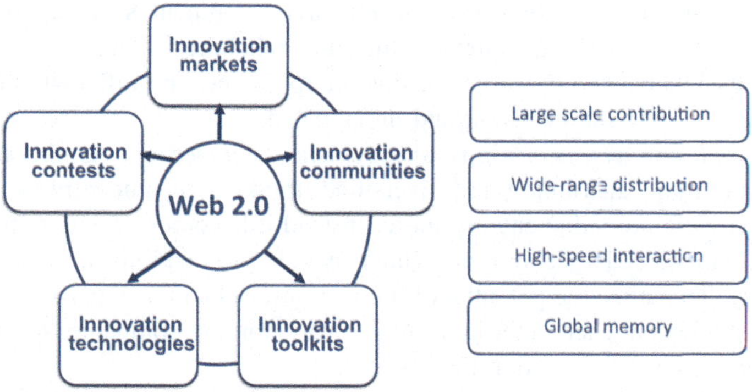

Fig. 5.2 Tools for open innovation and their effects

The different types of tools evolved more or less separately over the last decades, targeted to specific innovator groups. Today, they are frequently combined to create more sophisticated solution environments which can be adapted easily to the requirements of different participants. In this sense, they can be understood as basic building blocks to facilitate a growing number of new approaches to open innovation. To combine these building blocks wisely requires a better understanding of their modes of operation, but also of the tensions that open innovation creates in all kinds of value creation settings. These tensions are manifold and of utmost importance for leading innovation. Only a few can be sketched in the following.

Single versus team innovators: The existing conflict between individual and team innovation is intensified in open innovation. Additional and globally dispersed innovators (individuals as well as teams) are integrated in organizational innovation activities. Furthermore, the community of innovators, as an increasingly internet-based innovative community, reaches a more and more dominant status in open innovation. However, a loose interconnection of individuals and groups of individuals established to be common in innovation activities across boundaries of organizational units and organizations as a whole.

Local versus global range: The question between local pooling versus global dispersion of innovation activities was long regarded as an 'either-or' question. Several successful cases clearly proof that prerequisites are given to bundle and complementarily combine global creativity and innovation potential of many dispersed participants with local strengths. This step is enabled by the presented tools. However, handling these tools successfully and to gain competitive advantage remains as a core challenge for strategy and management.

Evolution versus revolution: Whilst the opinion prevailed for a long time that outside innovators could only contribute to continuous (evolutionary) innovations, many examples show that they support both continuous and discontinuous (revolutionary) innovations. Organizations have to deal with the predominant challenge of how to design, implement and strategically incorporate open innovation.

Professional versus amateur: Open innovation does not make experts obsolete, but it challenges them to accommodate their own practices to the contributions made by others. While the times are over in which industry experts could decide on their own about the future direction of technical development, such experts still play a vital role for securing the infrastructure and reliability of innovative solutions provided by others. The dynamics of this interplay between different stakeholders still remains widely intransparent and unregulated.

Single solution engineering versus systems engineering: As the innovation process gets more distributed and diversified, thinking of innovative solutions as single devices is not sufficient any more. Instead it becomes important to work on the development of larger systems which include new additions, but also legacy technology. To realize the potential of innovations, old and new parts of the system have to complement each other in a suitable way, forcing companies to think much more about systems integration than before.

The development story of Apple's iPod in combination with iTunes is an impressive example of discontinuous innovation that influences individual as well as affects existing boundaries of business sectors or even changes economic rules of whole industries, with various consequences for the interaction between all stakeholders that are involved. It is also a paradigmatic example of a successful and strategically oriented combination of efforts from different contributors in an overall corporate innovation strategy. We can currently see that such combinations cease to be an exception—they are increasingly becoming common practice.

Acknowledgments A very warm thank you to all inside and outside innovators who are part of our ongoing innovation research journey! This article bundles knowledge from the research projects "Open-I: Open Innovation *within* Organizations", funded by the federal ministry of research and education (BMBF) and the European social fund (ESF) (support code 01FM07053 and 01FM07054), the service-innovation initiative of Fraunhofer IIS, and draws on earlier work (e.g. Moeslein and Bansemir 2011; Moeslein et al. 2012; Moeslein 2013; Fritzsche 2017).

Bibliography

Bullinger, A.C., Moeslein, K.M.: Innovation contests – where are we?. AMCIS 2010 Proceedings. Paper 28 (2010). http://aisel.aisnet.org/amcis2010/28

Berger, C., Moeslein, K., Piller, F., Reichwald, R.: Co-designing modes of cooperation at the customer interface: learning from exploratory research. Eur. Manage. Rev. **5**(2), 70–87 (2005)

Chesbrough, H.: Open innovation: where we've been and where we're going. Res. Technol. Manage. **55**(4), 20–27 (2012)

Chesbrough, H.: Open Innovation: The New Imperative for Creating and Profiting from Technology. Harvard Business School Press, Boston (2003)

Fritzsche, A.: Open innovation and the core of the engineer's domain. In: Newberry, B., Michelfelder, D., Zhu, Q. (eds.) Philosophy and Engineering: Exploring Boundaries, Expanding Connections. Springer, Dodrecht (2017)

Fritzsche, A.: Dancing the device: a translational approach to technology. In: Pitt, J., Shew, A. (eds.), The Routledge Companion to Philosophy of Technology. Routledge, New York (2016)

Gershenfeld, N.: FAB—The coming revolution on your desktop—from personal computers to personal fabrication, Basic Books, Cambridge (2005)

Haller, J.: Open evaluation: integrating users into the selection of new product ideas. Springer Science & Business Media (2013)

Haller, J.B.A., Bullinger, A.C., Moeslein, K.M.: Innovation contests—an IT-based tool for innovation management. Bus. Inf. Syst. Eng. **3**(2), 103–106 (2011)

Huff, A., Moeslein, K., Reichwald, R. (eds.): Leading Open Innovation. MIT Press, Cambridge (2013)

Huizingh, E.K.: Open innovation: state of the art and future perspectives. Technovation **31**(1), 2–9 (2011)

Kahney, L. (2004). Inside Look at Birth of the IPod. Wired, 21 Apr 2004

Lakhani, K.R., Jeppesen, L.B., Lohse, P.A., Panetta, J.A.: The value of openness in scientific problem solving, Harvard Business School Working Paper No. 07-050, Cambridge (2007)

Mahr, D., Lievens, A.: Virtual lead user communities: drivers of knowledge creation for innovation. Res. Policy **41**(1), 167–177 (2012)

Moeslein, K.: Open innovation: actors, tools and tensions. In: Huff, A., Moeslein, K., Reichwald, R. (eds.) Leading Open Innovation, pp. 69–85. MIT Press, Cambridge (2013)

Moeslein, K., Bansemir, B.: Strategic open innovation: basics, actors, tools and tensions. In: Hülsmann, M., Pfeffermann, N. (eds.) Strategies and Communications for Innovations, pp. 11–23. Springer, Berlin (2011)

Moeslein, K., Huff, A., Reichwald, R.: Corporate open innovation: learning from longitudinal exploratory research. Guest editorial preface, special issue. Int. J. Knowl.-Based Organ. **2**(1), i–x (2012)

Naik, H.S., Velamuri, V.K., Möslein, K.M.: Simplifying solution space: a multiple case study on 3D printing toolkits. In: Proceedings of the European Conference on Information Systems (ECIS), Istanbul, Turkey (2016)

Neyer, A.-K., Bullinger, A., Moeslein, K.: Integrating inside and outside innovators: a sociotechnical perspective. R&D Manage. **39**(4), 410–419 (2009)

Reichwald, R., Piller, F.: Interaktive Wertschöpfung - Open Innovation, Individualisierung und neue Formen der Arbeitsteilung. Gabler, Wiesbaden (2006)

Rosenfeld, R., Servo, J.: Business and Creativity. Futurist **18**(4), 21–26 (1984)

Trott, P., Hartmann, D.A.P.: Why 'open innovation' is old wine in new bottles. Int. J. Innov. Manag. **13**(04), 715–736 (2009)

von Hippel, E.: Democratizing Innovation. MIT Press, Cambridge (2005)

West, J., Salter, A., Vanhaverbeke, W., Chesbrough, H.: Open innovation: the next decade. Res. Policy. **43**(5), 805–811 (2014)

Chapter 6
Cognitive Diversity of Top Management Teams as a Competence-Based Driver of Innovation Capability

Michael Hülsmann, Meike Tilebein, Philip Cordes and Vera Stolarski

Abstract In order to gain and maintain innovation capability, organizations have to adapt their profiles and processes to perpetually changing environmental conditions. However, the resulting need for a high degree of flexibility, which includes avoiding an information undersupply by being stable but inflexible, entails the risk of an information overload. Therefore, a balance between an organization's flexibility and its stability is needed. Top Management Team (TMT) cognitive diversity seems to constitute a promising resource, which under certain circumstances can be turned into an organizational competence, allowing for a high but stable level of organizational flexibility. Employing insights from complexity theory and adopting agent-based simulation is suggested as a further research method in order to deduce underlying causal interrelations.

6.1 Introduction: Strategizing in a World of Variety and Change

Permanent changes in so-called "real-time economies" have amplified the characteristics of markets as worlds of variety and change (Tapscott 1999; Siegele 2002). Beside others, this is due to fast and constitutional developments in information and communication technologies in the last 10 up to 20 years, which contributed to a drastic shift from an industrial to an information age of the global society

M. Hülsmann (✉)
University of Bayreuth, Bayreuth, Germany

M. Tilebein
University of Stuttgart, Stuttgart, Germany

P. Cordes
Jacobs University, Bremen, Germany

V. Stolarski
European Business School, Wiesbaden, Germany

N. Pfeffermann and J. Gould (eds.), *Strategy and Communication for Innovation*,
DOI 10.1007/978-3-319-49542-2_6

(Ottens 2003). Additionally, strategic management of organizations is more and more confronted with multiple, intertwined, sometimes contradictious, and mostly competing demands articulated by worldwide stakeholders and resource holders (Müller-Christ and Hülsmann 2003). In consequence, organizations like companies and their strategic management have to cope with an increasing complexity and dynamic in their relevant environments (Hülsmann and Berry 2004).

For that reason, the bases on which decisions in management are rendered are increasingly characterized by imperfect information (introduced into the management literature by Simon 1972). Organizations face the challenge that the amount of external information, emerging from their complex and dynamic environments, might exceed their information-processing capacities (Hülsmann et al. 2008). These capacities however are necessary in order to develop ideas, concepts, and practices that improve the respective company's product or service characteristics that are perceived as new and valuable by any stakeholder of the respective organization, in other words for the innovation capability (see, e.g., Rogers 2003). Gaining and maintaining innovation capabilities requires the ability to adapt the company's profile and processes to the perpetually changing environmental conditions (e.g., shifting customer demands). A lack of this ability as an effect that emanates from a too low information-processing capacity might lead to a so-called lock-in situation (Schreyögg et al. 2003). This includes a vital risk to the functionality and robustness of the locked organization (Hülsmann and Wycisk 2005b).

From a complexity-science-based perspective organizations can be regarded as Complex Adaptive Systems (CAS). They consist of autonomous, heterogeneous, interactive, and learning elements and coevolve with their environments. Thereby, they exist in a so-called melting zone between the edge of chaos and the edge of order (Wycisk et al. 2008; based on Holland 2002; Kauffman 1993). According to Wycisk et al. (2008), international supply networks are one example for such a CAS, which leads to the term Complex Adaptive Logistics Systems (CALS). Hence, the management of such a CALS in particular, but as well as of systems (e.g., a network of companies) and single organizations (e.g., companies) in general, are claimed to deal with increasing complexity and dynamics and the endangerment of lock-in-situations.

The challenge lies in two opponent requirements: On the one hand organizations need as much as possible information to be processed in order to react flexibly to changing environmental demands. On the other hand the information inflow has to be on a manageable level (Hülsmann et al. 2008). In this means it is essential to understand how companies can adjust the level and quality of their capabilities and capacities with regard to the accessing, acquiring, and processing of knowledge and information to external requirements. Thus, a key factor for successful strategic management, aiming at the organization's long-term survivability by gaining and maintaining the innovation capability, even under complex and dynamic conditions, seems to be the organization's adaptivity, which is defined by Hülsmann (2008) as the ability to keep the balance between an organization's flexibility and its stability (Hülsmann et al. 2008).

This article addresses this challenge by focusing on an approach that aims at an increase of organization's flexibility, without losing sight of the limitations of a flexibility spillover. According to Top Management Team (TMT) cognitive diversity research the heterogeneity of a team potentially contributes to organizational flexibility (Allen 2001; Akaishi and Arita 2002; Hülsmann et al. 2008; Stolarski and Tilebein 2009) by leading to an increased knowledge pool and inducing an extensive range of perspectives (Kauer et al. 2007; Pitcher and Smith 2001). The competence-based view, in which perspective flexibility can be seen as a competence itself, provides an adequate theoretical framework for an examination of cognitive flexibility as a source for flexibility as a competence on an organizational level (Sanchez 2004). However, former empirical diversity research does not deliver data of how and under which conditions cognitive diversity will reach its full potential as a driver for organizational flexibility. Additionally, it is not clarified yet from which degree on diversity endangers an organization's stability.

The overarching objective of this article is to analyze theoretically possible contributions of TMT cognitive diversity to organizational adaptivity and to point out approaches, elucidating the issue further. Therefore, several subgoals emerge: On a descriptive level, the need for organizational adaptivity and thus for a high but limited level of organizational flexibility will be specified. Furthermore, a description of flexibility as an organizational competence and TMT cognitive diversity as a possible approach to develop this competence shall be given. On an analytical level, the cause and effect chains between TMT cognitive diversity and the respective organizational flexibility and hence adaptivity shall be examined. Finally, implications for the management of organizations as well as for further research shall be deduced.

For that reason, the article proceeds as follows: In section two, potential effects of organizational flexibility in relation to a company's information-processing capability will be outlined, in order to show that a certain degree of flexibility is essential for an organization's innovation capability. Section three describes flexibility from a competence-based view, because TMTs are characterized by individual competencies that might lead in the following to organizational competences. Therefore, diversity management for TMTs will be emphasized as a special kind of the management of flexibility as an organizational competence. With recourse to these findings, section four introduces TMT diversity as a potential driver of organizational flexibility and proposes agent-based simulation as an alternative research approach, in order to derive deeper insights regarding the specific interdependencies between diversity and the flexibility of organizations. Section five sums up the findings and illustrates the attended limitations of this research as well as further research requirements.

6.2 Locked Organizations: Limitations to Rationality-Based Decision-Making in Strategic Management

According to Dörner (2001) a complex system (e.g., an organization or an organization network) can be seen as "(...) the existence of many interdependent characteristics in a section of reality (...)". The more interrelations among the elements within a system as well as between a system and its surrounding environment exist, the more information is potentially available (Dörner 2001; Malik 2000). Burmann (2002) gives the example of new global sales opportunities for companies, evolving from the vast amount of information on new products and technologies, which is offered via Internet (Burmann 2002). The term "dynamics" describes the accelerated variation of the system status (e.g., the Internet) over time (Coyle 1977; Probst and Gomez 1989). Applied to the mentioned example, dynamics could be understood as the permanently altering and available information on the Internet. In this case, the elements (pieces of information) themselves change and thus the relations between them and other organizations and their elements alter as well.

Hence, in order to obtain the organization's abilities to react to timely demands, they have to deal with complex information that are perpetually changing. In other words, organizations have to be capable to adapt to changes in their relevant environments so as to ensure their survivability. A system's adaptivity, in turn, helps to keep the balance between a high level of flexibility and a system's stability (Hülsmann et al. 2008). Whereas the former can be achieved by processes of system opening (Hicks and Gullett 1975; Garavelli 2003), the latter can be achieved by processes of system closure (Luhmann 1973, 1994). Both processes are opposite action alternatives, from which an organization's management has to choose in individual situations. Thus, the management is confronted with perpetual decision-making situations regarding the system's endowment with information, which is necessary for the management to render their decisions rationally, respectively, nearly rationally (Hülsmann and Wycisk 2005b).

Processes of system openings are based on mutual interrelations (Luhmann 1973), as they allow the system to communicate with the environment. Thereby it sustains the existential exchange process of resources (Staehle 1999; Böse and Schiepek 1989). While the system is in an open state, it is able to absorb some of the complexity within the environment and thereby the needed resources (Hicks and Gullett 1975). System openings are needed to gain an adequate amount of information for making preferably rational decisions. The higher the degree of complexity in the environment, the higher is by trend the degree of necessary information to find a solution for a certain problem. By system openings the decision maker therefore allows for more complexity that has to be absorbed by the system (Hülsmann et al. 2008), while the ability of processing this amount of complexity remains on a constant level. At the same time, the management has to take into account the dynamics of information and the risk of an information overload caused by system openings (e.g., Hülsmann 2005; Gebert and Boerner 1995; Gharajedaghi 1982). Furthermore, because flexibility is enabled by a system's ability to open its borders for required resources (e.g., information) (Hicks and Gullett 1975; Garavelli 2003), the system structures can alter permanently in the course of aiming at maintaining the system's ability to meet the demands of its relevant environment. In order to avoid system boundaries which diffuse too much, the degree of flexibility has to be balanced by processes of stabilization.

Hence, processes of system closures are required as a compensating means (Luhmann 1973, 1994). Not every single piece of information adds to the decision-making process, wherefore the flow of information has to be cut at some point to enable a timely decision. Furthermore, no social system (e.g., organizations) is able to absorb the whole complexity of its environment. Hence, the problem of bounded rationality is immanent in every decision-making situation, which implies that managers are forced to base their decisions on incomplete information (Simon 1972). For this reason systems have to select the information they process, confining themselves to those parts which are still manageable by the system and necessary to solve specific problems (Luhmann 1994). In consequence, processes of system closure must be incorporated in a decision-making process in order to select information in terms of quality and quantity with regard to its contribution to the company's innovation capability.

Therefore, in order to provide an organization, respectively, a system such as a CALS with a preferably high degree of adaptivity, its management has to find the optimum balance between system openings and system closures. In turn, this leads to perpetual decision-making situations. The management has to provide the system on the one hand with a preferably high degree of flexibility, but on the other hand it has to assure its stability. This can be seen as one of the major tasks and challenges in the management of such systems like organizations or organization networks.

If an organization does not succeed in balancing the underlying contradictory objectives, the management's ability to render decisions as rationally as possible might be restricted (Hülsmann and Berry 2004). The quantity and quality of information needed for a total rational choice in a complex and dynamic environment cannot be met by any system's capability to access, acquire, or process information. Too much system opening, respectively, flexibility, might therefore lead to an information overload. Too much system closure, respectively, stability, might lead in contrast to an undersupply of information (Hülsmann et al. 2008). Both decrease the system's abilities in adapting to the changing and challenging demands of the environment and therewith to be innovative. If this is permanent and not only a temporary incident the respective system or organization can become a so-called "locked organization" (Schreyögg et al. 2003). This phenomenon is based on path dependencies that develop in the course of repetitive actions of a system's single elements, fortified by feedback loops (David 1985). Actions that have been carried out in the past can therewith decrease the current range of possible action alternatives. Path dependency theory refers to this phenomenon as "history matters" (Arthur 1989; David 1994). As a result, a dysfunctional and suboptimal situation with a limited choice of possible decisions might occur

(Schreyögg et al. 2003). Processes of locking can therewith reduce an organization's capability to cope with the complexity of its environment by decreasing its ability to continue its exchange of vital resources with its environment. Neither can it identify all necessary resources (e.g., information about new products, new trends or innovations) from the offered mass in the environment nor evaluate or integrate them. As a consequence, the organization cannot respond to the demand of the environment (e.g., products of the company which potential buyers need) in time, quality, quantity, or place. This lack of flexibility could result in a disequilibrium which in turn could lead to negative environmental responses in terms of the required resources (e.g., through a lower volume of sales the company could loose its market shares). In the worst case, a lock-in situation may result in an organizational collapse (Hülsmann and Wycisk 2005b).

Therefore, the questions arise how to avoid a lock-in situation (ex-ante) and how to cope with an existing lock-in situation (ex-post). With recourse to the above stated dependencies between organizational flexibility and the risk of a lock-in situation, two challenges can be deduced: The increase of an organization's flexibility itself, as well as the increase of an organization's ability to cope with a high degree of flexibility. In other words, it is necessary to implement organizational flexibility on a level as high as possible, but low enough to ensure that the system's stability is not endangered, in order to gain and maintain the innovation capability.

6.3 Organizational Flexibility by Competence-Based Management

Subsequent to the above-mentioned need for organizational flexibility on a high but stable level, the question arises, how the organizational system's structure (e.g., of a company or of an organizational network of companies) can be created in order to allow a maximum level of flexibility without loosing the system's stability. One possible approach applying to the flexibilization of organizations is competence management (Sanchez 2004). According to the competence-based view, flexibility can be seen as a competence itself (Sanchez 2004; Krüger and Homp 1997) and can be understood as a basic demand of the organization structure, allowing it to form and apply competences and to refine their arrangement if necessary (Hülsmann et al. 2006). Important thereby is the differentiation between individual competencies and organizational competences (Müller-Martini 2008). In order to provide organizations with the required adaptiveness so as to secure a sustainable development and survival in dynamic, complex, and highly competitive environments (Hülsmann and Wycisk 2005a), flexibility is not only necessary on the individual levels. Rather it has to be implemented as a competence on an organizational level, wherefore the competence-based perspective offers an appropriate theoretical framework (Sanchez 2004).

According to Sanchez et al. (Sanchez and Heene 1996, p. 8; Sanchez 2004, p. 521) competences can be described as "(…) the ability to sustain the coordinated deployment of assets in ways that help a firm achieve its goals." In the theory of the competence-based view a firm is seen as a learning organization that builds and deploys assets, capabilities, and skills to achieve strategic goals (Hamel and Heene 1994).

Five different "modes" of competences have been defined by Sanchez (2004):

(1) cognitive flexibility to imagine alternative strategic logics;
(2) cognitive flexibility to imagine alternative management processes;
(3) coordination flexibility to identify, configure, and deploy resources;
(4) resource flexibility to be used in alternative operations; and
(5) operating flexibility in applying skills and capabilities to available resources.

Competence modes 3–5 are highly dependent on and limited by the competence modes 1 and 2. The cognitive flexibility, in turn, provides an appropriate connectivity to the approach of cognitive diversity. Therefore, this article focuses on the first two competence modes.

Competence mode 1 reflects an organization's ability to imagine different strategic ways of creating value in a certain market. Hence, it is highly dependent on the individual competencies of the organization's top managers to perceive the organization's relevant environment. Competence mode 2 reflects an organization's ability to imagine different processes in order to achieve the organization's strategic goals, which are, in turn, dependent on competence mode 1. In analogy to competence mode 1, this competence again highly depends on the individual competencies of an organization's top managers (Sanchez 2004). Although each meaning of the competence modes refers therewith to a different kind of flexibility, they all react toward a changing environment (e.g., changing markets or new technologies). Hence, they contribute, up to a certain point, to an organization's adaptivity. Therewith, the risk of a lock-in situation can be reduced. In consequence, the question arises, how organizations can achieve the competence "cognitive flexibility" regarding their strategic logics and their management processes. The latter are the preconditions for developing coordination, resource, and operating flexibility in order to enable organizations to avoid as well as to cope with lock-in situations as described above and hence, to gain and maintain innovation capability.

6.4 Design of Cognitive Diversity of TMT—A Key Issue of Strategic Organization?

According to the upper echelon perspective individual characteristics of TMT members like values, personalities, and experiences take influence on strategic choices and the overall performance of a firm (Hambrick 2007). TMTs constitute the aggregate informational entity, which is responsible for strategic decisions regarding the whole system's, respectively, organization's or company's future behavior and thus its development (Hambrick et al. 1996).

Speaking in terms of the competence-based view, TMT members' individual competencies might contribute to the TMT goals. In conjunction, individual competencies can constitute a resource, which composes an organizational competence under certain circumstances (Müller-Martini 2008). Focusing on TMT characteristics and composition (Hambrick et al. 1996), TMT diversity research contributes to the clarification of these circumstances.

TMT diversity is defined as the extent to which a top management team is heterogeneous with respect to its members' demographics or cognitions (Simons et al. 1999). As cognitive diversity, accounting for deep-level differences like cognitions, beliefs, and attitudes, is much more potent in group interactions than surface-level demographic differences (Harrison et al. 1998) this article focuses on cognitive differences opposed to surface-level demographic differences, e.g., age or gender. Cognitive differences are defined as distinctions in individual information processing, which comprises the way a person perceives and uses information (Lord and Maher 1990). Similarly, team-level information processing describes the way a team perceives and uses information (Akgün et al. 2006; Hambrick 2007). From a competence-based perspective, the information-processing capability of individuals as well as of teams can be regarded as a precondition to imagine alternative strategic logics and management processes. In other words, it constitutes a precondition for the development of cognitive flexibility, which is, in turn, a precondition for the competence organizational flexibility (Sanchez 2004). In this context the questions arise, in what way cognitive TMT diversity contributes to the creation of such an organizational competence and how this competence again contributes to organizational adaptivity, respectively, how does it help to avoid or to cope with organizational lock-ins and to foster an organization's innovation capability.

In empirical diversity research it is often assumed that cognitive diversity potentially enhances a team's information-processing capabilities, e.g., in means of coding and sorting out information or considering alternative solutions to a problem (Akgün et al. 2006). This again provides several assets relevant to organizational flexibility, e.g., allowing for decision-making based on less imperfect information. The basic idea underlying this assumption is that diversity broadens a team's knowledge pool. The heterogeneity of team members with regard to, e.g., knowledge, experience, and educational background enables them to produce a wide range of different ideas and to approach problems and tasks from different perspectives, resulting in enhanced problem-solving quality, creativity, and innovation (e.g., Cox and Blake 1991; Thomas and Ely 1996). In addition to these effects, taking place inside the organization, diversity may lead to a better acquisition and assimilation of new external knowledge (Jansen et al. 2005) and thereby again can contribute to an organization's flexibility (Akaishi and Arita 2002; Allen 2001).

However, in analogy to the risk of an information overload and the resulting need for a balance between flexibility and stability, diversity does not proof beneficial by all means. A number of studies revealed risks related to high degrees of team diversity (e.g., Miller et al. 1998; Cronin and Weingart 2007). Accordingly, communication and cooperation barriers tend to arise in heterogeneous teams, hampering information processing. In this case diversity leads to rigidity, counteracting the potential benefits it may have otherwise and confronting an organization with further risks in addition to the difficulties in today's business environments.

Apparently, within the field of diversity research there is no consensus on diversity effects, respectively, under which circumstances diversity emerges as a risk or a resource (see, e.g., overviews in Kilduff et al. 2000; Van der Vegt and Bunderson 2005). Although there seems to be a high potential, there are no definite implications for how to manage a diverse team successfully, respectively, how to unlock organizations by diversity management.

However, numerous researchers agree that unmeasured moderator, mediator, intervening, independent, or dependent variables contribute most predominantly to the observed inconsistency of empirical studies (e.g., Carpenter 2002; Hambrick et al. 1996; Pitcher and Smith 2001). Further there is evidence that diversity is a multifaceted construct, whose facets interact with team processes in converse ways (Pelled et al. 1999). Aside from the complexity of the subject matter, the underlying processes of information processing in diverse teams are highly dynamic and take influence on team performance in the course of time (Harrisson et al. 1998; Picher et al. 2001).

As Stolarski and Tilebein (2009) point out, classical empirical methods are not capable of comprising the complex and highly dynamic relationships, accompanying diversity effects. For example, numerous potentially relevant moderator and mediator variables, taking influence on the "cognitive diversity"—"team information processing" relationship must remain unconsidered (Stolarski and Tilebein 2009). Apparently, this calls for an alternative research approach and research method, which takes account of all relevant variables and of the dynamics evolving between the variables (Stolarski and Tilebein 2009). Thus, it will be possible to clarify TMT diversity effects and consequently to explore TMT diversity as a driver for organizational flexibility.

Complexity research provides illuminating insights regarding the relationship between diversity, information-processing capability and adaptivity of systems. Thus a basic understanding of the complex and dynamic mechanisms, which underlie diversity effects, can be derived from complexity theory (Tilebein 2006a). Complexity theory, which has its roots in systems theory, evolutionary biology, game theory, and information science, looks for common principles of CAS. The dynamics of these systems can be either chaotic, locked in a stable state, or "on the edge of chaos" (Kauffman 1993).

Although chaotic systems are highly dynamic, they cannot respond to external impulses in a timely and adequate manner, comparable with organizations, being too flexible. Systems with a lack of flexibility on the other hand always reproduce their prior states wherefore they are not capable of adapting to changing conditions. Apparently, systems with too much flexibility and systems lacking flexibility are not able to cope in an adequate manner with the information provided by their environment, which, in turn, can lead to lock-in situations and hinder the organization to be innovative.

The edge of chaos is a balanced state where a system shows both stable and changing characteristics. In this state, a system reaches a maximum in information-processing capability, and it can display spontaneous order and absorb external disturbances (Langton 1992; Kauffman 1993; Wolfram 1994; Tilebein 2006b). This again reflects organizational adaptivity as described in Sect. 2.

The different types of dynamics in CAS are based on information-processing procedures performed by the interrelated system elements, which are called agents, and on the information flows within the network of agents (Holland 1995). The more interaction partners each agent has, i.e., the higher the network density is the more information can be passed on through the network (Boisot and Child 1999). In order to be adaptive, CAS have to combine this information transfer capacity with adequate information storage capacities, which strongly relates to the idea of balancing system opening and system closure that was introduced in Sect. 2.

This analogy is acknowledged by results from agent-based simulation models, used in complexity theory (Law and Kelton 1991). For example, there is evidence from very simple agent-based models, Kauffman's Boolean Networks that an optimized fit between network density and diversity leads to adaptivity of the system. In contrast, networks with low diversity and low density might lead to locked systems, while highly dense networks between highly diverse agents might result in chaos. Lowering or rising diversity can optimize information-processing capacity of those extreme systems and thus allow for adaptivity again (Kauffman 1993).

Agent-based simulation in general strives to model individual behavior and interactions of a system's micro-level entities, called agents (Klüver et al. 2003). Most important concerning the current issue agents possess schemes determining their individual goals and their actions, including their behavior toward and the information exchange with other agents (Dooley 2002; Klüver et al. 2003). This allows for simulating, e.g., the emergence of cooperation in social groups, the communication of beliefs and aspirations, the negotiation with or without resorting to conflict, and the coordination of activities (Schieritz and Milling 2003; Wooldridge 2002). Additionally agents are able to interact not only among themselves, but also with their environment. In other words, they absorb the information provided by their environment and adapt to them (Wooldridge 2002).

To sum up, in complexity research agent-based simulation is an established approach to examine diversity effects in regard to information-processing capability and flexibility, respectively, stability of a system. Recently management science is beginning to realize and discuss on a broader basis the vast potential of simulation methods for experimental theory building (Davis et al. 2007; Hazy 2007).

6.5 Conclusions

In competitive environments of increasing variety and change strategic management has to face the risk of "locked organizations" (Schreyögg et al. 2003), which includes vital risks to the organization's operational reliability and robustness (Hülsmann and Wycisk 2005b) and to its innovation capability. Unlocking an organization from this risky state and reconstructing, respectively, maintaining its ability to cope with volatile and diverse environmental demands requires organizational adaptivity, which in turn requires a high, but stable level of organizational flexibility (Hülsmann et al. 2008). From a competence-based perspective the capability to imagine alternative strategic logics and management processes is an essential precondition for organizational flexibility, respectively, cognitive flexibility (Sanchez 2004).

The upper echelon view acknowledges a major role of management team characteristics in strategic decision-making and flexibility. Accordingly, one of the drivers of organizational flexibility seems to be TMT cognitive diversity. As shown many times in TMT diversity research, diversity leads to an increased knowledge pool, inducing an extensive range of perspectives (Kauer et al. 2007; Pitcher and Smith 2001). Thus diversity potentially enhances the flexibility of the organization, which might allow for adaptivity of the system (Allen 2001; Akaishi and Arita 2002).

To sum up, TMT diversity might constitute an organizational competence, which fosters the innovation capability. But the circumstances, which enable organizations to utilize this resource and develop an organizational competence, have not been identified by empirical research yet. While prior research has shown that top management team diversity affects strategic flexibility and performance, there is no consensus on the mechanisms underlying these effects.

Hence, strategic management should be aware of cognitive diversity as a potential driver of unlocking organizations. Adjusting the heterogeneity of a management team, e.g., via appointing new members to the team or by providing similar, respectively, dissimilar information to the team members, in order to allow for organizational adaptivity could be a first, practical approach. However, regarding the contradictory empirical results described above, caused by, e.g., unmeasured moderator and mediator variables changing TMT diversity by way of trial in order to optimize organizational flexibility appears rather hazardously.

Accordingly, it is proposed to use agent-based models to clarify how and under which conditions diversity will reach its full potential as a driver for innovation capability. Referring to the shortcomings of empirical diversity research described above, agent-based simulation appears to be an appropriate approach to cope with the complexity and dynamic of intertwined variables, which contribute predominantly to the inconsistent results in empirical research.

Regarding the huge amount of TMT diversity studies, each focusing on different variables, which moderate the relationship between cognitive diversity and information processing (Stolarski and Tilebein 2009), it becomes clear that simulation models in management research will have to be more complex than those originally used in complexity theory, e.g., Boolean Networks. This renders more complex agent-based systems appropriate tools for modeling diversity and information processing in management teams as they are able to model aspects like, heterogeneity, autonomy, openness to the environment and communication (Bandtke 2007).

Thus, it seems promising to develop and apply agent-based systems, which are capable of modeling decision processes in TMTs in order to ascertain the conditions under which TMT diversity contributes to the innovation capability of organizations.

Bibliography

Akaishi, J., Arita, T.: Misperception, communication and diversity. In: Standish, R.K., Abbass, H.A., Bedau, M.A. (eds.) Artificial Life VIII. MIT Press, Sydney (2002)

Akgün, A.E., Lynn, G.S., Yilmaz, C.: Learning process in new product development teams and effect on product success: a socio-cognitive perspective. Ind. Mark. Manag. **35**(2), 210–224 (2006)

Allen, P.M.: What is complexity science? Knowledge of the limits to knowledge. Emergence **3**(1), 24–42 (2001)

Arthur, W.B.: Historical events [1989] competing technologies, increasing returns, and lock-in by historical events. Econ. J. **99**(394), 116–131 (1989)

Bandte, H.: Komplexität in Organisationen. GWV Fachverlag, Wiesbaden (2007)

Boisot, M., Child, J.: Organizations as adaptive systems in complex environments: The case of China. Organ. Sci. **3**, 237–252 (1999)

Böse, R., Schiepeck, G.: Systemische Theorie und Therapie. Asanger, Heidelberg (1989)

Burmann, C.: Strategische Flexibilität und Strategiewechsel als Determinanten des Unternehmenswertes. Deutscher Universitäts-Verlag, Wiesbaden (2002)

Carpenter, M.A.: Implications of strategy and social context for the relationship between top management heterogeneity and firm performance. Strateg. Manag. J. **3**, 275–284 (2002)

Cox, T.H., Blake, S.: Managing cultural diversity: implications for organizational competitiveness. Acad. Manage. Executive **3**, 45–56 (1991)

Coyle, R.G.: Management System Dynamics. Wiley, London (1977)

Cronin, M.A., Weingart, L.R.: Representational gaps, information processing, and conflict in functionally diverse teams. Acad. Manage. Rev. **32**(3), 761–773 (2007)

David, P.A.: Economics of QWERTY [1985] clio and the economics of QWERTY. Am. Econ. Rev. **75**, 332–337 (1985)

David, P.A.: Why are institutions the 'carriers of history'?: path dependence and the evolution of conventions, organizations and institutions. Struct. Change Econ. Dyn. **5**, 205–220 (1994)

Davis, J.P., Eisenhardt, K.M., Bingham, C.B.: Developing theory through simulation methods. Acad. Manag. Rev. **2**, 480–499 (2007)

Dooley, K.: Simulation research methods. In: Baum, J. (ed.) Companion to Organizations. Blackwell, London (2002)

Dörner, D.: Die Logik des Misslingens: Strategisches Denken in komplexen Situationen. Rowohlt, Reinbek near Hamburg (2001)

Garavelli, A.C.: Flexibility configurations for the supply chain management. Int. J. Prod. Econ. **8** (2), 141–153 (2003)

Gebert, D., Boerner, S.: Manager im Dilemma—Abschied von der offenen Gesellschaft. Campus-Verlag, Frankfurt a, M. et al (1995)

Part II
Communicative Perspectives on Innovation

Chapter 7
Pairwise Communication for Innovation at Work

Martin J. Eppler and Lawrence McGrath

Abstract In this chapter, we argue for the overlooked potential of pairwise communication for innovation efforts in organizations. We review the research that documents that pairs outperform other team constellations when it comes to idea development and refinement. We describe how to setup and optimize such dyadic communication so that it can contribute to organizational innovation. In the chapter, we give readers a strategy to leverage teams of two for innovation efforts, and we show how managers can enable pairwise communication. Cornerstone of the chapter is a concise and instructive typology of creative pairs as well as guidelines for their setup, management, and transition. We conclude the chapter with an outlook on future research on dyadic communication for innovation.

7.1 Introduction: The Overlooked Potential of Pairs for Creativity and Innovation

Thomas Edison is widely recognized as one of the single greatest inventors of the modern era. However, he is perhaps just now beginning to receive recognition for his greatest invention—modern innovation management. "Edison is in reality a collective noun and means the work of many men" stated Edison's long-time engineering assistant Francis Jehl (Hargadon 2003, p. 93). Edison's inventiveness was the result of close pairwork with collaborators such as Nicholas Tesla and his second wife Mina. Until recently, the innovation management field has largely overlooked the power of pairwork. Today, an increasing body of work is beginning to demonstrate the high value of pairs for ideation (coming up with new inventions and refining them) and innovation (bringing them to commercial viability).

M.J. Eppler (✉) · L. McGrath
University of St. Gallen (HSG), Blumenbergplatz 9, 9000 St. Gallen, Switzerland
e-mail: martin.eppler@unisg.ch

L. McGrath
e-mail: l.mcgrath.lm@gmail.com

N. Pfeffermann and J. Gould (eds.), *Strategy and Communication for Innovation*,
DOI 10.1007/978-3-319-49542-2_7

Why have pairs been overlooked as innovation catalysts? Osborn's popular brainstorming technique (Osborn 1957) emerged in 1957, and sections of the scientific community immediately condemned it (Taylor et al. 1958). This led the scientific community to a preoccupation with the relative creative merits of working alone or in groups of approximately four to twelve people for four decades (Bouchard et al. 1974; Campbell 1968; Chatterjea and Mitra 1976; Dunnette et al. 1963; Graham and Dillon 1974; Gurman 1968; Mullen et al. 1991; Paulus et al. 1995; Renzulli et al. 1974; Stroebe and Diehl 1994; Sutton and Hargadon 1996; Vroom et al. 1969). While preoccupied with these polar opposites, most academics ignored the smallest possible innovation group—two. Below we review why innovation can profit from teams of two, how to best setup pairs for creativity and innovation and what guidelines to follow when doing so. We also offer a strategy of how to harness the power of pairwork for innovation as an individual in the final section of the chapter.

Our review of creative pairs (known as dyads in the literature) begins in 1960. It was then that Cohen, Whitmyre, & Funk found cohesive, trained dyads to be creatively efficient and effective. Their study positioned cohesion as the key creativity success factor for dyads. On the heels of Cohen, Whitmyre & Funk's article, Janis' influential work on groupthink (Janis 1971) began emerging. Within academia, Janis' popular book (1972) established groupthink as the enemy of creative problem solving. Cohesion became synonymous with groupthink, and creativity research on dyads largely ceased. Large group brainstorming remained beloved with practitioners and became part of the popular vernacular, while researchers continued to find evidence of solo idea generation's relative advantages (Bouchard et al. 1974; Chatterjea and Mitra 1976; Graham and Dillon 1974; Renzulli et al. 1974).

In contrast to the mainstream, the pooling of individual ideation efforts became the academic gold standard. Two years after the second edition of Janis' book Groupthink was published (Janis 1982), German researchers Pape & Bölle conclusively demonstrated higher idea count among ad hoc, untrained dyads than pooled individuals (Pape and Bölle 1984). This research went virtually unnoticed, and academics continued in vain to attempt to sway practitioners to substitute solo ideation for large group ideation.

Creativity research on dyads slowed to a trickle, but did not stop. Between Cohen et al. (1960) study, and groupthink's dominance of discussions, the renowned creativity researcher Torrance found significant ideation benefits for dyads. Torrance was the author of the world's most-used creativity test—the Torrance Tests of Creative Thinking (TTCT). Torrance found higher task persistence, participant perceptions of enjoyment, originality of expression and stimulation in dyads than lone ideators (Torrance 1970, 1971). Torrance also found that dyads exhibit more cognitive flexibility than lone ideators—dyads covered more categories of thought (Torrance 1971). There was more variety in their ideas. Such cognitive flexibility is a valuable creative mechanism (Nijstad et al. 2010) and a major component of the original TTCT (Kim 2006). Moreover, dyad's access to cognitive flexibility sits in complete discordance with groupthink's homogeneity of thought.

Unfortunately Torrance's work on dyads' creative value found little resonance. This is curious when one considers his stature among creativity researchers. That is, until we note that much of Torrance's work focused upon the identification and encouragement of talented youth. Management researchers tend not to place much stock in pediatric research. However, the two studies mentioned above both used an adult sample. Torrance (1970) even compared the results of adults to those of children, and found that the originality effect of dyads to be more pronounced in an adult sample. Be that as it may, it appears that dyads may finally be getting the acknowledgement they deserve from creativity and innovation researchers.

Numerous dyadic advantages in creative pursuits are simply functions of the number two. The creativity literature commonly references three major impediments to creative processes: Freeriding, production blocking, and evaluation apprehension (Diehl and Stroebe 1987).

Freeriding can be defined as letting others pay the price for attaining the group's objective (Kerr and Bruun 1983). Freeriding increases with group size (Petty et al. 1977). Thus the smallest possible group—two—is that with the lowest freeriding. In pairs, the work rate of each individual is completely transparent to both participants. Social mechanisms of responsibility thus reduce freeriding to a minimum in pairs (Petty et al. 1977).

Production blocking refers to "the rule that only one group member speaks at a time" (Diehl and Stroebe 1987, p. 498). As only one can produce a contribution, the production of all others in the group is blocked. Larger groups have more production blocking. It is logical that the length or the probability of delay likely increases with group size (Bouchard and Hare 1970; Diehl and Stroebe 1987). Thus, the smallest possible group size—two—is that with the lowest production blocking. One need only wait for the partner to finish contributing, and then individuals may contribute, or even begin a direct dialogical exchange.

Evaluation apprehension is the fear of being judged negatively for sharing an idea (Rosenthal and Rosnow 2009). It has long been recognized as a potent ideation disruptor (Camacho and Paulus 1995; Collaros and Anderson 1969; Girotra et al. 2010; Henchy and Glass 1968; Kavadias and Sommer 2009; Paulus and Dzindolet 1993). Early creativity research showed that evaluation apprehension damaged creative processes by inducing individuals to completely withhold ideas (Collaros and Anderson 1969). Arguably even more deleterious is the finding that evaluation apprehension leads to production of dominant, pedestrian responses instead of subordinate, original responses to prompts (Henchy and Glass 1968). Such behavior fixates groups upon non-original solutions (Smith 1995). Evaluation apprehension increases with group size (Gallupe et al. 1992), thus groups of two suffer the least evaluation apprehension. Moreover, as dyads disallow the formation of minority and majority in- and out-groups (Moreland 2010), valuable disagreement and debate can freely develop.

As a unit, pairs themselves are useful creativity building blocks within an organization. Not only does evaluation apprehension reduce within pairs, but pairwork enables pairs to more confidently interact with parties external to the pair (McGrath 2015b). Dew and Hearn (2009) found pair-based groups to solve problems equally well as individually based nominal groups. Both these groups outperformed traditional interactive groups (Dew and Hearn 2009). Lowered evaluation apprehension certainly aided the problem-solving efforts of pair-based groups. In further dyad research, Parks and Cowlin (1996) found that groups accepted unverifiable information and integrated it into conversation when it was contributed by pairs, but not single individuals. This effect was pronounced when the group was familiar with the topic area (Parks and Cowlin 1996). Therefore, pairs are both more likely than individuals to deliver original, unorthodox messages (McGrath 2015b), and have them accepted by a group (Parks and Cowlin 1996). This is of great consequence for creative problem solving and innovation when one considers that highly original new ideas cannot be verified, and much closed innovation continues to take place among groups of experts (Herzog and Leker 2010).

The numerical functions and pair level benefits mentioned above have profound creativity effects, but how can creative pairs be managed for innovation? Which kind of pairs should managers foster in order to use pair work for their innovation efforts? Our typology in the next section and subsequent guidelines will address these questions.

7.2 Typology of Creative Pairs

This section presents a useful and clear classification of creative pairs, which instructively assists innovation managers in using pairs as a creativity management tool. Our classification uses our own observation and experiments, as well as the distinctions documented in Shenk (2014). We distinguish the following six creative pair constellations:

Pair #1: The Dreamer and the Doer
Pair #2: The Mentor and the Mentee
Pair #3: The Hero and the Sidekick
Pair #4: The Rivals
Pair #5: The Paradox Pair
Pair #6: The Leader and the Manager

Each class of pair will be expanded upon below. This typology was conceptualized for relational pairs within organizations, yet applies equally well to ad hoc task pairs. With the exception of the leader and the manager, the pairings apply equally well regardless of whether members are intra-organizational or external. Guidelines for pair setup, management, and transition follow below.

7.2.1 The Dreamer and the Doer

The label dreamer and the doer originates with Shenk (2014). This is a high-potential creative pair existing of an enthusiastic stargazer and an implementation specialist. In many ways, this pair mirrors the typical view of the division between creative and routine work. The former is often glorified while the latter is seen as pedestrian. Truth be told—both need each other.

Dreamers are often starters of new trains of thought. They restlessly scout and make suggestions. However, they seem to skip through activities without an implicit interest in implementation. Doers are detail-oriented realists. They are often indispensable and bear a heavy workload in teams. Yet, many of these workhorses may struggle to inspire others and see possibilities beyond the track of project milestones. The Virgin Group's Richard Branson and Joshua Bayliss are an illustrative corporate example.

Dreamers and Doers have been responsible for a number of humanity's creative breakthroughs. Shenk (2014) points to the radioactivity work of Pierre and Marie Curie, as well as Wilbur and Orville Wright's gift of flight to civilization. When we bear in mind what creativity is, we get to the crux of dreamer/doer pair's potency—and how to leverage them.

The widely accepted definition of creativity is process output that is simultaneously original and feasible (Diehl and Stroebe 1987; Dean et al. 2006). That is, both novel and able to be realized. The dreamer brings novelty to this equation. The doer brings feasibility. To make a synergistic dreamer/doer pair, both parties should be able to fully express their strengths.

Briggs and Reinig (2010) add a third dimension to the standard definition of creativity—instrumentality. Besides being original and feasibly, creative ideas must be useful. This is typically an overlooked hygiene factor, yet instrumentality as creativity delivers the management insight below.

By virtue of their role as initiator in the creative project process, dreamers set the direction for projects. Managers of dreamer/doer pairs thus need to ensure that dreamers are completely informed of, and aligned with detailed organizational objectives. This will ensure the instrumentality of the solution the pair provides. Managers should give dreamers an absolute free hand on how to reach objectives—but they should give dreamers a prioritized list of concrete objectives. Competent dreamers may then use tools such as the synergy map (Eppler and Pfister 2011) to simultaneously achieve multiple objectives through a combinatory solution (Koestler 1964).

Much research shows that creativity is highly iterative (Brown 2008)—this partnership is predetermined to develop its own iteration spirals. Much to the chagrin of the doer, the dreamer will periodically reimagine and restart the project from a different tack. Managers should encourage this dynamic, providing the project remains focused on its objectives and uses resources efficiently. The doer may find the need for such iterations difficult to swallow. To avoid frustration, managers must make doers understand the iterative nature of creative progress and the sunk cost fallacy associated with their previous efforts. The dreamer tends to be a classic creative who requires space. At early project stages it is advisable to employ the doer in a shielding role. This serves the dual purpose of protecting the dreamer from distraction and preventing unnecessary effort on the doer's behalf. Eventually the dreamer and doer will become equally visible as the joint fruits of their labor emerge.

7.2.2 The Mentor and the Mentee

Many organizations already employ mentoring to acclimatize and transfer knowledge to new hires. However, mentor/mentee pairs offer a powerful creative problem-solving perspective. Mentor/mentee pairs can be particularly adept at solving interconnected problems. Experience not only flows down from long-term to new employees—both educate each other.

To illustrate the value and dynamics of mentor/mentee pairs, we will examine the experience of Hartford Insurance—a division of The Hartford, a Fortune 500 company founded in 1810. Hartford Insurance was faced with an aging workforce, low retention among young employees and difficulties in digitally reaching customers. In response, Hartford Insurance paired senior managers and junior employees into mentoring relationships. The juniors then mentored the senior managers on social media. The result was improved campaign planning by senior managers, advancement of the company intranet, and higher engagement among junior employees.

Mentoring is more than knowledge transfer. Values may also be transferred—or knowledge and values may both be challenged. A mentoring relationship may thus provide employees with refreshing new perspectives—and allow them to make otherwise unlikely cognitive connections. Although symmetry of knowledge leads to symmetry of learning and mentees will acquire more in learning situations, mentors will also attain some new knowledge (Kozlov and Grosse 2016). Often, the most powerful knowledge a mentor can transfer to a mentee is the knowledge structure of interconnecting fields. This allows the mentee to understand where their knowledge overlaps—or not—with that of their mentor, or others. Moreover, managers should not forget that the opportunity to act as a mentor or mentee is typically appreciated as a compliment by both parties, and thus provides motivational impetus.

Mentees with the empowerment to persist in unconventional directions can be particularly valuable to organizations. Such mentees will push through the comfort zone of the mentor and bring the pair into new solution spaces. The expertise of the mentor can then be focused upon an entirely new question statement. The mentee essentially serves as a problem reframing device for the mentor (Chi et al. 1989; Renkl 1997). Solving these new problems will push the mentor to the limits of his or her capabilities. There is likely to be resistance from the mentor, and managers will need to reiterate the rewards of solving newly defined problems. Regular, radical reframing may indicate that a mentee has graduated from the mentor/mentee relationship and will thus require empowerment to retain his or her capabilities within the organization.

An explicit transactive memory system (Wegner 1986), or understanding of who knows what in the relationship, is typically useful but perhaps no more so than in mentor/mentee pairs. Concept maps are a useful tool for visualizing knowledge and quickly creating an awareness of who holds which knowledge (Eppler, 2006; Schreiber and Engelmann 2010).Concept maps help groups to solve problems more quickly and accurately (Engelmann and Hess 2010). In domains where the mentee holds higher knowledge levels than the mentor, an awareness thereof may even see the existing mentor/mentee relationship flipped. Such domains are areas in which the mentor may be able to substantively reframe question statements and raise their knowledge level. Transparency of knowledge helps individuals to define roles (Chi et al. 2004), reduce conflicts (De Dreu and Weingart 2003), and learn from each other (Kozlov and Grosse 2016). Each participant should create a concept map before their collaboration begins, and these maps may also serve as useful pairing tools for management.

7.2.3 The Hero and the Sidekick

Modern intra-organizational creativity requires both talent and effort behind the scenes and in the spotlight. Engineers and developers typically work behind the scenes to make impossible customer demands possible. Project managers and assistants do the more public work of negotiating innovation stage gates, presenting project milestones, and networking with external stakeholders. When innovation processes are divided into fixed front- and backstage roles, we speak of a hero and sidekick pair. Here the sidekick has relative prominence to the hero, and thus need not be completely behind the scenes. Moreover, both these roles are interdependent and of equal value to the entire creativity and innovation process.

Hero and sidekick pairs are relatively common in history. History has often attributed the joint creativity and perseverance of two or more to one single individual. This is referred to as the lone genius creativity myth (Montuori and Purser 1995). The current academic environment reinterprets a number of lone geniuses as masterful hero collaborators in hero/sidekick pairs. Perhaps chief amongst these is Thomas Edison. Edison is well known for managing a team of inventors called "muckers" who worked on innovations ranging from motion pictures to lighting. Edison appears to have been the first modern innovation manager—and Edison was the hero. Edison's primary sidekick was his second wife, Mina Edison—the daughter of a farm machinery inventor. As typical of this epoch, Mina Edison was wife, business confidant, and home executive. While Thomas Edison headed his empire, Mina Edison bolstered the Edison reputation by managing public relations activities such as the social calendar. The hero and the sidekick worked in complete symbiosis.

In modern hero and sidekick pairs, positions emerge more out of personality and cultural variables than archaic gender roles. Role alignment is vital in hero/sidekick pairs. Partners should recognize the strengths of their counterpart in progressing the pair as a complementary team. Moreover, heroes in particular should work to ensure that "differential recognition" (Farrell 2001, p. 255) from external parties does not undermine the creative relationship.

Cultural differences are also a factor in front- and backstage tendencies. Individuals from some cultures with low levels of individualism and high levels of uncertainty avoidance may absolutely prefer to be sidekick to a hero. South Korea and Taiwan are examples of such cultures (Hofstede 2016). Highly individualistic cultures that do not avoid uncertainty are Australian and Canadian cultures (Hofstede 2016). Individuals from the latter cultures are more likely to perform well in front stage roles. Individual personality traits, independent of culture, may also play a part in preferences for front- or backstage roles.

Some individuals may prefer the sidekick role due to the evaluation apprehension discussed above. In theory, such individuals should outgrow their timidity after decision makers repeatedly accept their ideas, especially if they are accepted over other competing ideas (Cottrell 1972). For team members carrying excessive evaluation apprehension loads, a sidekick role may be a welcome relief and an ideal training ground in which to develop confidence.

It is not overly important whether evaluation apprehension, cultural differences, or personality traits determine the front or backstage preferences and performances. What is absolutely crucial to effectiveness is that the reasons for occupying both roles are open and transparent to both dyad members (Varvel et al. 2004). Understanding and tolerance are vital. Frank private discussions early in the relationship will pay later dividends for hero and the sidekick pairs. Such intimate conversations may be usefully based upon the results of reference objects such as Briggs–Myers personality test results, cultural dimensions (Hofstede et al. 2010), or cultural metaphors (Gannon 2011).

Exactly as the lone genius myth long prevailed throughout history, it no doubt exists within contemporary organizations. Managers should let apparent lone geniuses continue to profit from their sidekick, while ensuring that the sidekick receives resources to master his or her craft.

7.2.4 The Rivals

The creatively productive combination of rivals is one of the most theoretically attractive, yet practically difficult endeavors in pairwise management. Research on conflict in small groups shows that creativity is one of the few areas that can actually benefit from friction (Nemeth and Ormiston 2007). The full expression of a multiplicity of perspectives allows full solution space exploration (Briggs and Reinig 2010) and works against groupthink (Janis 1982). This is particularly valuable at the question framing stage of creativity processes (Verganti 2016). The stormy rivalry relationship of Bill Gates and Larry Ellison is evidential of rivalry driving progressive innovation. Further interpersonal communication research finds that conflict leads to a rapid refinement of concepts through the mechanism of "improvised enrichment" (Hewing 2013, p. 16). Improvised enrichment refers to an individual's spontaneous ideation in response to an unforeseen criticism. Research points to the value of such mechanisms in promoting originality in solutions (McGrath 2015a). Improvised enrichment is an invaluable idea refinement mechanism in dyads with creative objectives.

Long-term relational dyads with an undercurrent of rivalry will benefit from bouts of improvised enrichment. However, the time scale for response to criticism must be adapted to the heat of the rivalry. Dyadic conflicts can be classified as disagreements, cool conflicts, and hot conflicts (Grove 1991); and will likely move through all three classes in the course of a creative rivalry pairing. Disagreement is healthy and the sparring of socially competent rivals will trigger new thought patterns and refine ideas. Disagreement is a goal state for rivalry pairs. Rivals in cool and hot conflicts require time and breathing space separate from one another, especially those in hot conflicts. The functional management of cool creative conflicts begins with the establishment of procedures to ensure conflict does not inadvertently escalate into a hot conflict. Cooling-off periods not only lower the risk of inadvertent conflict escalation, but also allow individuals time to reflect upon the best response to new perspectives and input from rivals. An absolute key for a functional rivalry pairing is thus the level of interpersonal communicative competence and reflective self-awareness levels of both parties involved (Grove 1991; Schön 1983).

Mastery and performance goals are a valuable tool for managers of rivalry pairs. Performance goals refer to an individual's wish to competitively demonstrate competencies relative to others—such as one's manager. Mastery goals are the desire to understand a task, acquire new knowledge, and develop abilities (Darnon et al. 2007). Dyads in general, and rivals in particular, should be encouraged to pursue mastery goals. Mastery goals are absolutely in line with the intrinsic motivation so central to creative performance (Amabile 1997; Harackiewicz et al. 1998), and have been linked to promoting and sustaining self-regulated learning increases (Pintrich 1999). In the pursuit of mastery goals, individuals search for challenges and persist at tasks even in the face of difficulties. Individuals with mastery goal orientations adapt to develop their performance (Harackiewicz et al. 1998). Mastery goal orientation goes hand in hand with disagreement and improvised improvisation.

The interpersonal dynamics between rivals may eventually balance into a state of convivial disagreement, or even metamorphosis into another type of creative pair. Should this reconciliation occur, then research suggests that the creative performance of the incohesive rivals transitioned into camaraderie will be even higher than consistently cohesive pairs (Triandis et al. 1965). Visual tools such as the iteration spiral (Eppler et al. 2014) can aid in keeping conflict at the cool, productive level of a disagreement (Mengis and Eppler 2006).

7.2.5 The Paradox Pair

Paradox pairing is an intensive twist on the classical teambuilding approach of competency gathering. This pairing is a combination of polar opposites into a self-reinforcing creativity system. This category covers all creatively productive pairings of opposites—traditionalist versus iconoclast, rural versus urban, simplifier versus complicator—it is thus broad.

What paradoxical pairings have in common is that the partners simultaneously push and balance each other. The contradiction in characters drives an upward spiral (Andriopoulos and Lewis 2009). Farrell (2001) uses two influential social reformers of the nineteenth century women's rights movement—Elizabeth Cady Stanton and Susan B. Anthony as an example of paradoxical pairing. Stanton was a gifted writer but a poor manager; and Anthony was her diametrical opposite. In their joint striving to achieve the common goal of women's suffrage together, the two drive each other forward. In the case of Stanton and Anthony, this was apparently through pushing each other to draw on their strengths and express their eccentricities. In other paradoxical pairs, progress is achieved through partner's strengths and weaknesses balancing each other out. The success of Apple, with Steve Jobs leading the commercial side and Steve Wozniak leading the technical is exemplary balancing in a paradox pair.

A prime example of balance in a paradox pair is the provocateur and the charmer. The provocateur draws outlandish influences into the pair's discussion, and uses them to bait his or her counterpart. The charmer has the means and composure to calmly integrate these impulses into the joint solution. In the next iteration of the solution, the provocateur will likely warp the charmer's integrative solution into its most challenging version. This process of collection, integration, and transformation can reveal useful permutations on the original input. In their encounters with external parties, the provocateur delights in pushing boundaries and gaining attention—the charmer smoothes these wrinkles out. Memorable impact with positive overtones is thus achieved. Hollywood abounds with such scriptwriting duos.

There are numerous subtypes of paradox pairs. For example, Goal-directed versus resource-oriented thinkers, niche experts versus generalists, optimists versus pessimists, and intuition versus data-driven thinkers. The first step in managing these pairs is forming them. Managers may need to draw out and identify individuals' unique personality and skill dimensions before forming paradox pairs. Mastery goals will again give better results than performance goals (Darnon et al. 2007). Members of paradox pairs need room to balance and extend each other. Micromanagement is unlikely to succeed with paradox pairs (White 2010). Managers will be better served by delegating problems to paradox pairs and supporting them where necessary. Management needs to give an initial problem statement, then empower and trust the paradox pair to solve the problem—or a reframed version of the problem (Blair 1992).

A critical point is that the actual functioning of paradox pairs requires testing before mission-critical commitments. Just because individuals have paradoxical qualities, this does not mean that they will synergistically draw them out or work well with a contrasting personality. Many personal and contextual variables are at play (Grove 1991). The partner's interpersonal skills are likely to be tested. Limited scope test projects and prototyping are useful for seeing how pairs actually work together. Moreover, these can help cement the partnership via quick wins

(Appelbaum et al. 2012; Kotter 1996). Alternatively, Verganti describes a "speed dating process" (2016, p. 94) whereby partners briefly discuss visualizations of ideas and decide themselves if they could further work together.

This is a pairing that requires much management trust despite its relatively visible individual members. Visualizations that draw on the strengths of the two partners can be tremendously useful to these pairs. For example, linear and lateral thinkers can benefit from the explicitness of paths to success (Eppler and Pfister 2011) and goal-directed versus resource-oriented thinkers are advised to use brainswarming to combine goals and resources into a new solution (McCaffrey and Pearson 2015).

7.2.6 The Leader and the Manager

Leaders are one of the most powerful predictors of workplace creativity (Mumford et al. 2002).

Role clarity is imperative for a functional leader/manager pair. The quintessential quality of a leader are that he or she has followers—who are inspired to follow the leader in the direction he or she creates through calculated risk taking (Bass 1990; Chapman 1989; Kotter 1990; Yukl 1989). By definition, the manager in this pair is thus an inspired follower. The role of the manager is continual planning, organizing, supervising, and controlling of resources to achieve objectives aligned with the direction set by the leader (Nebeker and Tatum 2002). In essence, managers produce predictability and order while leaders produce potential for disruptive change and chaos (Kotter 1990). Unlike the dreamer/doer pair, this is a vertical, hierarchical arrangement (Liden and Graen 1980). The leader is in charge.

A clearly defined leader/manager pair is extremely powerful. Clear definitions allow both parties to focus upon the relative strengths that enable them in their respective position. 2016 examples of creative leader/manager pairs are Tesla's Elon Musk and JB Straubel and Oracle's Larry Ellison and Mark Hurd. As the leader has an inherently senior position to the manager, the leader should put extra effort into inspiring the manager to exhibit certain behaviors. Perhaps chief among these is the encouragement to establish differentiated relationships with each member of their workplace team, instead of simply using an identical interaction style with every employee (Liden and Graen 1980). Higher quality relationships between workplace team leaders and their team members are characterized by extended resources, confidence, support, autonomy, and decision-making latitude (Dansereau et al. 1975; Graen and Uhl-Bien 1995). Lower quality relationships—which the manager may also choose to employ in some instances—exhibit the contrary. These relationships, or exchanges, may be of high or low quality—as the manager sees fit. Reciprocal high-quality relationships have been associated with innovative behavior, bissociative problem solving, and creative work involvement (Sanders et al. 2010; Scott and Bruce 1998; Volmer et al. 2012)—and creativity in general (Hammond et al. 2011). The leader's establishment and maintenance of a close creative leader/manager dyad is already an exemplary model of this social exchange behavior.

Both leaders and managers should note that job design plays a moderating role in creative work involvement (Volmer et al. 2012). Granting job autonomy boosts creative work involvement. Involvement increases creative performance through a range of mechanisms (Carmeli and Schaubroeck 2007). Therefore, managers in particular should note that team members empowered to determine the pace, sequence, and methods for accomplishing tasks will display higher creative performance. Employees who experiment with task completion variables—and take responsibility for their results—will be more involved, and more creative. This organizational capability will allow leaders latitude in the risks they engage.

Unfortunately confusion between the roles of leaders and managers abounds. This is largely due to a small overlap in their roles (Kotter 1990). The manager must inspire his team to follow at a workplace level, while the leader must also maintain resources. Leaders and managers can thus learn from each other, and opportunities for effective exchange of mutually useful knowledge, values, and abilities are crucial to leader/manager pairs (Euler and Hahn 2014). Visually supported face-to-face experience sharing is the optimal method for this. Decision trees, goal charts, and synergy maps are helpful instruments (Eppler and Pfister 2011). Leaders of entire organizations have a number of managers following them, and will thus need to carefully select which they want to engage in a close creative leader/manager pair. Innovation and marketing managers are obvious candidates, yet this type of pairing is appropriate in any field where innovation is of exceptional strategic value to an individual organization. This is ultimately a leadership choice.

Leaders and managers are found exclusively within organizations. However, the dynamics of the other pairings apply whether parties are internal or external to an organization. In the case of collaboration with external parties such as customers, one will only have partial information on the external party. It is therefore wise for managers to understand the internal party as well as possible and make an educated guess at their most suitable external partner. Employing rivalry pairing is inadvisable in an internal/external pairing as the lack of mutual dependency and objectives will render this pair ineffective at best, and destructive at worst.

7.3 Guidelines for Communication in Creative Pairs

General guidelines for leveraging dyads to produce and implement highly instrumental, original, and feasible ideas are provided below.

Dyads' single communication link makes them efficient and effective communicative units, but also makes them fragile (Moreland 2010). Keeping this communication link clear and open is vital.

Dyads, especially new dyads, typically operate in line with the norm of reciprocity (Gouldner 1960; Sober and Wilson 1998). The norm of reciprocity predicts positive reactions to favorable treatment and negative reactions to unfavorable treatment.

Conflict has three levels: disagreement, cool conflict, and hot conflict (Grove 1991). Conflict and criticism are necessary for creativity (Nemeth and Ormiston 2007), especially problem framing (Verganti 2016).

In ideation, dyads should aim to surface and combine multiple lines of thought (Koestler 1964).

Visualization facilitates a full exchange and discussion of perspectives. Specialized visual tools are useful in particular situations. For example, concept maps for surfacing who knows what about which topics or the iteration spiral for keeping conflicts cool (Eppler 2006; Schreiber and Engelmann 2010; Eppler et al. 2014).

Elaborative dialogue, or building on other's ideas, improves idea quality—particularly early elaborative comments (Berg 2014). Elaborative dialogue can be encouraged by apparently unfinished visualizations (McGrath 2015a) and the acronym IER (Idea, Elaborate, Respond) (Hausmann 2006).

"Test fast, fail fast, adjust fast" (Peters 1991). Use short feedback iterations to test and learn. Dyads should not be afraid to rapidly build and test low fidelity prototypes with potential users.

Creativity pairs need not be in constant, uninterrupted contact. Time apart is useful for gathering new experiences and insights. Moreover, time alone facilitates the "creative worrying" needed to pull a problem apart and the incubation needed to distance oneself from commonplace solutions (Olton 1979).

Creativity pairs tend to form, then actively produce towards an objective before naturally dissolving once their objective is accomplished. Personality, distracting life events, conflicting expectations, promotion, and betrayal are the five primary causes of workplace relationship deterioration (Sias et al. 2004). Providing a new objective before dissolution can renew the lifecycle of a creative pair (Tzokas and Saren 2004).

To actively disengage from a creativity relationship unworthy of renewal, employees commonly use indirect behavioral communication tactics (Baxter 1982). Most common are: Avoidance of nonwork conversation topics, nonverbal cues, and avoidance of nonworkplace socialization (Sias et al. 2004).

7.4 A Strategy for Transitioning Pairs from Creativity to Innovation

The creative advantages of pairwork are clear. But how does dyadic creativity translate into concrete innovation? Being that the implementation of creative ideas is a widely accepted definition of innovation (Baer 2012), the answer lies in the translation of ideas into reality. Maier and Schmidt's (2015) knowledge maturing model, depicted below in Fig. 7.1—a version slightly adapted for dyads, provides a

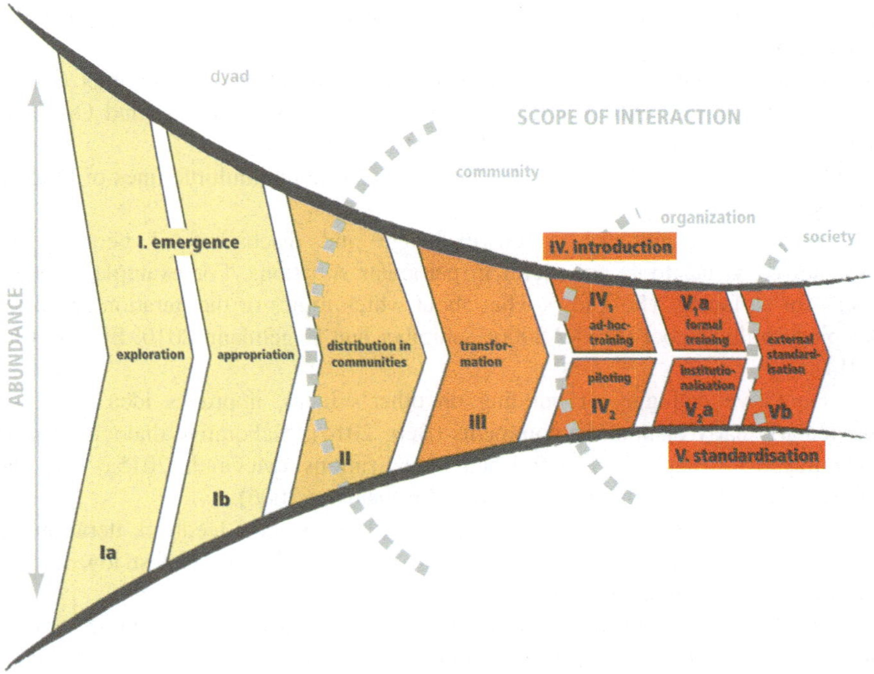

Fig. 7.1 The knowledge maturing model (Maier and Schmidt 2015)

useful basis. This funnel process sees new knowledge transition through emergence, community distribution, transformation, introduction into organizations before standardization in societies. This chapter has thus far covered emergence and appropriation.

New ideas often do not survive sharing beyond the pair. The stage of community distribution often suffocates ideas. Many dyads instinctively know this. As the point of further idea distribution approaches, emotions may become heated as the dyad seeks to pitch the best possible version of its idea. Visualization has proven useful for productively focusing attention and preventing disagreements from heating into disruptive hot conflict (Grove 1991; Mengis and Eppler 2006). An ideal tool for cooling conflicts and refining ideas is the iteration spiral, which leads dyads through a controlled confrontation based around their solution (Eppler et al. 2014).

Once ideas enter a community, dyads are more effective than individuals during the delicate distribution phase. Dyads are more robust champions of ideas than individuals (McGrath 2015b). On the management side, research shows that innovation managers personally struggle with terminating individuals' high involvement projects—because they do not wish to demoralize individual project heads (Daly et al. 2012). Such accommodating behavior sees the costs of ill-fated projects multiply before their inevitable late termination or failure. Dyadic project ownership has less personal attachment. This makes it easier for innovation managers to confront dyads and terminate projects with limited potential. Dyadic ownership thus prevents individual pet projects from unnecessarily draining organizational resources.

Should new knowledge survive community distribution, it then enters the transformation phase. Here new knowledge is codified into documents and artifacts. Maier and Schmidt find that "gardening" or "collaborative efforts to improve structure and quality of artifacts and their organization" are imperative for transitioning knowledge from distribution to transformation (Maier and Schmidt 2015, p. 370). Dyads' persistence (Torrance 1970), reflective and elaborative capabilities (McGrath 2015a) will benefit this iterative process. Once codified, new knowledge is piloted and integrated into ad hoc training over the width and breadth of the organization. The centrality of dyads to training and educational settings is well recognized (Euler and Hahn 2014). Small teams, such as dyads, typically carry out piloting in many organizations. The standardization phase consists of formal training and institutionalization. The latter is heavily colored by internal standardization in line with external standards such as ISO 9001. Dyads of intra-organizational and external representatives would facilitate this alignment process.

7.5 A Personal Strategy for Pairwork

The pairing typology presented above can also be used to manage one's own creative interactions and relationships. Moreover, individuals can use these roles to simultaneously manage more than one creative relationship. A manager who has the resources to make a new hire, may choose to reflect upon his or her own core long-term behavior and hire a complementary employee. For example, those who identify themselves as dreamers should bring doers into their team. We present a few guidelines for personal pairwork below.

7.5.1 The Dreamer and the Doer

This pairing requires a great deal of honesty with oneself about character. One must identify whether one is naturally exceptional at one of the dreamer/doer roles. If so, identify and leverage counterparts to counterbalance strengths.

7.5.2 *The Mentor and the Mentee*

Regardless of one's position in a formal or informal mentoring relationship, the encouragement of debate draws the value out of multiple perspectives. Mentoring partners should be both emboldened and listened to.

7.5.3 *The Hero and the Sidekick*

Protagonists require backstage support. Supportive stage managers need partners for visibility. Ensure that both partners contribute to mastering the task and both receive recognition.

7.5.4 *The Rivals*

The engagement of a rival as a sparring partner can fuel highly rewarding rapid idea iteration. In private, propose concepts to the rival and use them to refine ideas. Be sure to elicit detail and rationale for criticism. For maximum benefit, balance listening with adaptive defense of the idea.

7.5.5 *The Paradox Pair*

Identify your core personal and creative qualities. What is the exact opposite of these qualities and who has them?

7.5.6 *The Leader and the Manager*

Leaders and managers are formally defined, yet role overlap can blur the lines. Are you and your counterpart acting in line with your roles? What can you learn from your counterpart? Do your counterpart's competencies allow you to stretch yourself in your role, for example by exploring more high potential risks or further optimizing resource use?

7.6 Conclusion and Outlook for Future Research on Dyadic Communication for Innovation

Pairs are unique amongst groups and teams (Moreland 2010). A pair is the smallest possible group or team. This chapter has detailed the multitude of cognitive and communicative advantages pairwork has over lone work. Pairwork derails the three major deriders of creativity—freeriding, production blocking, and evaluation apprehension while simultaneously taking advantage of interpersonal dialectics. The chapter has shown how the simultaneous robustness and flexibility of pairs lends itself to both creativity and the resulting innovation processes. Pairs are powerful building blocks for innovation, and the reader now has the communicative tools to classify and leverage pairs for creativity and innovation.

Contemporary organizations continue to build themselves from individuals. They have done so for millennia. Organizational components of all types and sizes —from ad hoc project teams to divisions—are based upon the consolidation of individuals into units. They need not be. Organizations can build themselves from pairs—as Farrell proposes that successful creative circles of associates do (Farrell 2001). This "dyadic organizing" (Graen and Scandura 1987) approach is widely adopted in the leadership and relationship marketing fields, where at least dyad member is highly visible. However, the interweaving of dyadic networks within organizations is less widely studied. This perspective raises some methodological challenges, yet modern social network methodologies may well be equipped to strip away the complexity and examine organizational networks through a dyadic lens.

Bibliography

Amabile, T.M.: Motivating creativity in organizations: on doing what you love and loving what you do. Calif. Manage. Rev. **40**(1), 39–58 (1997)

Andriopoulos, C., Lewis, M.W.: Exploitation-exploration tensions and organizational ambidexterity: managing paradoxes of innovation. Organ. Sci. **20**(4), 696–717 (2009)

Appelbaum, S.H., Habashy, S., Malo, J.L., Shafiq, H.: Back to the future revisiting Kotter's 1996 change model. J. Manage. Dev. **31**(8), 764–782 (2012)

Baer, M.: Putting creativity to work: the implementation of creative ideas in organizations. Acad. Manage. J. **55**(5), 1102–1119 (2012)

Bass, B.M.: Bass and Stogdill's handbook of leadership: theory, research & managerial applications, 3rd edn. The Free Press, New York (1990)

Baxter, L.A.: Strategies for ending relationships: two studies. W. J. Commun. **46**(3), 223–241 (1982)

Berg, J.M.: The primal mark: how the beginning shapes the end in the development of creative ideas. Organ. Behav. Hum. Decis. Process. **125**, 1–17 (2014)

Blair, G.M.: The art of delegation. Eng. Manage. J. **2**(4), 165–169 (1992)

Bouchard, T.J., Hare, M.: Size, performance, and potential in brainstorming groups. J. Appl. Psychol. **54**, 51–55 (1970)

Bouchard, T.J., Drauden, G., Barsaloux, J.: Brainstorming procedure, group size and sex as determinants of the problem solving effectiveness of groups and individuals. J. Appl. Psychol. **59**, 135–138 (1974)

Briggs, R.O., Reinig, B.A.: Bounded ideation theory. J. Manage. Inf. Syst. **27**(1), 123–144 (2010)

Brown, T.: Design thinking. Harvard Bus. Rev. **86**(6), 84–92 (2008)

Camacho, L.M., Paulus, P.B.: The role of social anxiousness in group brainstorming. J. Pers. Soc. Psychol. **68**(6), 1071–1080 (1995)

Campbell, J.: Individual versus group problem solving in an industrial sample. J. Appl. Psychol. **52**, 205–210 (1968)

Carmeli, A., Schaubroeck, J.: The influence of leaders' and other referents' normative expectations on individual involvement in creative work. Leadersh. Quart. **18**(1), 35–48 (2007)

Chapman, E.N.: Leadership. Prentice Hall, Englewood Cliffs, NJ (1989)

Chatterjea, R.G., Mitra, A.: A study of brainstorming. Manas **23**, 23–28 (1976)

Chi, M.T.H., Bassok, M., Lewis, M.W., Reimann, P., Glaser, R.: Self-explanations: how students study and use examples in learning to solve problems. Cogn. Sci. **13**, 145–182 (1989)

Chi, M.T.H., Siler, S., Jeong, H.: Can tutors monitor students' understanding accurately? Cogn. Instr. **22**, 363–387 (2004)

Cohen, D., Whitmyre, J.W., Funk, W. H.: Effect of group cohesiveness and training upon creative thinking. J. App. Psych. **44**(5), 319–322 (1960)

Collaros, P., Anderson, L.: Effect of perceived expertness upon creativity of members of brainstorming groups. J. Appl. Psychol. **53**(2), 159–163 (1969)

Cottrell, N.B.: Social facilitation. In: McClintock, C.G. (ed.) Experimental social psychology, pp. 185–237. Holt, Rinehart and Winston Inc, New York (1972)

Daly, J.A., Sætre, A.S., Brun, E.: Killing mushrooms: the realpolitik of terminating innovation projects. Int. J. Innov. Manage. **16**(05), 1250024 (2012)

Dansereau, F., Graen, G., Haga, W.J.: A vertical dyad approach to leadership within formal organizations. Organ. Behav. Hum. Performance **13**, 46–78 (1975)

Darnon, C., Butera, F., Harackiewicz, J.M.: Achievement goals in social interactions: learning with mastery vs. performance goals. Motiv. Emot. **31**(1), 61–70 (2007)

De Dreu, C.K.W., Weingart, L.R.: Task and relationship conflict, team performance, and team member satisfaction: a meta-analysis. J. Appl. Psychol. **88**, 741–749 (2003)

Dean, D.L., Hender, J.M., Rodgers, T.L., Santanen, E.L.: Identifying quality, novel, and creative ideas: constructs and scales for idea evaluation. J. Assoc. Inf. Syst. **7**(10), 649–699 (2006)

Dew, R., Hearn, G.: A new model of the learning process for innovation teams: networked nominal pairs. Int. J. Innov. Manage. **13**(04), 521–535 (2009)

Diehl, M., Stroebe, W.: Productivity loss in brainstorming groups: toward the solution of a riddle. J. Pers. Soc. Psychol. **53**(3), 497–509 (1987)

Dunnette, M.D., Campbell, J., Jaastad, K.: The effect of group participation on brain-storming effectiveness for two industrial samples. J. Appl. Psychol. **47**, 30–37 (1963)

Engelmann, T., Hess, F.W.: How digital concept maps about the collaborator's knowledge and information influence computer-supported collaborative problem solving. Int. J. Comput.-Support. Collaborative Learn. **5**, 299–319 (2010)

Eppler, M.J.: A comparison between concept maps, mind maps, conceptual diagrams, and visual metaphors as complementary tools for knowledge construction and sharing. Inf. Visual. **5**(3), 202–210 (2006)

Eppler, M.J., Pfister, R.: Sketching at Work. St. Gallen, Switzerland: = mcminstitute (2011)

Eppler, M.J., Hoffmann, F., Pfister, R.A.: Creability: Gemeinsam Kreativ. Schäffer-Poeschel Verlag, Stuttgart, Germany (2014)

Euler, D., Hahn, A.: Wirtschaftsdidaktik, 3rd edn. Haupt UTB, Stuttgart, Germany (2014)

Farrell, M.P.: Collaborative circles: friendship dynamics and creative work. The University of Chicago Press, Chicago, IL (2001)

Gallupe, R.B., Dennis, A.R., Cooper, W.H., Valacich, J.S., Bastianutti, L.M., Nunamaker, J.F.: Electronic brainstorming and group size. Acad. Manage. J. **35**(2), 350–369 (1992)

Gannon, M.J.: Cultural metaphors: their use in management practice as a method for understanding cultures. Online Readings Psychol. Cult. **7**(1), 4–16 (2011)
Girotra, K., Terwiesch, C., Ulrich, K.: Idea generation and the quality of the best idea. Manage. Sci. **56**(4), 591–605 (2010)
Gouldner, A.W.: The norm of reciprocity: a preliminary statement. Am. Sociol. Rev. **25**, 161–178 (1960)
Graen, G.B., Scandura, T.A.: Toward a psychology of dyadic organizing. Res. Organ. Behav. **9**, 175–208 (1987)
Graen, G.B., Uhl-Bien, M.: Development of leader-member exchange (LMX) theory of leadership over 25 years: applying a multi-level multi-domain perspective. Leadersh. Quart. **6**, 219–247 (1995)
Graham, W.K., Dillon, P.C.: Creative supergroups: Group performance as a function of individual performance on brainstorming tasks. J. Social Psychol. **93**, 101–105 (1974)
Grove, T.G.: Dyadic Interaction: choice and change in conversations and relationships. Wm. C. Brown Publishers, Dubuque, IA (1991)
Gurman, E.B.: Creativity as a function of orientation and group participation. Psychol. Rep. **22**(2), 471–478 (1968)
Hammond, M.M., Neff, N.L., Farr, J.L., Schwall, A.R., Zhao, X.: Predictors of individual-level innovation at work: a meta-analysis. Psychol Aesthetics Creativity Arts **5**, 90–105 (2011)
Harackiewicz, J.M., Barron, K.E., Elliot, A.J.: Rethinking achievement goals: When are they adaptive for college students and why? Educ. Psychol. **33**(1), 1–21 (1998)
Hargadon, A.: How breakthroughs happen: the surprising truth about how companies innovate. Harvard Business School Press, Boston, MA (2003)
Hausmann, R. G. M.: Why do elaborative dialogs lead to effective problem solving and deep learning. Paper presented at the 28th Annual Meeting of the Cognitive Science Society, pp. 1465–1469, Vancouver (2006)
Henchy, T., Glass, D.: Evaluation apprehension and the social facilitation of dominant and subordinate responses. J. Pers. Soc. Psychol. **10**(4), 446–554 (1968)
Herzog, P., Leker, J.: Open and closed innovation–different innovation cultures for different strategies. Int. J. Technol. Manage. **52**(3/4), 322–343 (2010)
Hewing, M.: Merits of collaboration with potential and current users in creative problem-solving. Int. J. Innov. Manage. **17**(03), 1340009 (2013)
Hofstede, G. (2016). Country comparison. Retrieved from: http://geert-hofstede.com/countries.html. Accessed 10 Mar 2016
Hofstede, G., Hofstede, G.J., Minkov, M.: Cultures and organizations: software of the mind, 3rd edn. McGraw-Hill, Columbus, OH (2010)
Janis, I.L.: Groupthink. Psychol. Today **5**(6), 43–46 (1971)
Janis, I.L.: Victims of groupthink: a psychological study of foreign-policy decisions and fiascoes. Houghton Mifflin, Boston, MA (1972)
Janis, I.L.: Groupthink: psychological studies of policy decisions and fiascoes, 2nd edn. Houghton Mifflin, Boston, MA (1982)
Kavadias, S., Sommer, S.C.: The effects of problem structure and team diversity on brainstorming effectiveness. Manage. Sci. **55**(12), 1899–1913 (2009)
Kerr, N.L., Bruun, S.E.: Dispensability of member effort and group motivation losses: free-rider effects. J. Pers. Soc. Psychol. **44**(1), 78–94 (1983)
Kim, K.H.: Can we trust creativity tests? A review of the torrance tests of creative thinking (TTCT). Creativity Res. J. **18**(1), 3–14 (2006)
Koestler, A.: The act of creation. Penguin Books, New York, NY (1964)
Kotter, J.P.: A force for change: how leadership differs from management. The Free Press, New York (1990)
Kotter, J.P.: Leading change. Harvard Business School Press, Boston, MA (1996)
Kozlov, M.D., Grosse, C.S.: Online collaborative learning in dyads: effects of knowledge distribution and awareness. Comput. Hum. Behav. **59**, 389–401 (2016)

Liden, R.C., Graen, G.: Generalizability of the vertical dyad linkage model of leadership. Acad. Manage. J. **23**, 451–465 (1980)
Maier, R., Schmidt, A.: Explaining organizational knowledge creation with a knowledge maturing model. Knowl. Manage. Res. Pract. **13**(4), 361–381 (2015)
McCaffrey, T., Pearson, J.: Find innovation where you least expect it. Harvard Bus. Rev. 83–89 (2015)
McGrath, L.: The power of provisionality: an experimental evaluation of idea appearance in online knowledge creation. Paper presented at the 15th international conference on knowledge technologies and data-driven business, pp. 20–28. Graz (2015a)
McGrath, L.: When pairing reduces scaring: the effect of dyadic ideation on evaluation apprehension. Int. J. Innov. Manage. **19**(4), 1550039-1–1550039-35 (2015b)
Mengis, J., Eppler, M.J.: Seeing versus arguing the moderating role of collaborative visualization in team knowledge integration. J. Univers. Knowl. Manage. **1**(3), 151–162 (2006)
Montuori, A., Purser, R.E.: Deconstructing the lone genius myth: toward a contextual view of creativity. J. Humanistic Psychol. **35**, 69–112 (1995)
Moreland, R.L.: Are dyads really groups? Small Group Res. **41**(10), 251–267 (2010)
Mullen, B., Johnson, C., Salas, E.: Productivity loss in brainstorming groups: a meta-analytic integration. Basic Appl. Soc. Psychol. **72**(1), 3–23 (1991)
Mumford, M.D., Scott, G.M., Gaddis, B., Strange, J.M.: Leading creative people: orchestrating expertise and relationships. Leadersh. Quart. **13**, 705–750 (2002)
Nebeker, D.M., Tatum, B.C.: Understanding organisational processes and performance. In: Lowman, R.L. (ed.) Handbook of Organisational Consulting Psychology, pp. 668–691. Jossey-Bass, San Francisco, CA (2002)
Nemeth, C.J., Ormiston, M.: Creative idea generation: harmony versus stimulation. Eur. J. Soc. Psychol. **37**(3), 524–535 (2007)
Nijstad, B.A., De Dreu, C.K., Rietzschel, E.F., Baas, M.: The dual pathway to creativity model: creative ideation as a function of flexibility and persistence. Eur. Rev. Soc. Psychol. **21**(1), 34–77 (2010)
Olton, R.M.: Experimental studies of incubation: searching for the elusive. J. Creative Behav. **13** (1), 9–22 (1979)
Osborn, A.F.: Applied Imagination. Scribner's, New York (1957)
Pape, T., Bölle, I.: Einfallsproduktion von Individuen und Dyaden unter "Brainstorming"-Bedingungen: Replikation einer Studie und allgemeine Probleme eines Forschungsgebietes. Psychologische Beitrage **26**, 459–468 (1984)
Parks, C.D., Cowlin, R.A.: Acceptance of uncommon information into group discussion when that information is or is not demonstrable. Organ. Behav. Hum. Decis. Process. **66**(3), 307–315 (1996)
Paulus, P.B., Dzindolet, M.T.: Social influence processes in group brainstorming. J. Pers. Soc. Psychol. **64**(4), 575–586 (1993)
Paulus, P.B., Larey, T.S., Ortega, A.H.: Performance and perceptions of brainstormers in an organizational setting. Basic Appl. Soc. Psychol. **17**(1–2), 249–265 (1995)
Peters, T.: Thriving on Chaos: handbook for a management revolution. HarperPerrenial, New York (1991)
Petty, R.E., Harkins, S.G., Williams, K.D., Latane, B.: The effects of group size on cognitive effort and evaluation. Pers. Soc. Psychol. Bull. **3**(4), 579–582 (1977)
Pintrich, P.R.: The role of motivation in promoting and sustaining self-regulated learning. Int. J. Educ. Res. **31**(6), 459–470 (1999)
Renkl, A.: Learning from worked-out examples: a study on individual differences. Cogn. Sci. **21**, 1–29 (1997)
Renzulli, J.S., Owen, S.V., Callahan, C.M.: Fluency, flexibility, and originality as a function of group size. J. Creative Behav. **8**(2), 107–113 (1974)
Rosenthal, R., Rosnow, R.L.: Artifacts in behavioral research. Oxford University Press, New York (2009)

Sanders, K., Moorkamp, M., Torka, N., Groenveld, S., Groenveld, C.: How to support innovative behaviour? The role of LMX and satisfaction within HR practices. Technol. Investment **1**, 59–68 (2010)

Schön, D.A.: The reflective practitioner: how professionals think in action. Basic Books, New York (1983)

Schreiber, M., Engelmann, T.: Knowledge and information awareness for initiating transactive memory system processes of computer-supported collaborating ad hoc groups. Comput. Hum. Behav. **26**(6), 1701–1709 (2010)

Scott, S.G., Bruce, R.A.: Following the leader in R&D: the joint effect of subordinate problem-solving style and leader-member relations on innovative behavior. IEEE Trans. Eng. Manage. **45**, 3–10 (1998)

Shenk, J.W.: Powers of two: finding the essence of innovation in creative pairs. Houghton Mifflin Harcourt Publishing Company, New York (2014)

Sias, P.M., Heath, R.G., Perry, T., Silva, D., Fix, B.: Narratives of workplace friendship deterioration. J. Soc Pers. Relat. **21**(3), 321–340 (2004)

Smith, S.M.: Getting into and out of mental ruts: a theory of fixation, incubation, and insight. In: Sternberg, R.J., Davidson, J.E. (eds.) The Nature of Insight, pp. 229–252. The MIT Press, Cambridge, MA (1995)

Sober, E., Wilson, D.S.: Unto Others: The Evolution and Psychology of Unselfish Behavior. Harvard University Press, Cambridge, MA (1998)

Stroebe, W., Diehl, M.: Why groups are less effective than their members: on productivity losses in idea-generating groups. Eur. Rev. Soc. Psychol. **5**(1), 271–303 (1994)

Sutton, R.I., Hargadon, A.: Brainstorming groups in context: Effectiveness in a product design firm. Adm. Sci. Quart. 685–718 (1996)

Taylor, D.W., Berry, P.C., Block, C.H.: Does group participation when using brainstorming facilitate or inhibit creative thinking? Adm. Sci. Quart. 23–47 (1958)

Torrance, E.P.: The influence of dyadic interaction on creative functioning. Psychol. Rep. **26**, 391–394 (1970)

Torrance, E.P.: Stimulation, enjoyment, and originality in dyadic creativity. J. Educ. Psychol. **62** (1), 46–48 (1971)

Triandis, H.C., Hall, E.R., Ewen, R.B.: Member heterogeneity and dyadic creativity. Hum. Relat. **18**(1), 33–55 (1965)

Tzokas, N., Saren, M.: Competitive advantage, knowledge and relationship marketing: where, what and how? J. Bus. Ind. Mark. **19**(2), 124–135 (2004)

Varvel, T., Adams, S.G., Pridie, S.J., Ruiz Ulloa, B.C.: Team effectiveness and individual Myers-Briggs personality dimensions. J. Manage. Eng. **20**(4), 141–146 (2004)

Verganti, R.: The innovative power of criticism. Harvard Bus. Rev. 88–95 (2016)

Volmer, J., Spurk, D., Niessen, C.: Leader-member exchange (LMX), job autonomy, and creative work involvement. Leadersh. Quart. **23**, 456–465 (2012)

Vroom, V.H., Grant, L.D., Cotton, T.W.: The consequences of social interaction in group problem-solving. Organ. Behav. Hum. Perform. **4**, 77–95 (1969)

Wegner, D.M.: Transactive memory: a contemporary analysis of the group mind. In: Mullen, B., Goethals, G. (eds.) Theories of Group Behavior, pp. 185–208. Springer, New York (1986)

White, R.D.: The micromanagement disease: symptoms, diagnosis, and cure. Public Pers. Manage. **39**(1), 71–76 (2010)

Yukl, G.: Managerial leadership: a review of theory and research. J. Manage. **15**(2), 251–290 (1989)

Chapter 8
Communication Model Design Innovation—Authentic Open Innovation-Culture

Nicole Pfeffermann

Abstract Innovation is the most important driver of growth. In the digitalized information age, the way of innovating, however, has changed and attaining a better understanding of communication—as a critical factor supporting successful innovation—is exceedingly decisive to strategic success. This chapter answers the question: How can companies develop information communication designs to successfully carry on stakeholder dialogs for enhancing corporate innovation? The objective of this chapter is to present an agile method for communication planning and emphasize the importance of dialog communication in multisensory worlds to shape an authentic open innovation-culture. After presenting a theoretical approach of innovation communication management, this chapter focuses on communication planning and introduces an agile communication method related. In addition, the frame of this chapter *Visual and Scent-based Communication* describes a new form of communication in the digitalized information age.

8.1 Introduction

Innovations can represent a valuable, rare and inimitable resource of a corporation and, thus, lead to strategic competitiveness according to the resource-based view (e.g. Barney 1991; Grant 1991; Hall 1992, 1993). A deeper look at the facts and numbers related to innovation show, however, an increased rate of different types of innovations caused by a tremendous organizational pressure to continuously reinvent themselves in order to seize entrepreneurial opportunities and operate in highly connected innovation ecosystems. In fact, there is a pressing need to rapidly offer innovative customer solutions involving *prosumers* and partners in open processes.

The digitalized information age, an 'era of world shrinking, time shrinking, and constant, instantaneous contact' (Brumfield et al. 2008: xviii), places also signifi-

N. Pfeffermann (✉)
Pfeffermann Strategy and Research, Im Medienhafen,
D-40221 Düsseldorf, Germany
e-mail: np@npfeffermann.de

N. Pfeffermann and J. Gould (eds.), *Strategy and Communication for Innovation*,
DOI 10.1007/978-3-319-49542-2_8

cant demands on organizations to successfully innovate in terms of discovering new ways in information and communication to capture stakeholder attention, to valuably interact with various known and unknown knowledge-empowered stakeholders via new channels and platforms (applications) to build reputation and brand awareness, and to positively influence stakeholder adoption, idea dissemination and innovation diffusion.

The first priority of managers has to become, then, to get a better understanding of information & communication management for innovation in the digitalized information age. It is neither simply using a social media platform nor publishing a new business idea via a crowdfunding platform. Connecting with human beings and carrying on an interactive dialog with stakeholders in communities cannot result from sending out message upon message, impression upon impression; this strategy may instead trigger information overload and avoidance. Companies 'have overzealously exhausted traditional forms of communication, achieving poor results from print, TV, radio, and online advertising' and 'inadvertently ignore or have not yet become enlightened to this secret, ancient tool of communication; the silent key that unlocks the inner recesses of our mind, breaking through our fortified barriers with its eloquently muted message'. (Brumfield et al. 2008: xx). Purposefully managing *visuals* and this 'silent key' *scents* and, therefore, addressing human senses, can lead to an open dialog communication.

Visuals can be used to offer a means to overcome an individual's information overload and to communicate new product/innovation characteristics (Esch and Michel 2008; Kroeber-Riel 1993).

> Visual communication is the 'communication through visual aid [and]... includes art, signs, typography, drawing, graphic design, illustration, colour, and electronic resources [...] to explore the idea that a visual message with text has a greater power to inform, educate or persuade a person' (Martin 2008: 1).

Relating to innovation, the use of ***visual communication*** also plays a central role in complex information transmission (Kroeber-Riel and Weinberg 2003) and imagery communication can be used to trigger direct and indirect image associations, such as free associations as an instrument for innovative product launches (Esch and Michel 2008; Morgan and Welton 1992). People usually tend to trust their emotions linked to what they already think they know, when they are introduced to unfamiliar ideas, conceptions, practices, etc.—innovation understood as anything perceived as new—or should follow new paths to create and adopt innovations. Visual communication can transmit information to trigger emotions and create a world of experiences for consumers in marketing (Esch and Michel 2008; Kroeber-Riel and Esch 2004). Regarding the implementation visual stimuli require strategic planning in order to be effective (Esch and Michel 2008).

In addition to visual communication, though, there is yet another means to communicate, targeted at an individual's primary senses, using smell. The sense of smell is the slowest, most emotional and primitive human sense—the 'silent' key. Although it is the oldest human sense, it is only in recent years that scholars have studied the phenomenon ***scent*** and its psychological effects (Herz 2010; Krishna

2010; Rempel and Esch 2008) and find that 'the sense of smell is the quickest way to tap into someone's emotional responses and memory [, which] makes scents an excellent tool for many different kinds of communication' (Brumfield et al. 2008: xv). To touch the emotions of consumers, innovative companies have understood that 'the simple element of scent is in itself a very dynamic form of information, so they are studying, testing, and implementing ways to use scent to communicate more effectively with consumers' (ibid., p. xx; see also Knoblich et al. 2003).

> ***Scent communication*** can be understood as interactions through the aid of scent stimuli to address an individual's sense of smell and alter recognition, mood and behaviour with positive effects for an organization.

If practical, physiological and psychological factors are considered in scent communication, 'an ambient fragrance that is emotionally and thematically associated to a product [or innovation] should be able to alter perception, cognition, and behaviour with positive consequences for revenue' (Herz 2010: 103). However, as with visual communication, the implementation of scent communication has to be strategically planned in sensory communication to achieve positive effects.

Consequently, both visual communication and scent communication for innovation play a vital role in the digitalized information age but implies strategic planning to be effective. The question, then, should be posed: **How can companies develop information communication designs to successfully carry on stakeholder dialogs for enhancing corporate innovation?** This chapter provides an answer to this question for forward-thinking companies and start-ups, who restructure communication management to capture stakeholder attention, to valuably interact with various known and unknown stakeholders to build reputation and brand awareness, and to positively influence innovation adoption and diffusion, which in turn leads to competitive advantage and business growth in the long run. The objective of this chapter is to present an **agile method for communication planning and emphasize the importance of dialog communication in multisensory worlds to shape an authentic open innovation-culture**.

8.2 Communication Planning

8.2.1 Innovation Communication Management: A Theoretical View

From a theoretical view, the following three management concepts and corporate communication frameworks represent a basis to integrate innovation communication in corporate communication:

- Competing value framework for corporate communication (CVFCC) (Belasen 2008)
- Strategic management concept (Grant 2008)

- St. Gallen Management Model and extended versions of it (Bleicher 1991, 1999; Müller-Stewens and Lechner 2001)

These concepts and frameworks are selected because they provide a differentiated, balanced view (four perspectives) and focus on value creation, resources/structures, stakeholders/transactions and strategic management, which are essential aspects of innovation communication.

Figure 8.1 shows the **elements of the theoretical management approach of innovation communication (ICOMM)**:

- Four main system perspectives based on the CVFCC by Belasen (2008).
- Change/reconfiguration & innovation (transform), knowledge & value creation (perform); regulation & standards/policies (conform); integration & profile/positioning (reform).
- Basic framework of the linkage between a company's structures/resource base and its stakeholders/environment adapted from Grant (2008): structures/resource base; strategy and goals; stakeholder and transactions.
- Three horizontal levels based on the St. Gallen Management Model by Bleicher (1991, 1999): Structures level; activities level; and behaviour level for strategic management and operational management; the normative management could be added but here it is understood as an integral factor in corporate communication management.

Beginning with the ***ICOMM activities*** dimension, illustrated in the centre of Fig. 8.1, the theoretical management approach of innovation communication consists of strategic ICOMM management, operational ICOMM management and ICOMM performance measurement. From the top, these levels are indirectly influenced by the corporate vision and strategy including corporate philosophy and values, corporate communication strategy and goals, and directly from the ICOM strategy and goals. From the bottom, the dynamics of three types of markets (resource markets, communication markets and sales markets) and several environmental factors have an impact on *ICOMM activities*. For instance, new legal and political requirements in transport logistic chains can affect ICOMM in logistic companies as well as stakeholder adoption of an innovation (Daschkovska et al. 2010).

On the left-hand side, the ***structures and resource base*** dimension encompasses three basic elements related to the three levels of ICOMM:

(1) The organizational structure and management system dimension in strategic ICOMM management, such as a web-based management system;
(2) The process-oriented dimension in operational ICOMM management, such as cross-functional information processes of ICOMM; and
(3) The communication controlling system incl. ICOMM controlling.

The ***stakeholder and transaction*** dimension is shown on the right hand of the *ICOMM activities* dimension in Fig. 8.1. Different stakeholder behaviours regarding various transactional procedures of information transmission between an

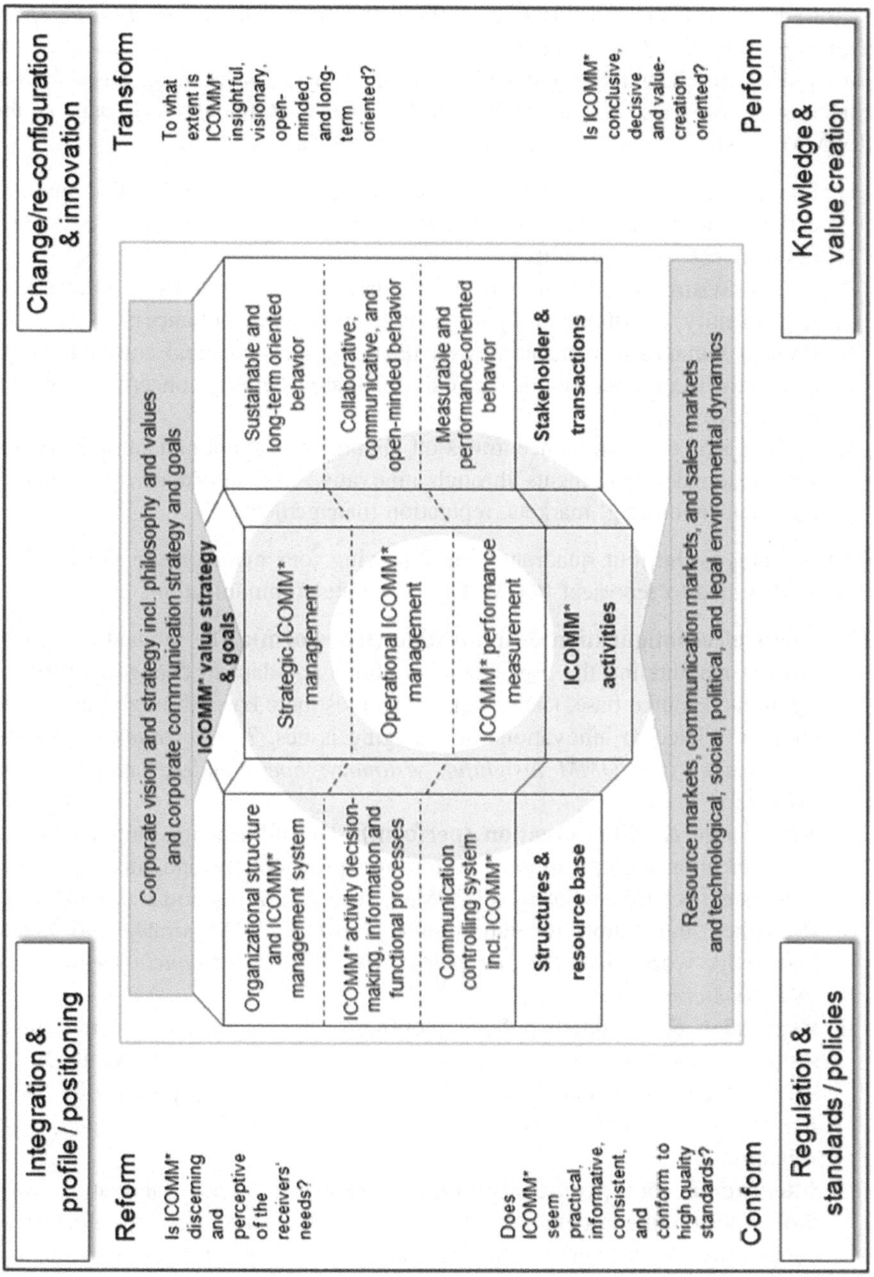

Fig. 8.1 A theoretical approach to innovation communication management (ICOMM). *Source* Adapted from Belasen (2008), Bleicher (1991, 1999)

organization and its stakeholders are enumerated in this dimension. For example, on the operational ICOMM management level employees should be communicative, open-minded and collaborative to execute, monitor and coordinate ICOMM.

Based on the CVFCC by Belasen (2008), ***four main communication perspectives represent the four basic quadrants*** for ICOMM that reflect sociological paradigms (Burrell and Morgan 1979) and serve as a fundamental basis for the construct of corporate communication; in this context, a fundamental basis for the ICOMM construct. These four communication perspectives are:

(1) Functionalism: concentrates on the process and measurement of communication performance, roles and behaviours (e.g. external image, goals, strategy, performance, accountability);
(2) Interpretivism: concentrates on the regularization of systems of interactions (e.g. identity, coordination, symbolic convergence, compliance);
(3) Radical humanism: concentrates on relational-inter-personal communication and human communication (e.g. culture, shared beliefs, concerns of human resources);
(4) Radical structuralism: concentrates on alignment of communication activities with external requirements through innovative and informative systems of communication (e.g. markets, reputation management).

According to the four quadrants, the following four main system perspectives can be identified to represent ICOMM in corporate communication:

(1) **Change/reconfiguration & innovation (transform):** The innovative system aims at maintaining the organizational ability to adapt to change and reconfigure its resource base, such as the human resource base, through knowledge creation related to innovations or emerging issues. *The key question is: 'To what extent is ICOMM insightful, visionary, open-minded, and long-term oriented?!'*
(2) **Knowledge & value creation (perform):** The information/knowledge and value creation system aims at performing productively and meeting stakeholders expectations related to ICOMM, such as information transparency to strengthen innovation reputation and global ICOMM strategy to create knowledge worldwide. *The key question is: 'Is ICOMM conclusive, decisive and value-creation oriented?!'*
(3) **Regulation & standards/policies (conform):** The regulative system aims at supporting the flow and dissemination of structured administrative ICOMM, such as the use of ICOMM standards and policies. *The key question is: 'Does ICOMM seem practical, informative, consistent, and conform to high quality standards?!'*
(4) **Integration & profile/positioning (reform):** The integrative and profile/positioning system aims at providing formal structures and information communication network platforms within an organization or network and creating opportunities to exchange regarding the alignment of corporate strategy/goals and receivers' needs in Open Innovation projects or in general in the Open Innovation economy (see new stakeholder demands (Davenport

et al. 2006), such as the building of stakeholder trust through the informative transparency (de la Fuente Sabate and de Quevedo Puente 2003) in formal structures, which can lead to innovation reputation (positioning). *The key question is: 'Is ICOMM discerning and perceptive of the receivers' needs?!'*

To understand ICOMM in the four-system perspective, the interdependencies of all perspectives have to be considered; effective ICOMM can only be realized through a balance among the four-system perspective based on the requirements of effective corporate communication (see Belasen 2008).

8.2.2 *Communication Planning: A Practical View*

Communication planning is of expanding interest due to the: (1) ubiquitous of information in highly dynamic and complex environments; (2) new understanding of knowledge-empowered stakeholder dialogs on idea exchange platforms (crowdsourcing/co-creation); (3) explosion of start-up businesses and entrepreneurial marketing activities resulting in skew distributions of firms; (4) increased demand in engagement hubs, innovation ecosystems and open innovation initiatives; and (5) need for digital transformation in many companies.

8.2.2.1 Strategy-Communication Approach

Communication facilitates each strategizing process in a different stage in order to translate ideas into execution, accomplish goals and meet the requirements for renewal and transformation in the digitalized information age. Related to the five strategy approaches (Reeves et al. 2015), the five communication approaches/roles linked to strategy approaches are, as illustrated in Fig. 8.2:

	Communication Role	Strategizing Process / Mantra	Strategy Approach
1	ENABLER	Analyze → Plan → **Execute**	Classical
2	SELECTOR	Experiment → **Select** → Scale up → Iterate	Adaptive
3	VISIONARY	**Envision** → Realize → Exploit	Visionary
4	INFLUENCER	**Influencing** → Collaborating → Orchestrating → Co-evolving	(Re)Shaping
5	DESIGNER	Focusing → Freeing up → **Transiting to…**	Renewal

Fig. 8.2 Strategy-communication approach. *Source* Adapted from Reeves (2015)

- **ENABLER:** In the classical view, communication is also understood in a traditional way of being supportive in the execution phase to coordinate all communication activities [secondary activity in Porter's value chain].
- **SELECTOR:** In the second approach, communication has the pivotal role to select the right idea after the experimentation phase has ended. For instance, the full-spectrum method of critical thinking could be a means to dig deeper and get a better understanding of ideas from a wider, more powerful view, which supports the selection process of ideas and effectively communicating with others in figuring out solutions to complex problems (Collison et al. 2000; McCurry et al. 2013).
- **VISIONARY:** Imagine being an entrepreneur; visionary communication is key to envision the future and to create the New. Enhancing the engagement with external stakeholders, hence, can help to shape and predict an environment. For example, scenario planning is a powerful tool for mapping the future and communication assists in painting a colourful picture of the future (Pfeffermann and Breuer 2013).
- **INFLUENCER:** In the exotic 'collectively-reshaping-an-industry' environment, communication is the ability to influence in order to get the results you want. Communication is a competence in collaborative innovation. It is simply impossible to 'control' an innovation ecosystem and the dynamic process of collectively reshaping an industry. But communication is key to positively influence and make the first move, such as coming up with a disruptive solution.
- **DESIGNER:** Finally, the fifth approach focuses on the renewal environment, the harsh conditions in the survival mode. In this context, communication design is key in the transition phase, for instance, in digital transformation projects to enable creation processes.

Consequently, it is crucial for project managers and innovation & communication managers to understand the instruments and activities in each strategizing process and to effectively plan instruments and activities to positively impact each strategizing approach and strategy-communication collage on the company-wide level.

8.2.2.2 Communication Plan

First, it is important to understand the difference between a communication plan and communication model. As an enabler, a communication plan is used to accomplish a specific goal. The communication plan is a practical tool for developing and managing communication instruments and activities from a project management perspective. The communication plan usually consists of different sections as follows:

- **Goal:** The first step of developing a communication plan is to identify and set up overall goal and goal statements for specific outputs or outcomes. It is important in order to determine resources and provide future directions for an idea or project.

- **Strategy:** After defining the 'end point', the outputs or outcomes of communication activities, an overall strategy is required to clearly define a pathway and better select communication instruments and activities. It also helps to prioritize activities, find the right collaborative partners, and build up a community. This section can also include a 360° stakeholder analysis to understand the recipients' (stakeholder) specifications, for instance, their interaction behaviour and use of communication channels.
- **Instruments:** Having a goal and strategy is a first step. Assigning both strategic and tactical-operational instruments is critical for accomplishing the goal. This section describes specific instruments to ensure alignment to strategy in order to realize the impact their involvement has on the achievement of goals.
- **Activity Roadmap:** This section encompasses a communication activity roadmap and schedule for a variety of integrated communication activities related to an instrument and target group. It also includes responsibilities and milestones for each activity.
- **Resources:** In addition to developing a communication activity roadmap, resources have to be identified for consistent and effective communication. That includes to provide an overview of resources needed to achieve communication goals.

8.2.2.3 Agile Communication Planning Method

In the last years, innovation of business models has become increasingly important for many organizations. 'Business model innovation is a multi-stage process whereby organizations transform new ideas into improved business models in order to advance, compete and differentiate themselves successfully in their marketplace'. (Eppler and Hoffmann 2014, p. 5). Following the definition by Osterwalder et al. (2005) the key term 'business model' is understood as 'a conceptual tool that contains a set of elements and their relationships and allows expressing the business logic of a specific firm' (Eppler and Hoffmann 2014, p. 5). Linked to communication, 'business model innovation affects and concerns various, if not all stakeholders inside and outside the firm' (ibid., p. 4), which means it is impossible to ignore the information-interaction design in a transformation process of a business model or development process of a new business model. In other words, every organization needs to focus on new information-interaction designs for mastering innovating business models and transforming the business logic of a firm.

Based on the definition of a business model, a *communication model is a conceptual tool that contains a set of elements and their relationships and allows expressing an information-interaction logic of a specific firm (business model).* It is a (visual) description of the relationships among strategic and tactic-operational instruments to (1) translate strategy into execution and achieve overall goal and

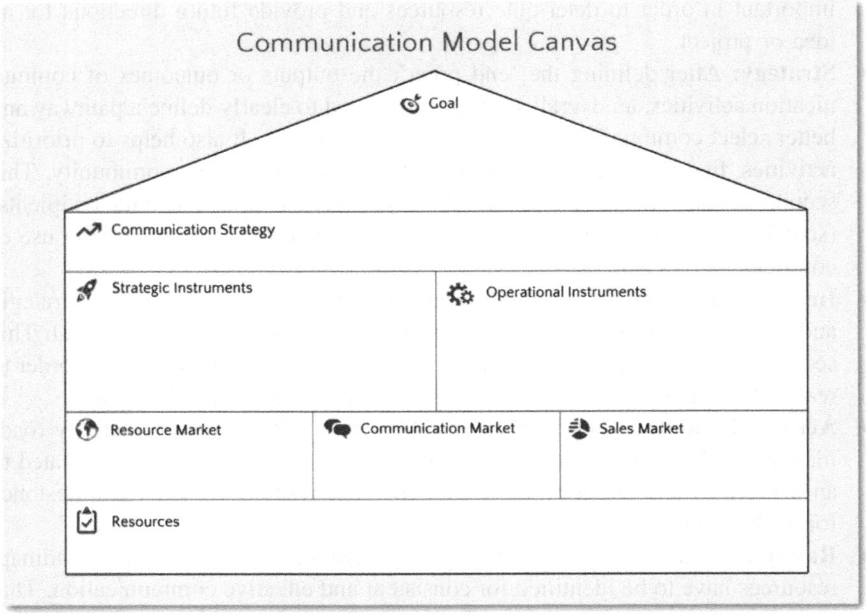

Fig. 8.3 Communication model canvas. *Source* By the author

goal statements (value a company offers) and (2) address different market structures to create value in terms of activating and modifying knowledge schemata, intensifying corporate reputation and trustful stakeholder relationships (legitimacy) and improving the management of strategic assets (e.g. strategic collaboration, innovation community, patents). In fact, a communication model design is key to define the success of any transformation process and development process of (new) business models.

According to research, resource fluidity is a core challenge for business model innovation (Eppler and Hoffmann 2014; Christensen and Raynor 2000; Doz and Kosonen 2010; Leonard-Barton 1992; Zott and Amit 2010). Consequently, it is an essential task for managers to focus on **resources and resource markets** because it builds 'the foundation for the whole idea generation and later implementation' (Eppler and Hoffmann 2014, p. 7). Furthermore, information communication supply and demands are the driver for connecting with human beings and carrying on an interactive dialog with stakeholders in communities (**communication markets**). Related to the business model, a broad range of information communication channels and activities help to create, deliver and market the company's value to prosumers in order to generate revenue streams (**sales markets**).

Communication model innovation as is the multi-stage process whereby organizations develop and advance their information-interaction design to enable business model transformation and new business model development as well as to proactively create value in terms of shaping corporate culture, knowledge and relationship capital (resource base) for gaining competitive advantage and co-evolving open innovation system (growth zone).

As suggested by Eppler and Hoffmann, a visual solution, and in particular the canvas approach by Osterwalder and Pigneur (2009), can be effective to overcome challenges in the multi-stage process of transforming and creating (new) business models. This chapter suggests using the communication model canvas to better build, reflect and present (lean) communication models, as shown in Fig. 8.3. The visual solution is derived from previous contributions in the first edition and second edition of this volume (innovation communication toolbox). The communication model canvas was tested in start-up coaching sessions in practice; however, there is still a need to better understand the multi-stage process of information-interaction design related to strategy approach, business model innovation, and resource market, communication market and sales market.

8.3 Authentic Open Innovation-Culture

A better understanding of the drivers of innovation and culture has a great strategic importance for any system because it leads to superior performance and growth. The most important driver of innovation and culture is dialog communication:

- **Innovation dialog**: Storytelling and framing are key instruments of innovation communication. Narratives can drive idea exchange, learning and engagement. Framing is a method to modify and activate knowledge schema for grasping new ideas and positively influencing innovation adoption.
- **Learning dialog**: Co-creation is a new crowd-powered innovation tool for gaining insights and learning by selecting and refining new ideas or understanding underlying needs of consumers (Kröper et al. 2014). The dialog has changed in the prosumer age and, hence, new communication instruments and activities facilitate stakeholder interaction and mindful listening.
- **Multisensory dialog**: Our senses have an impact on awareness and cognitive skills. It is still a necessity to feel, touch and hear to better connect with others and engage in innovation processes. Diving into different 'worlds', which means discovering novelties with human senses, can enhance knowledge acquisition, creation and collaborative innovation.

As far as multisensory dialog is concerned, visual and scent communication represent a state-of-the-art means of communicating innovation, which can be designed and strategically used to capture stakeholder attention and create emotional ties.

The different uses of scents in marketing are (Morrin 2010):

(1) **Primary or secondary product attributes** (e.g. perfumes, room deodorizers and added-value of products through distinguishable scents);
(2) **Advertising and sales promotion** (use of scent communication, e.g. PoS; public relations; trade fairs, etc.);
(3) **Ambient scents** (e.g. more favourable stores, hotels, retail rooms, etc.; and product evaluation for an increase in sales revenues); and
(4) **Signature scents** (e.g. unique combinations of scents as a potential competitive differentiator).

In particular, the third and fourth uses are growing fields of interest in scent marketing because of the positive effects on sales revenues and competitive differentiations (Morrin 2010).

Positive effects of odour impacts on an individual's behaviour

- If a congruent scent is correctly used in a specific retail environment, purchase behaviour and sales were found to increase (Herz 2010; Rempel and Esch 2008; Spangenberg et al. 2006)
- Positive associations with scents can trigger specific positive or negative emotions that have a direct influence on an individual's mood and behaviour (Ehrlichman and Bastone 1988, 1992; Herz 2010; Rempel and Esch 2008)
- Branding-related advantages of scent communication are as follows (Rempel and Esch 2008): strong emotional ties to brands; improved learning and efficiency in branding; improved clear, attractive and active mental imagery of brands; and consumer disposition to buy and pay a price premium

Negative effects of odour impacts on an individual's behaviour

- When an odour is below the level of perceptual detection there are no behavioural or psychological consequences (Herz 2010)
- A lack of attention to odours can lead to a decrease of the odour detection ability (Herz 2010; Plailly et al. 2008; Zelano et al. 2005)
- The phenomenon of cross-adaptation can affect the recognition of specific odours, such as the differentiation of scents after several samples (Herz 2010)

From a literature review of the effects of visual and scent stimuli on consumer behaviour in (sensory) marketing, theoretically driven implications for communicating innovation can be deduced.

Visual innovation communication

- Using framing in visual communication to create a '*frame of reference*' can improve the understanding of an innovation (Huck 2009)
- Imagery communication offers a means for transmitting complex information (Kroeber-Riel 1993; Kroeber-Riel and Weinberg 2003), through such means as videos, exhibitions and animations for radical innovations

- Free associations and imagery analogies in systematic interrelations can support introducing new products and services (Esch and Michel 2008; Morgan and Welton 1992), such as innovation clusters

Scent-based innovation communication

- Congruent scents of innovative products, services, issues, etc., or scents in stores, showrooms, etc., can lead to higher attraction and mental imagery, which in turn tend to positively affect innovation adoption through mental application (for mental application see Rogers 2003)
- Ambient fragrance (Herz 2010; Morrin 2010), which is emotionally and thematically associated to an innovation, can positively change the perception and thus tends to influence an individual's decision-making to adopt an innovation
- Scent communication can be used to introduce new products through the improvement of learning (Brumfield et al. 2008; Rempel and Esch 2008), through positive associations and changes to knowledge schemata to influence innovation adoption

Hence, both visual and scent-based innovation communication can influence stakeholder attention and enhance an individual's mental application through mental imagery, through strong emotional recalls and by tapping into learning processes. This positively affects innovation adoption through the activation and modification of schemas (for schema theory: Bartlett 1932; Brewer and Nakamura 1984; Rumelhart and Norman 1977, 1988; Waldmann 1990; in a communication context: Bruhn 2009; Esch 2006; Kroeber-Riel 1993).

From the operational management view, a combination of visual and scent-based innovation communication can be implemented in the following areas:

Exhibitions: Exhibitions represent a communication means that offers a dialog platform for organizations and stakeholders (e.g. the broader public, employees or customers) for inter-personal discussions regarding the presented innovations. A mixture of visual design elements and scent-based communication technologies can be used to transmit complex information related to innovation, innovation clusters, context issues of innovations and the innovative capability of an organization or collaborative network.

Showrooms/shows/cinemas/live entertainment: Showrooms and shows are communication platforms for interactions among organizations and their stakeholder groups. Brumfield et al. (2008) also identify several possibilities to apply scent-based communication, such as in stores, theatres, cinemas, concerts and hotels because all possibilities 'share the common intention to elicit emotion from the viewer' (Brumfield et al. 2008: 247; see also Drobnick 2009; Knoblich et al. 2003). For example, product placement of perfumes in scent cinemas can effectively communicate the emotional positioning of perfumes in movies (Knoblich et al. 2003).

Trade fairs/displays: The tool *trade fairs* are commonly used in marketing to present new products and services as well as the innovative capability of an organization. Knoblich et al. (2003) mention the implementation of scent communication in a trade fair, for instance, to trigger a positive emotional mood for a product (innovation) or the complete presentation of a corporation. Among other things, creative displays can be implemented as a combination of visual stimuli and scent stimuli (e.g. expected or unexpected scent) in order to attract visitor attention and create interest in a product or services (Knoblich et al. 2003). According to Knoblich et al. (2003) the communication tool *displays* used in combination with scent communication can positively influence the emotional product/innovation's positioning. This development can lead to new advertising and film formats and new means to communicate innovative brands as well as innovations.

Creative workshops/seminars: 'Among other cognitive benefits, aroma can increase comprehension, learning, and recall'; thus education is a 'natural application for scent' (Brumfield et al. 2008: 255), and new creative techniques can be developed to facilitate learning in seminars based on scent-based communication effects on memory and recall functions. Moreover, a combination of visual and scent communication can create new associations in innovation workshops and information events.

8.4 Conclusion

After presenting a theoretical approach to innovation communication integrated into existing management concepts and frameworks, this book chapter introduced an agile, visual instrument for communication planning: The communication model canvas. The frame of this chapter was given by the introduction of visual communication and scent-based communication as state-of-the-art instruments in the digitalized information age including implications and use cases for visual and scent-based innovation communication.

The communication model canvas is a visual, practical and strategic-entrepreneurial instrument to better build, reflect and present communication models. Although this instrument was used in start-up coaching sessions, it is required to further do research on how to systematically transform and develop communication models (communication model innovation) in order to create value through communication and enhance corporate innovation. Nevertheless, it can be stated that an agile, responsive information-interaction design, using the communication model canvas method, can support changing the winning game in the digitalized information age.

As for future research, scent-based innovation communication on the corporate level is an emergent theme and scholars can examine 'whether and under what conditions [...] behaviours do indeed tend to result from scent [communication] efforts' (Morrin 2010: 77). Future research can also investigate the effects and limitations of the presented instrument for communication model design and how communication instruments and activities can change different strategizing processes and impact authentic open innovation-culture.

Bibliography

Barney, J.B.: Firm resources and sustained competitive advantage. J. Manage. **17**, 99–120 (1991)

Bartlett, F.C.: Remembering. A study in experimental and social psychology. Cambridge University Press, Cambridge, UK (1932)

Belasen, A.T.: The Theory and Practice of Corporate Communication. A Competing Values Perspective. Sage, Thousand Oaks, CA (2008)

Bleicher, K.: Das Konzept integriertes Management. Campus Verlag, Frankfurt am Main, Germany (1991)

Bleicher, K.: Das Konzept Integriertes Management. Campus Verlag, Frankfurt am Main, Germany (1999)

Brewer, W.F., Nakamura, G.V.: The nature and functions of schemas. In Wyer, R. S Jr., Srull, T. K (eds.) Handbook of Social Cognition, vol. 1, pp. 119–160. Lawrence Erlbaum Associates, Hillsdale, NJ (1984)

Bruhn, M.: Integrierte Unternehmens- und Markenkommunikation: Strategische Planung und operative Umsetzung, 5th edn. Schäffer-Poeschel, Stuttgart, Germany (2009)

Brumfield, C.R., Goldney, J., Gunning, S.: Whiff. The Revolution of Scent Communication in the Information Age. Quimby Press, New York (2008)

Burrell, G., Morgan, G.: Sociological Paradigms And Organizational Analysis: Elements of the Sociology of Corporate Life. Ashgate, Brookfield, VT (1979)

Christensen, C., Raynor, M.E.: Meeting the challenge of disruptive change. Harvard Bus. Rev. **78** (2), 66 (2000)

Collison, G., Erlbaum, B., Haavind, S., Tinker, R.: Facilitating Online Learning: Effective Strategies for Moderators. Atwood Publishing, Madison, WI (2000)

Daschkovska, K., Pfeffermann, N., Meyer-Larsen, N., Hülsmann, M.: Adoption framework for innovations in logistic transport chains. In Kersten, W., Blecker, T., Lüthje, C. (eds.) Pioneering Solutions in Supply Chain Management. A Comprehensive Insight into Current Management Approaches, pp. 347–364. Erich Schmidt Verlag, Berlin (2010)

Davenport, T.H., Leibold, M., Voelpel, S. (eds.): Strategic Management in the Innovation Economy. Strategy Approaches and Tools for Dynamic Innovation Capabilities. Publics Publishing and Wiley-VCH, Erlangen (2006)

de la Fuente Sabate, J.M., de Quevedo Puente, E.: The Concept and Measurement of Corporate Reputation: An Application to Spanish Financial Intermediaries. Corp. Reputation Rev. **5**(4), 280–301 (2003)

Doz, Y.L., Kosonen, M.: Embedding strategic agility: a leadership agenda for accelerating business model renewal. Long Range Plann. **43**(2/3), 370–382 (2010)

Drobnick, J. (2009). Sense and *reminiSCENT*: performance and the essences of memory. Can Theatre Rev. 137/Winter 2009, 6–12

Ehrlichman, H., Bastone, C.: Affect and memory: effects of pleasant and unpleasant odors on retrieval of happy and unhappy memories. J. Pers. Soc. Psychol. **55**, 769–779 (1983)

Ehrlichmann, H., Bastone, C.: The use of odor in the study of emotion. In: Van Toller, S., Todd, G.H. (eds.) Fragrance: The Psychology and Biology of Perfume, pp. 143–159. Elsevier, London (1992)

Eppler, M.J., Hoffmann, F.: Strategies for business model innovation: challenges and visual solutions for strategic business model innovation. In: Pfeffermann, N., Minshall, T., Mortara, L. (eds.) Strategy and Communication for Innovation, pp. 3–14. Springer Verlag, Heidelberg (2014)

Esch, F.-R. (2006). Wirkung integrierter Kommunikation. Ein verhaltenswissenschaftlicher Ansatz für Werbung, (4th ed.). Wiesbaden, Gabler, Germany

Esch, F.-R., Michel, M.: Visuelle Reize in der Kommunikation. In: Bruhn, M., Esch, F.-R., Langer, T. (eds.) Handbuch Kommunikation: Grundlagen, innovative Ansätze, praktische Umsetzungen, pp. 713–734. Gabler Verlag, Wiesbaden (2008)

Grant, R.M.: The resource-based theory of competitive advantage. Implications for strategy formulation. Calif. Manage. Rev. **33**, 114–135 (1991)
Grant, R.M.: Contemporary Strategy Analysis, 6th edn. Blackwell, Oxford (2008)
Hall, R.: The strategic analysis of intangible resources. Strateg. Manage. J. **13**, 135–144 (1992)
Hall, R.: A framework linking intangible resources and capabilities to sustainable competitive advantage. Strateg. Manage. J. **14**, 607–618 (1993)
Herz, R.: The emotional, cognitive, and biological basics of olfaction. Implications and considerations for scent marketing. In Krishna, A. (ed.), Sensory Marketing: Research on the Sensuality of Products, pp. 87–107. New York, Routledge (2010)
Huck, S.: Innovationskommunikation in den Arenen der Medien. Campaigning, Framing und Storytelling. In Zerfaß, A., Möslein, K.M. (eds.) Kommunikation als Erfolgsfaktor im Innovationsmanagement. Strategien im Zeitalter der Open Innovation. Gabler, Wiesbaden, Germany (2009)
Knoblich, H., Scharf, A., Schubert, B.: Marketing mit Duft, 4th edn. R. Oldenbourg Verlag, München, Germany (2003)
Krishna, A. (ed.): Sensory Marketing: Research on the Sensuality of Products. Routledge, New York (2010)
Kroeber-Riel, W.: Bildkommunikation: Imagerystrategien für die Werbung. Vahlen, München, Germany (1993)
Kroeber-Riel, W., Weinberg, P.: Konsumentenverhalten, 8th edn. Vahlen Verlag, München, Germany (2003)
Kroeber-Riel, W., Esch, F.-R.: Strategie und Technik der Werbung. Verhaltenswissenschaftliche Ansätze. (6th ed.). Stuttgart, Germany: Kohlhammer Verlag (2004)
Kröper, M., Bilgram, V., Wehlig, R.: Empowering members of a brand community to gain consumer insights and create new products: the case of vorwerk thermomix research community. In: Pfeffermann, N., Minshall, T., Mortara, L. (eds.) Strategy and Communication for Innovation, pp. 415–426. Springer Verlag, Heidelberg (2014)
Leonard-Barton, D.: Core capabilities and core rigidities: A paradox in managing new product development. Strateg. Manage. J. **13**, 111 (1992)
Martin, P.: Visual Communications. Global Vision Publishing House, New Dehli, India (2008)
McCurry, D., Knight, P., Meiers, M.: Retrieved at http://qct.edu.au/pdf/Research%20Periodicals/QCTResearchDigest2013-9.pdf, 8. July 2016 (2013)
Morgan, J., Welton, P.: See What I Mean. An Introduction to Visual Communication. Edward Arnold, London (1992)
Morrin, M.: Scent Marketing. An Overview. In Krishna, A (ed.) Sensory Marketing: Research on the Sensuality of Products, pp. 75–85. Routledge, New York (2010)
Müller-Stewens, G., Lechner, C.: Strategisches Management. Wie strategische Initiativen zum Wandel führen. Der St.Galler Management Navigator. Stuttgart. Germany: Schäffer-Poeschel (2001)
Osterwalder, A., Pigneur, Y.: Business model generation. A handbook for visionaries, game changers, and challengers. Wiley, Hoboken, NJ (2009)
Osterwalder, A., Pigneur, Y., Tucci, C.: Clarifying business models: Origins, present, and future of the concept. Commun. Assoc. Inf. Syst. (CAIS) **16**(1), 1–25 (2005)
Pfeffermann, N., Breuer, H.: Scenario-based learning architectures as a management tool. In: Brem, A., Viardot, E. (eds.) Evolution of Innovation Management—Trends in An International Context. Palgrave Macmillan, Hampshire, UK (2013)
Plailly, J., Howard, J.D., Gitelman, D.R., Gottfried, J.A.: Attention to odor modulates thalamocortical connectivity in the human brain. J. Neuroscience **28**, 5257–5267 (2008)
Reeves, M., Haanaes, K., Sinha, J.: Your strategy needs a strategy: how to choose and execute the right approach. Harvard Business Review Press, Boston, MA (2015)
Rempel, J.E., Esch, F.-R.: Olfaktorische Reize in der Kommunikation. In: Bruhn, M., Esch, F.-R., Langer, T. (eds.) Handbuch Kommunikation: Grundlagen, innovative Ansätze, praktische Umsetzungen, pp. 285–315. Gabler Verlag, Wiesbaden (2008)
Rogers, E.M.: Diffusion of Innovations, 5th edn. The Free Press, New York (2003)

Rumelhart, D.E., Ortony, A.: The representation of knowledge in memory. In: Anderson, R.C., Spiro, R.J., Montague, W.E. (eds.) Schooling and the Acquisition of Knowledge, pp. 99–135. Erlbaum, Hillsdale, NJ (1977)

Rumelhart, D.E., Norman, D.A.: Representation in memory. In Atkinson, R.C., (ed.) Stevens' Handbook of Experimental Psychology: Vol. 2 Learning and cognition (2nd ed., pp. 511–587). Wiley, New York (1988)

Spangenberg, E.R., Sprott, D.E., Grohmann, B., Tracy, D.L.: Gender-congruent ambient scent influences on approach and avoidance behaviors in retail store. J. Bus. Res. **59**, 1281–1287 (2006)

Ulrich, H., Krieg, W.: *Das St. Galler Management-Modell*. Bern: Paul Haupt, updated version in: Ulrich, H. (2001). *Gesammelte Schriften*, Vol. 2, Paul Haupt, Bern (1972)

Waldmann, M.R.: Schema und Gedächtnis: das Zusammenwirken von Raum- und Ereignisschema beim Gedächtnis für Alltagssituationen. Roland Asanger Verlag, Heidelberg, Germany (1990)

Zelano, C., Bensafi, M., Porter, J., Mainland, J., Johnson, B., Brenner, E., et al.: Attentional modulation in human primary olfactory cortex. Nat. Neurosci. **8**, 114–120 (2005)

Zott, C., Amit, R.: Business model design: An activity system perspective. Long Range Plann. **43** (2/3), 216–226 (2010)

Chapter 9
Open Innovation: Enhancing Theory and Practice by Integrating the Role of Innovation Communication

Ellen Enkel, Annika Dingler and Carsten Mangels

Abstract Firms aim to foster open forms of innovation by collaborating with external partners, e.g., suppliers, customers or firms operating in foreign industries, in order to gain a competitive advantage. At the same time, there is no unique role model of how to realize the potential of open innovation in terms of systematically addressing the external resources. Based on survey data and multiple case study researches, we suggest that more attention should be directed toward a strategic application of innovation communication. We draw on open innovation literature and relational based view to investigate how distinct innovation communication assists a holistic (in terms of breadth and depth) open innovation approach. Our findings illustrate that firms can foster open innovation visibility, partner acquisition, and relationship management through systematic innovation communication. Furthermore, the results reveal how distinct aspects such as timing of communication during the innovation process, the communication channels and corresponding target groups constitute the idiosyncratic open innovation approach. The content of communication, however, is mostly similar to that of firms with a restrained open innovation approach. Accordingly, the findings of our study enhance understanding of how firms can benefit from distinct innovation communication in terms of open innovation.

9.1 Introduction

In times of ubiquitous chances, firms are more and more challenged by phenomena like converging industries. This phenomenon refers to blurring industry boundaries and the challenge of value propositions, products, services, and innovation converging (Bröring and Leker 2007; Hacklin et al. 2013). Accordingly, industry convergence is a development that forces many firms to commercialize technology

E. Enkel · A. Dingler (✉) · C. Mangels
Lehrstuhl Für Innovationsmanagement, Zeppelin Universität, Am Seemooser Horn 20, 88045 Friedrichshafen, Bodensee, Germany
e-mail: a.dingler@zeppelin-university.net

N. Pfeffermann and J. Gould (eds.), *Strategy and Communication for Innovation*,
DOI 10.1007/978-3-319-49542-2_9

and knowledge outside their main markets. It leads to faster innovation cycles and, in that vein, a lack of resources (ibid). This emphasizes the relevance for firms to systematically opening their innovation activities and investing in collaboration with external partners to innovate.

Open innovation as a term introduced by Chesbrough (2003) hereby describes the systematic approach of firms to involve external partners in their innovation processes. In exploiting open innovation, firms, e.g., aim to overcome long product development cycles or reduce risks by sharing it with partners (Cheng and Huizingh 2014; Herzog and Leker 2010). However, it is not an occasional task to introduce corresponding changes and activities into a firm's existing innovation processes but a holistic approach to collaborate with external partners (Enkel et al. 2009). Since open innovation became a dominant innovation management design, research and practice discovered that it is not a simple task to find the right partner or idea at the right time and to engage in collaboration. Additionally, it becomes very resource consuming to search and address every possible source and partner. Therefore, the question arises how ideas and partners are able to find you as a firm pursuing the open innovation approach. Open innovation as a cooperative approach with various communicative interfaces makes innovation communication a constitutive element (Zerfaß 2009) as an open innovation process not just comes with the challenge of exchanging details on technological details but also with the need of proper coordination, communication and general exchange of relevant information (Hauschildt and Salomo 2011).

In the following chapter, we elaborate the strategic importance of innovation communication and how firms can adapt their approaches in terms of enhancing their capacities to network with potential partners. Specifically, the question of how firms can benefit from open innovation visibility and why it holds high potential for firms to communicate their innovation capabilities and processes is discussed.

9.2 Innovation Communication

The opening of a firm's innovation activities generates multiple communicative transition points (Beckett and Hyland 2011) which have to be managed throughout the innovation process. While innovation communication in line with the innovation process itself is of a mainly operative nature, its utilization as a central initiation and steering element for partnerships and cooperation is of strategic relevance (Pfeffermann et al. 2008).

Accordingly, the innovation process in most firms is no longer a linear and mostly internally performed process but a cooperative one comprising various iterative steps (Zerfaß and Huck 2007a). As a consequence, innovation management is not a simple coordination of the internal R&D departments anymore, but the additional coordination of external partners, their linking to internal departments and the management of this innovation network (Möslein 2010; Zerfaß and Ernst 2008). This leads to the understanding of innovation as a social construct consisting of processes of adjustment and participation of different actors (von Busse 2005) which, in turn, leads to a distinct meaning and task of innovation communication. Following Zerfaß and Huck (2007b, p. 848), innovation communication is a "systematically planed, performed and evaluated communication of innovation aiming to generate understanding and trust for the innovation as well as positioning the corresponding organization as innovator." Besides its process-oriented character, innovation communication hereby also refers to the potential addressees as it includes the positioning of a firm as being innovative. Literature on innovation communication identifies a broad range of goals including the addressing of internal and external partners or potential partners (Möslein 2009). This emphasizes the relevance of innovation communication to foster a complaisant innovation culture (Zerfaß and Huck 2007a) and a strategic knowledge management (Zerfaß 2009) as well as the enabling of connecting one's innovation activities to the market, building a high innovation reputation (Fink 2009) and the integration of different stakeholders in the innovation process (Möslein 2009). Attended with those goals is a broad range of addressees including internal as well as external actors (Zerfaß 2009). This notion of innovation communication is of the category persuasive information politics (Zerfaß and Ernst 2008) aiming to inform and positively influence different stakeholders and make the open innovation approach visible and understandable to them.

9.3 Open Innovation Visibility

Framing theory describes that subjectively observed conditions, which are influenced by the information provided, cause cognitive structures which, in turn, influence the observation of reality (Goffman 1974). As rationality of addressed actors is limited and availability of information is incomplete (Simon 1979), innovation communication is able to set certain frames and therefore influence observations and actions of the actors addressed (Pfeffermann et al. 2008). In the context of open innovation, this approach aims for the perception of a firm's innovation capability, in particular open innovation capability, as well as its internal innovation culture (Gassmann et al. 2009; Trautmann and Enkel 2014). In that vein, innovation communication aims to foster internal absorption and, more importantly, a high perception of the firm's innovation capability in order to appear highly attractive for potential partners (Dollinger et al. 1997). Based on the thereby arising possibilities of innovation communication, firms can enhance their open innovation visibility and networking potential to build new partnerships.

Focusing on internal addressees, innovation communication can be facilitated to build an open innovation culture when it comes to acceptance and assimilation of innovation (Zerfaß and Huck 2007a). Such utilization of innovation communication is especially valuable for firms pursuing an open innovation approach as it is not just about opening up to the outside but also about reducing internal barriers and systematizing knowledge flows (Möslein 2010). In this vein, it supports the development of a firm's dynamic capabilities and accordingly its networking with different actors to generate and reorganize new knowledge (Pfeffermann 2011; Zaremba 2003). Also innovation communication can enable firms to renew and develop their resources and adapt external changes during their open innovation processes (Barreto 2010).

For a successful open innovation approach it is not sufficient to just generate an internal willingness, furthermore it is necessary to find external actors that are willing to agree to cooperation and to carve out complementary interests (Gassmann et al. 2010). As an instrument to increase demand for cooperation, open innovation visibility of a firm can foster the establishment of a high innovation reputation in the market (Pfeffermann 2011) (Fig. 9.1). By displaying the innovation capabilities, firms on the one hand emphasize their own interest in pursuing innovative activities and on the other hand they present a ticket of admission for potential partners (Rosenberg 1990). Furthermore, they communicate and demonstrate their open and collaborative approach of doing innovation. Firms hereby can illustrate that the external knowledge they search for is not a substitute for their own but an addition to their own resources and capabilities which illustrates the complementarity of the approach and the firms' capabilities to the potential partner (Mowery et al. 1996). In addition to a proactive approach to acquire new partners, innovation communication generating open innovation visibility attracts potential partners to approach the firm on their own initiative, thus create the ticket of admission for those partners (Mowery et al. 1996; Rosenberg 1990). In doing so, the potential partners become aware of potential cooperation options based on the firm's presented competencies and especially its willingness to cooperate.

Finally, by fostering the firms' dynamic capabilities, innovation communication not just enhances the internal acceptance and capacity to absorb innovation but also contributes to overcoming asymmetric information disposition at the market (Menon and Pfeffer 2003) and accordingly to inform (potential) partners about their open innovation approaches and search areas.

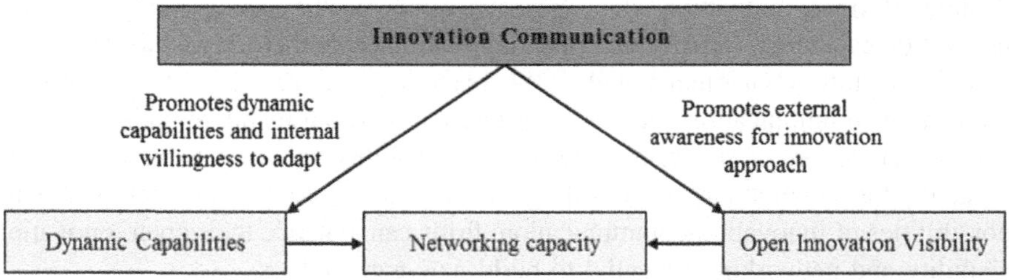

Fig. 9.1 Innovation communication as facilitator for a firm's networking capacity (illustration by authors)

9.4 Strategic Relevance of Innovation Communication

Focusing on open innovation, we have elaborated the various communicative interfaces and the strategic relevance of innovation communication in regard to firms' dynamic capabilities as well as their processes to overcome information asymmetries. Hence, innovation communication can be understood as a constitutive element of the innovation approach (Zerfaß 2009) to generate competitive advantages (see Fig. 9.2). This theoretical framework emphasizing the relevance of communicative activities to achieve competitive advantage and superior open innovation visibility leads to the analytical framework of innovation communication as the basis for our data.

9.5 Methodology and Data

Due to limited theoretical insights on the concept of open innovation visibility, we conducted an inductive, multiple case study approach (Eisenhardt 1989). Therefore, we do not aim to test an existing theory but to contribute to theory building. The main advantages of this interpretative and qualitative approach are the possibilities to generate and analyze the data in a contextualized way (Yin 2013) and to create—as a result—a stable access to social reality and the phenomenon under research.

By using theoretical sampling (Eisenhardt and Graebner 2007), we selected eight cases for the case study. By choosing the most representative, we further elaborate three of those cases as examples for this book chapter. The leading dimensions for case identification were (i) the implementation of a holistic and systematic open innovation approach and (ii) a strategic and integrative understanding of innovation communication. For the determination of the openness of innovation processes, we drew on data of an annual non-compulsory innovation survey, conducted by the "Dr. Manfred Bischoff Institute of Innovation Management of the Airbus Group" at Zeppelin University in Friedrichshafen, Germany. The data was analyzed by applying a concept developed by Laursen and Salter (2006) classifying innovation

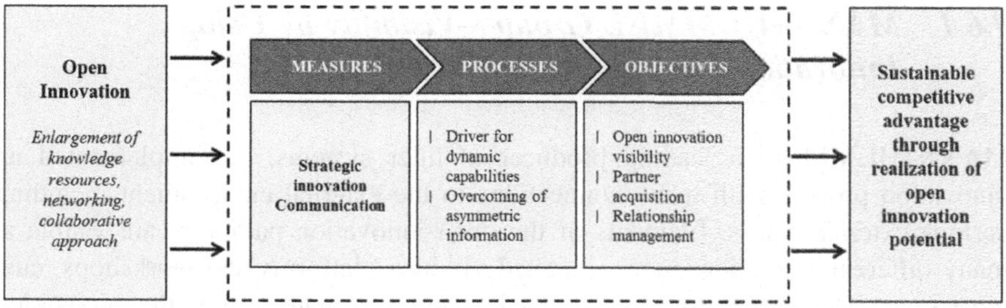

Fig. 9.2 Innovation communication as business capital to build competitive advantages (illustration by authors)

Table 9.1 Overview of the cases

Firm	Interview partner
Webasto SE	Innovation manager "Advanced Engineering & Product Strategy"
Mann+Hummel GmbH	Manager Corporate Strategy & Innovation
Sixt SE	Head of Business Development
Rehau AG & Co. KG	Head of Innovation Management
Philips Austria GmbH	Head of Advanced Development & Patents
T-Systems Multimedia Solutions GmbH	Head of Innovations T-Systems MMS
Outfittery GmbH	Head of Marketing Cooperation
Robert Bosch GmbH	Senior Manager IT-Technology

approaches regarding search breadth and search depth. Moreover, the degree of systematization of the open Innovation approach was analyzed utilizing the open innovation maturity framework by Enkel et al. (2011) (Table 9.1).

To address our research questions different data sources were selected for each of the cases. First, we conducted semi-structured interviews. Second, we analyzed publicly available information like press releases, web pages, and presentations. Third, we included data of the annual innovation survey to review the gained insights. The semi-structured interviews focused on the alignment of innovation management in general and the value creation of innovation communication in particular. Based on the interview data, we conducted a within case as well as cross case analysis (Eisenhardt and Graebner 2007). We used an open code system based on an iterative and continuous development process (Strauss and Corbin 2008). The results of the within case analysis were sent to the interview partners for confirmation aiming to increase the validity of the results. In general, this approach guarantees a high degree of transparency, builds up a definite research focus, and emphasizes the strict observance of scientific quality criteria.

9.6 Case Studies

9.6.1 MANN+HUMMEL Group—Visibility by Using Innovation Communication

MANN+HUMMEL, a leading producer of filter systems, has implemented an innovation process with many connections to the external environment including various external actors. Elements of the open innovation paradigm are visible at many different steps: Ideas are generated via idea platforms and workshops, customers are encouraged to give feedback and collaborate within the innovation process, suppliers are asked to participate in call for ideas and collaboration projects, and a spin-off for a cross-industry project has been established. Moreover, all these steps are strictly structured to ensure organized interaction with the different actors and to identify trends and needs.

This open approach of innovation management is aligned with a strategic approach of communication activities. The integrated character of communication becomes apparent in various internal and external communication activities. Innovation as one of the main pillars of the corporate strategy is reflecting the commitment of the shareholders and the board. Therefore, the corporate strategy map including innovation as competitive advantage is forming the starting point for the innovation communication. With standardized methods and approaches, the company is ensuring that innovation communication is executed on a regular base. Examples for such approaches are the idea management platform with idea campaigns, regular reviews and challenges of innovation roadmaps and corresponding portfolios, information about content and status of innovation projects in innovation meetings, journals and within Technology Days at customers.

While the firm aims to foster the internal innovation culture as well as customer relations, the creation of open innovation visibility can be considered as one of the main strategic objectives of innovation communication. The company wants to be perceived as one of the leading experts, as provider for innovative solutions and as a reliable partner for cooperation.

"*I think it is important to the external environment […] that the cooperation partners address us […], that they approach us actively and perceive us as experts knowing our innovative products and appreciate them.*"

In order to attain this objective, MANN+HUMMEL uses trade fairs as well as congresses as platforms to present its products and to place topics of interest. In that vein, the company pursues to increase its innovation reputation by presenting its innovativeness and its open innovation visibility.

As a result, innovation communication at MANN+HUMMEL is more than an operative measure—it is a strategic tool to enhance partner relations and to enrich the open innovation approach. In this context, the creation of open innovation visibility is considered as one of the driving factors to build up the groundwork for an ongoing cooperative approach of innovation.

9.6.2 T-Systems Multimedia Solutions—"Ambidextrous Innovation" as Guiding Principle

T-Systems Multimedia Solutions, one of Europe's largest companies for IT services, has innovation processes which rely on constant exchange and cooperation throughout the entire process stream. The guiding principle is called "Ambidextrous Innovation" meaning the innovative progress is implemented on the operational level by continuous and incremental progress as close as possible to the customers as well as on a more strategic level by pursuing radical changes and adjustments. This approach demonstrates the intention to meet the efficiency and requirements driven by the company's key performance indicator while, at the same time, ensuring a certain flexibility. In this systematic framework, the openness is especially emphasized by cooperation-based market research and idea generation, systematic and open feedback processes and numerous development partnerships.

The innovation communication of T-Systems Multimedia Solutions is structured into outside-in communication for the continuous integration of external actors and innovation marketing for the improvement of the perception of innovativeness. This kind of communication and the aim to be perceived as more innovative compared to other market participants addresses the final customer as well partners. To foster open innovation visibility and to address partners, company representatives participate as keynote speakers at numerous conferences and symposiums, while they also operate a proactive partner marketing. By facilitating these different communication channels, the firm wants to present its topics and products as well as its way of innovating as well as its concept of "Ambidextrous Innovation." The strategic focus is not only on the presentation of innovative services but also on the presentation of open approaches, best practices, and processes to foster the cooperative concept.

T-System Multimedia Solutions pursue the objective to generate a unique selling point by communicating its innovation approach. In accordance with the innovation approach, T-Systems Multimedia Solutions aims to establish a high degree of awareness for its innovation model by using strategic communication measures.

9.6.3 Outfittery GmbH—Personal Networks Instead of Institutional Ties

Outfittery GmbH is Europe's biggest Personal Shopping Service for men. The innovative business model is heavily connected to its environment. Many new ideas —on the operational but also strategic level—are generated by collaborative approaches. Even the creation of those ideas is more based on personal contacts than institutional ties, the innovation process integrates external actors and therefore can be considered as open. This applies equally to prestigious projects like the "3D men scan" as well as adaption's and innovations in its operational processes.

Comparable to this innovation approach, the innovation communication is based on many personal ties and networks, which enhance the transfer of knowledge beyond corporate borders. The firm pursues the objective to gain access to several start-up networks and platforms by innovation communication. In this context, it is important for Outfittery GmbH to underline the innovativeness to get a "ticket of admission" to these networks. This ticket is necessary, because respective exchange processes are based on mutual help. Every participant needs to offer specific knowledge and has to signal its openness for cooperation. In this context, the creation of open innovation visibility through interpersonal innovation communication is of strategic importance.

Outfittery GmbH aims for a general visibility of its innovativeness by using innovation communication. On the one hand, the innovativeness of the business model has to be communicated to a broad target group; on the other hand, this kind of communication can be considered as access key for different partner networks inside the start-up ecosystem.

9.7 Results

9.7.1 Communication Channels to Enhance Open Innovation Visibility

The data reveals different ways of enhancing open innovation visibility through communication. A classic way of innovation communication is to facilitate trade show appearances to present products and innovation. Such appearances comprise a persuasive information character as well as interactive elements as firms not only display products at their booth but also interact with visitors and thus can lead their attention toward methods, approaches, and processes leading to the results they display. Such as, for example, MANN+HUMMEL engaging in many congresses and trade shows. Not just to present their new products, but to place trends and other topics of interest.

Firms can similarly profit from attending and presenting at congresses. They can communicate their innovation activities, processes, and existing partnerships as a first step and contact potential partners attending the congress based on their presentation. Depending on the event and availability, firms can make use of their CEO to present as a keynote speaker or have employees of the innovation management represent the firm as an external figurehead.

A more sales and reputation oriented form of innovation communication is the implementation of roadshows or similar marketing activities such as Outfittery's "Men scanner," an innovation they promote on various occasions on order to display their innovative approach of selling men's clothes.

9.7.2 Addressees of Innovation Communication

In addition to distinct channels of innovation communication, the data illustrates differences in groups of addressees and corresponding approaches, especially in order to enhance open innovation visibility. Most relevance as integrated actors of innovation communication is assigned to the firms' customers which can be explained by the importance of customer-oriented communication and traditional innovation marketing activities themselves but also the intention of integrating the customer into the innovation process. Accordingly, firms implement idea generation platforms or integrate customers in predefined feedback loops during their innovation process. Moreover, firms address research institutes, universities, or colleges through their innovation communication in a similar way in order to integrate them in their innovation activities. Thereby, firms not only focus on the generation of added value during the innovation process, e.g., during the idea generation phase, but also aim to appear attractive for future employees. As an example, T-Systems is very present at universities: The company supervises projects, offers industry-sponsored PhDs, and gives lectures to build up strong ties with potential future employees and to enrich HR-marketing.

Existing innovation partners as well as potential innovation partners, i.e., other firms or other experts, are addressed by innovation communication through fairs, exhibitions, congresses, or similar events. Firms thus foster innovation-oriented nursing of partnerships as well as continuous exchange in terms of ongoing projects or the generation of new ideas. At the same time, innovation communication on a superior level serves the promotion of visibility by leading the attention toward a firm's innovation capabilities to be attractive to potential and future partners and to initiate new innovation cooperation.

9.7.3 Innovation Communication and Open Innovation Visibility

As elaborated earlier, communicative activities can lead to adjustment in partners' perception. Innovation communication that actively copes with inequality of information and knowledge distribution conducts partnerships during an innovation project not only on a process base but also contributes to open innovation visibility in general. The data illustrates that aside from communication involved in the innovation process itself, firms pursue a superior perceived innovativeness in the market through purposeful innovation communication.

In that vein, innovation communication activities aim for the dissolution of knowledge asymmetries to enable future partnerships. At first, the firm has to be perceived as expert in a specific area by potential partners. While the general innovation capability is therefore communicated, firms further emphasize the cooperative approach of their innovation project in order to highlight their openness of collaborating with others. Accordingly, the goal is not only to increase the firms' perception of innovativeness but also the firms' perception of openness in the innovation process. For instance, T-Systems does not only want to be perceived as innovative by customers, but also by partners. In order to attain this objective, the firm presents its "Ambidextrous Innovation" approach on many expert conventions.

The data also shows that the open innovation visibility construct is multidimensional. One determining dimension is the need of an adequate internal knowledge base creating a "ticket of admission" for partnerships. Another determining dimension is the presentation of the firms' innovation strategy leading potential partners' attention toward a high willingness to cooperate. For example, Outfittery emphasizes the attendance and especially an active participation in network events to present their competencies and initiate interaction and partnerships. Simply be becoming visible to others, a certain value in order to actively or passively initiate partnerships arises. To demonstrate potential partners a real win-win opportunity, it is conducive for firms to additionally highlight their knowledge and capabilities as complementary to the partners addressed. MANN+HUMMEL further emphasizes that a respective reputation in the market, in turn, also leads to external actors approaching the firm proactively with inquiries to cooperate. To foster external open innovation visibility, the example of T-Systems shows that presenting their innovation approach to actors outside the firm is beneficial in similar vein.

According to our study, open innovation visibility is both a necessary prerequisite to passively build a relationship with partners (potential partners approach the firm proactively) and a sufficient condition to actively build a relationship with partners (approach potential partners).

9.8 Discussion

Our results illustrate that innovation communication decisively contributes to realizing open innovation potential. A firm that pursues the open innovation approach has to position itself in the market as a credible partner willing to cooperate in order to actually initiate partnerships. One way of option for such a positioning is a respective innovation communication leading to enhanced open innovation visibility of a firm.

In line with literature, the study illustrates the role of innovation communication to manage existing partnerships and elaborates its role to acquire new partners by creating an open innovation visibility to external actors. Literature (Zerfaß and Huck 2007a) emphasizes the function of innovation communication to position the focal firm as being innovative. This positioning is supposed to lead to a respective innovation perception of the firm (Gassmann et al. 2009) and is supposed to ensure a ticket of admission (Rosenberg 1990) necessary for cooperation. Thus, the constitution of such open innovation reputation is a communicative element of persuasive information policy (Zerfaß and Ernst 2008).

While the data confirms the constitution of a respective reputation beneficial for proactive and necessary for passive acquisition of new partners, there are differences in the concept of innovation reputation deducted from literature and the focal concept of open innovation visibility. Difference lies in the strategic relevance as well as in the connectional breadth and depth. The data illustrates that a concept to enhance visibility is more than a matter of innovation reputation: The firms in our study consider a purposeful communication of their own innovation capacity toward a broad range of addressees as a central element of their innovation and communication strategy. In that vein, attention is created toward their innovation capabilities leading to interest in cooperation. Accordingly, this is not an element of persuasive information policy (Zerfaß and Ernst 2008) but a strategic element closely aligned with the actual innovation management and open innovation processes.

In that context, we are able to differentiate between the concept of innovation reputation and open innovation visibility, whereas the constructs differ especially regarding strategic understanding and addressees. Open innovation visibility can be defined on the basis of three dimensions: First, for the positioning of a firm as being innovative, open innovation visibility comprises the communication of the general innovation capability of a firm as a central element. The results of our study illustrate that firms promote their innovation capacity in the market by facilitating their capabilities as ticket of admission (Rosenberg 1990). Second, while there is only little focus in terms of the actors addressed in the concept of innovation reputation and customers are addressees as well as potential employees, in the concept of open innovation visibility a direct reference to the firms' innovation processes is produced. This is achieved through a purposeful communication of the processes and approaches during the development of the innovation. Thus, stakeholders are not only informed about the outputs of the firms innovation processes, i.e., the new products or services, but also the underlying processes, activities, and cooperative behavior. The determining difference lies in the communication of the result when it comes to innovation reputation and the communication of the result in combination with the underlying processes when it comes to open innovation visibility. Innovation communication in the context of an implemented open innovation approach accordingly generates a value in communicating the approach itself, which is not directly associated with the newly created product or services: Firms can thus position themselves as innovator and additionally create attention among potential partners and signalize their willingness to cooperate. This is how innovation communication affects networking capacity of a firm as well as future innovation activities.

The third dimension, after innovation capability and a process-oriented approach, is the aspect of visibility anchoring the concept on an action level and within innovation communication. Building on the elements concerning the content of communication, firms furthermore need to facilitate innovation communication to create visibility. The results display that especially fairs, exhibitions and congresses are beneficial to promote innovation capabilities in combination with the underlying approaches differentiating the channels of communication from innovation reputation in the same way as the actors' addressed—by including and building up reciprocal effects with the innovation processes.

The interplay of innovation communication and open innovation visibility can be summarized as follows (see also Fig. 9.3): The distinct characteristic of open innovation visibility in contrast to innovation reputation is communication that comprises the innovation itself as well as the processes and partners that were part of the innovation. Building on a strategic and interactive understanding of communication, open innovation visibility is to not only able to initiate partnerships but can generally enhance the networking capacity fostering a firm's dynamic capabilities which, in turn, creates a competitive advantage.

9.9 Conclusion

This chapter illustrates that in the context of the open innovation paradigm, innovation communication is not just relevant in terms of a process-oriented coordination but is also of strategic relevance. While internally it contributes to the innovation culture of a firm, innovation communication externally enhances open innovation visibility and increased innovation perception through potential partners. The strategic relevance evolves from the positive influence on a firm's networking capacity, resulting in a competitive advantage.

The novel concept of open innovation visibility is illustrated by the selected cases and leads to a distinction of the concept of a persuasive-oriented innovation communication approach. Specifically, the intention of firms to communicate their innovation capabilities and the implemented open approach to acquire partners and, in turn, foster the open innovation approach needs to be highlighted. Innovation communication accordingly is enhancing an active initiation of partnerships as it provides the requested ticket of admission. Furthermore, innovation communication is a necessary prerequisite to build partnerships initiated by the external partner.

Additionally, the data illustrates that firms especially facilitate the participation in trade shows, congresses, or symposia as channel of innovation communication in order to create visibility for their innovativeness and their approaches to innovate. Also, our study highlights the relevance of communicating innovation processes, approaches and partnerships in addition to the innovativeness itself in order to manifest one's open innovation visibility.

Communication is purposefully implemented to enrich existing open innovation processes and to realize the underlying potential. On behalf of open innovation visibility, it becomes clear that innovation communication is not just an operative tool but should be an integrative strategic element of a firm's innovation processes (Fig. 9.3).

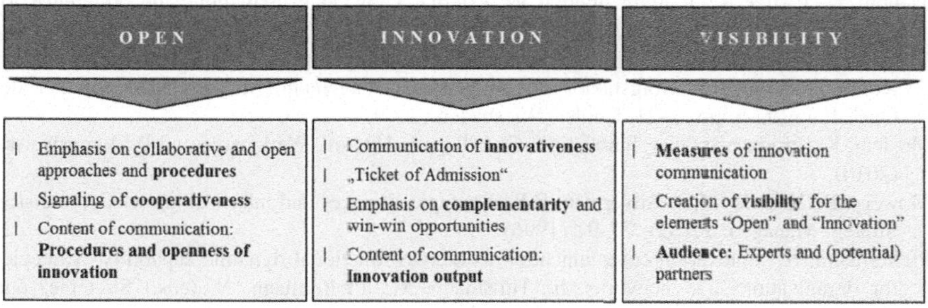

Fig. 9.3 The three dimensions of the open innovation visibility concept as an integrative strategic element of a firm's innovation processes (illustration by authors)

Bibliography

Barreto, I.: Dynamic capabilities: a review of past research and an agenda for the future. J. Manag. **36**(1), 256–280 (2010)

Beckett, R., Hyland, P.: Communications and learning structures that facilitate transfer of knowledge at innovation transition points. J. Sci. Commun. **10**(4), 1–8 (2011)

Bröring, S., Leker, J.: Industry convergence and its implications for the front end of innovation: a problem of absorptive capacity. Creativity Innov. Manag. **16**(2), 165–175 (2007). Doi:10.1111/j.1467-8691.2007.00425.x

Cheng, C.C.J., Huizingh, E.K.R.E.: When is open innovation beneficial? The role of strategic orientation. J. Prod. Innov. Manag. **31**(6), 1235–1253 (2014). Doi:10.1111/jpim.12148

Chesbrough, H.: The logic of open innovation: managing intellectual property. Calif. Manag. Rev. **45**(3), 33–58 (2003)

Dollinger, M.J., Golden, P.A., Saxton, T.: The effect of reputation on the decision to joint venture. Strateg. Manag. J. **18**(2), 127–140 (1997)

Eisenhardt, K.M.: Building theory from case study research. Acad. Manag. Rev. **14**(4), 532–550 (1989)

Eisenhardt, K.M., Graebner, M.E.: Theory building from cases: opportunities and challenges. Acad. Manag. J. **50**(1), 25–32 (2007)

Enkel, E., Gassmann, O., Chesbrough, H.: Open R&D and open innovation: Exploring the phenomenon. R&D Manag. **39**(4), 311–316 (2009)

Enkel, E., Bell, J., Hogenkamp, H.: Open innovation maturity framework. Int. J. Innov. Manag. **15** (06), 1161–1189 (2011)

Fink, S.: Strategische Kommunikation für Technologie und Innovationen *Kommunikation als Erfolgsfaktor im Innovationsmanagement*, pp. 209–225. Springer, Wiesbaden (2009)

Gassmann, O., Rumsch, W.-C., Rüetsche, E., Bader, M.A.: R&D reputation and corporate brand value. Res.-Technol. Manag. **52**(4), 16–19 (2009)

Gassmann, O., Enkel, E., Chesbrough, H.: The future of open innovation. R&D Manag. **40**(3), 213–221 (2010)

Goffman, E.: Frame Analysis: An Essay on the Organization of Experience. Harvard University Press, Cambridge (1974)

Hacklin, F., Battistini, B., Von Krogh, G.: Strategic choices in converging industries. MIT Sloan Manag. Rev. **55**(1), 65 (2013)

Hauschildt, J., Salomo, S.: Innovationsmanagement, vol. 5 (überarb., erg. und aktualisierte Aufl.). Vahlen, München (2011)

Herzog, P., Leker, J.: Open and closed innovation—different innovation cultures for different strategies. Int. J. Technol. Manag. **52**(3/4), 322–343 (2010). Doi:10.1504/ijtm.2010.035979

Laursen, K., Salter, A.: Open for innovation: the role of openness in explaining innovation performance among UK manufacturing firms. Strateg. Manag. J. **27**(2), 131–150 (2006)

Menon, T., Pfeffer, J.: Valuing internal vs. external knowledge: explaining the preference for outsiders. Manag. Sci. **49**(4), 497–513 (2003)

Möslein, K.: Innovation als Treiber des Unternehmenserfolgs. In: Zerfaß, A., Möslein, K. (eds.) Kommunikation als Erfolgsfaktor im Innovationsmanagement: Strategien im Zeitalter der Open Innovation, pp. 3–21. Gabler, Wiesbaden (2009)

Möslein, K. (Producer): Open Innovation: Grundlagen, Akteure, Werkzeuge und Wirkungsweisen (2010)

Mowery, D.C., Oxley, J.E., Silverman, B.S.: Strategic alliances and interfirm knowledge transfer. Strateg. Manag. J. **17**(S2), 77–91 (1996)

Pfeffermann, N.: Innovation communication as a cross-functional dynamic capability: Strategies for organizations and networks. In: Hülsmann, M., Pfeffermann, N. (eds.) Strategies and Communications for Innovations, pp. 257–289. Springer, Berlin (2011)

Pfeffermann, N., Hülsmann, M., Scholz-Reiter, B.: Framing innovations to grasp stakeholders' attention: a dynamic capability-based conception of innovation communication. Int. Grad. Sch. Dyn. Logistics **40** (2008)

Rosenberg, N.: Why do firms do basic research (with their own money)? Res. Policy **19**(2), 165–174 (1990)

Simon, H.A.: Rational decision making in business organizations. Am. Econ. Rev. **69**(4), 493–513 (1979)

Strauss, A.L., Corbin, J.: Basics of Qualitative Research: Techniques and Procedures for Developing Grounded Theory. Sage Publications, Thousand Oaks, CA (2008)

Trautmann, K., Enkel, E.: Success factors for strategic communication of corporate innovativeness for financial analysts. Int. J. Innov. Manag. **18**(01), 1450006 (2014)

von Busse, N.B.: Innovationen in der Unternehmenspraxis: Soziale Repräsentationen von Innovation und Innovationsprozessen. Deutscher Universitätsverlag, Wiesbaden (2005)

Yin, R.K.: Case study Research: Design and Methods. Sage Publications, Thousand Oaks, CA (2013)

Zaremba, A.J.: Organizational Communication: Foundations for Business & Management. South-Western Pub, Manson (2003)

Zerfaß, A.: Kommunikation als konstitutives Element im Innovationsmanagement. In: Zerfaß, A., Möslein, K. (eds.) Kommunikation als Erfolgsfaktor im Innovationsmanagement, pp. 23–55. Springer, Wiesbaden (2009)

Zerfaß, A., Ernst, N. (Producer): Kommunikation als Erfolgsfaktor im Innovationsmanagement. Ergebnisse einer Studie in deutschen Zukunftstechnologie-Branchen (2008)

Zerfaß, A., Huck, S.: Innovation, communication, and leadership: New developments in strategic communication. Int. J. Strateg. Commun. **1**(2), 107–122 (2007a)

Zerfaß, A., Huck, S.: Innovationskommunikation: Neue Produkte, Ideen und Technologien erfolgreich positionieren. In: Piwinger, M., Zerfaß, A. (eds.) Handbuch Unternehmenskommu nikation, pp. 847–858. Gabler Verlag, Wiesbaden (2007b)

Chapter 10
Scanalyse—A Case Study of the Role of Social Capital, Strategic Networking, and Word of Mouth Communication in the Diffusion of an Innovation

Tim Mazzarol, Peter Malone and Sophie Reboud

Abstract In this chapter, a longitudinal case study is examined of Scanalyse, a technology spin-out company from an Australian university that grew from start-up to a global market position within a decade and was then sold through trade sale to one of its main competitors. The case provides an example of how a small innovative firm assesses the value of the economic rent it can secure from an innovation and the strategic decision-making and "pivots" it must undertake to create value. It also demonstrates the key role played by social capital, strategic networking, and word of mouth communication in facilitating the commercialization process.

10.1 Introduction

This chapter examines the case of Scanalyse, a high-technology spin-out from an Australian university that was able to successfully commercialize its technology within a global market. The focus of this case study is on the roles played by social capital, strategic networking, and word of mouth (WOM) communication in assisting the firm to undertake this commercialization. It analyses the process through the conceptual framework of an entrepreneurial innovation value model (EIV), and demonstrates how both social capital as well as economic capital influenced the diffusion of the innovation. Relatively little work has been published in the academic literature about the way commercialization takes place in small innovative firms and how they capture value. The case brings together theories associated with the role played by entrepreneurial rents (Amit and Schoemaker 1993; Alvarez and Barney 2004; Alvarez 2007), and the resource-based view

T. Mazzarol (✉) · P. Malone
University of Western Australia, M263, 35 Stirling Highway,
Crawley, WA 6009, Australia
e-mail: tim.mazzarol@uwa.edu.au

S. Reboud
Groupe ESC Dijon Bourgogne, Burgundy School of Business, Dijon, France

N. Pfeffermann and J. Gould (eds.), *Strategy and Communication for Innovation*,
DOI 10.1007/978-3-319-49542-2_10

(RBV) of the firm (Barney 1991; Mahoney 1995; Collis and Montgomery 1995; Fahy and Smithee 1999). The case provides evidence of a process that entrepreneurial firms go through in the commercialization of an innovation, with specific reference to the influence of social capital, strategic networking, and WOM. This case offers a good illustration of the importance of early relationships built on trust, on the power of the social capital of the manager, and of the role that WOM can have when engaging lead customers to "co-create" the innovation with the firm.

10.2 Commercialization, Social Capital, Strategic Networks, and WOM

The concept of commercialization is poorly defined within the literature (Aarikka-Stenroos and Sandberg 2009). However, it can be broadly understood as a process involving both the sale and the adoption of new products, processes, and services generated from an investment in R&D and new product development (NPD). It typically involves marketing communications, sales, brand development and distribution, with social capital networks playing an important role (Lehtmäki et al. 2008; Partanen et al. 2008). If the sales side is in the hands of the firm, the adoption side is more difficult to monitor, as it depends on the perception of the customer. Such a process relies heavily on human relationships between different actors. They include: (i) the innovators in the firm undertaking the commercialization; (ii) the people they are usually seeking advice from; (iii) the targeted customers, and (iv) other actors potentially impacted by the innovation.

Commercialization is therefore a process that can benefit from the presence of a strong strategic network through which the firm's management team is able to secure access to information, suppliers, customers and providers of technology, skills and financial capital. It is particularly important for small firms that typically lack the necessary resources to proceed alone in the commercialization process (Alvarez and Busenitz 2001; Alvarez and Barney 2005). This strategic network, and the development of social capital to assist the firm to bring its innovation to market, should allow it to engage freely with lead customers, key suppliers, and other network actors to exchange information and secure positive word of mouth (WOM) (Brooks 1957; Arndt 1967; Dodson and Muller 1978).

Although commercialization is often viewed from a marketing or technical development perspective, the importance of social capital, interpersonal networking, and WOM communications should not be ignored when trying to understand the process (Czepiel 1974; Mazzarol 2013). Managers seeking to commercialize a new product or service should acknowledge this and focus on the generation of positive WOM within target market segments and influential decision-makers (Kawakami et al. 2013). If the rate of diffusion will be influenced by the nature of the innovation, and the characteristics of both the innovators and the environment in which they exist, it can be dramatically modified by positive or negative WOM (Mahajan et al. 1984). However, firms must take into account that such a process has to be built in advance, as the propensity of opinion leaders to provide positive WOM depends on their previous experience with the firm, on their commitment to it, and on their perception of the value offered by the innovation (see: Martilla 1971; Leonard Barton 1985; Valente and Davis 1999).

Marketing promotional activities that help to provide opinion leaders with early trials of the product or service, plus sufficient information to educate them as to its attributes and benefits are likely to boost their self-confidence in giving WOM (Rogers 1976, 1995). Moreover, this WOM will be better accepted by the receivers if the richness, strength, and valence of the message, as well as the credibility of the source and brand equity of the supplier are undisputable (Mazzarol et al. 2007; Sweeney et al. 2008, 2012). Building the message and gathering proofs will thus be critical activities.

Engaging customers, particularly lead customers, in a proactive dialogue aimed at reducing the perceived risk and uncertainties associated with the innovation is a key aspect of any commercialization marketing strategy (Mazzarol 2011). Customer education and collaboration in the NPD process builds trust, loyalty, and commitment that result in early adoption and the generation of positive WOM (Czepiel 1974; Mahajan et al. 1984; Kawakami et al. 2013). It is important that work on this process occurs long before the innovation actually reaches the market, and to also explore in advance who will be the most influential actors. Due to their resource constraints small firms are less likely to follow the same pattern of behavior in NPD and commercialization as that found in large firms; relying more on customer engagement and co-creation (Berends et al. 2014).

The dynamics of different markets and how disruptive the innovation is will also influence the nature of any marketing communications strategy (Corsaro et al. 2012). The more disruptive the innovation will be and/or the less innovative the market will be, the more convincing the evidence and explanations will have to be (Christensen et al. 2002; Sandberg 2002). The diffusion of innovation has been viewed as a social rather than an economic process (Rogers and Shoemaker 1971). Tarde (1903) observed that the diffusion of innovation is essentially a process of imitation by one individual of another. New ideas or practices are passed on from the originator of the invention to the imitator and then from imitator to imitator (Kinnunen 1996). Despite the influence of a range of communications media (e.g., mass media, social media), this social process still relies on the interrelationship between individuals and the capacity of the most influential to convince the more reluctant (Rogers 1976).

The success of an innovation is not so much its novelty, or even the merits of its technological sophistication or economic value. What is critical to its success is the ability for the invention to be imitated and adopted on a significant scale, and for this to be diffused as widely as possible (Grubler 2000). This continues to be the case as social media allows for the rapid diffusion of innovation at the global level (Hanna et al. 2011). The formal and informal network of relationships required to bring a product successfully to market form a key element in the successful diffusion of an innovation (van Hemert et al. 2013). Firms seeking to undertake the commercialization of innovations must create and sustain strong networks of lead customers and key suppliers in order to leverage knowledge and other resources (Arias 1995; Laperche and Liu 2013). Relationships here must be based on trust and a willingness to share ideas and assist with the co-creation of the new product or service (Conway 1995; Senker and Faulkner 1996; Echeverri-Carroll 1999). It will be the ability of each of the firm's managers, in particular the owner or senior leadership team responsible for the innovation's commercialization, to use the social network that will be decisive (Rothwell 1991; Steward and Conway 1996). This suggests that the innovative firm needs to possess not only the physical and human capital it needs for commercialization, but also the social capital (Cooke and Wills 1999; McGuirk et al. 2015).

10.3 Methodology

The case study is recognized as a valuable methodology for the development of theory (Eisenhardt 1989), and also a powerful method for understanding the behavior of small firms (Chetty 1996). Yin (2009) suggests that this methodology provides an appropriate approach for addressing research that focuses on 'how' and 'why' phenomena occur. It is also particularly relevant where research is seeking to understand contemporary issues, is exploratory in nature and the researchers have little control over the events being studied (Yin 2009). The Scanalyse case is a single, longitudinal case that saw the first interview with the firm's CEO in 2006 less than three years after its founding. A second interview took place in 2012 after the firm had grown significantly and developed its operations globally. Following that interview the business was acquired by one of its main competitors in a trade sale thereby providing a "full lifecycle" perspective of the business case. This single case analysis is an appropriate rationale for a research methodology where the data collection at the early and later stages of the firms lifecycle offer an opportunity to understand how the business evolves due to changes to its environment and internal decision-making (Yin 2009). It was the rationale for the longitudinal single case study of 60-year history of the Canadian entrepreneurial firm "Steinberg Inc." undertaken by Mintzberg and Waters (1982).

10.3.1 Case Selection and Data Collection

Scanalyse, the company is an exemplar case to study the process of commercialization, WOM diffusion, and the role of social capital within innovative start-up firms. It meets the criteria of theoretical case selection as detailed by Eisenhardt (1989) and provides the ability to develop the emergent theory. As noted above, the first interview took place with the firm's CEO in 2006. At this point the company was just two years old. The firm had just been formally established following its spin-out from a university. It was essentially transitioning out of being a research project within a university into an independent business located in separate facilities. Data was subsequently collected from a second interview with the CEO six years later in 2012. By this time the firm had grown significantly and was at the point where it was moving towards a trade sale. A key part of the data collection was the use of a diagnostic assessment questionnaire. This examined the CEO's perception of the value of the commercialization using a strategic forecast of the management's assessment of the economic rent that was to be anticipated from the innovation.

The assessment of anticipated rent was undertaken through a set of questions that evaluated the company's potential volume of sales over time, the rate of profit earned, and the length of innovation in the market (Mazzaro and Reboud 2005; Duhamel et al. 2014; Do et al. 2014). The questionnaire also examined the company's management of the NPD and commercialization process using a 40 item scale across four dimensions. The four dimensions covered: (1) likely market adoption and diffusion of the innovation; (2) the strength of 'isolating mechanisms' developed by the company; (3) ability to proceed given possession of the necessary technical, financial, and human resources required to fully commercialize the innovation and (4) the company's approach to the formulation of a coherent strategy for the commercialization of the innovation. This questionnaire was the basis of an in-depth interview with the company's CEO at both data collection points.

10.3.2 Coding and Analysis

The analysis of the case was undertaken with the NVivo qualitative software program that facilitates the drawing together, coding and organization of the data collected (Hoover and Koerber 2011). Additional interviews were also undertaken in the preparation of this chapter. Data consisted of interview transcription, field notes, documents from the company and other sources, as well as the innovation diagnostic assessment tools used in 2006 and 2012 interviews. NVivo provides a valuable tool for the coding and classification of qualitative data and is well suited to a case study analysis (Bazeley and Jackson 2013).

The coding and analysis was guided by the theoretical framework outlined in Table 10.1. This draws together the foundation literature relating to the identification of "innovation rent," how it is assessed by an entrepreneur or entrepreneurial firm during the process of commercialization, and the influence of social capital, strategic networking, and WOM communication for the diffusion of the innovation. The 2006 and 2012 interviews were undertaken by third-party research assistants who used a common case study protocol. As a result, the approach we took to the data analysis was to examine the data, coding it in NVivo into initial parent nodes defined by the theoretical categories listed in Table 10.1, and then grouping "emic" data derived from the CEO interviews and innovation diagnostic survey tool, with the "etic" data comprised of the field notes provided by the research assistants. Both kinds of data are important in order to obtain a comprehensive view of the phenomena being examined (Rovai et al. 2013).

A pattern matching approach was used with comparisons between the first and second interviews and a triangulation of the data against the theoretical categories (see Table 10.1) (Eisenhardt 1989; Yin 2009). This process continued with redundant nodes being removed and the remaining nodes nested as "tree nodes" into the operationalized categories outlined in Table 10.1, which grouped around the 5 main theoretical categories connected via layering (Rovai et al. 2013; Miles et al. 2014). In addition follow-up interviews were undertaken with the Scanalyse CEO to assist in the validation of the data and our conclusions (van de Ven and Poole 1990).

An additional method employed in the analysis was the critical incident technique (CIT), which is a procedure for collecting direct observations of human behavior so as to generate useful information for solving practical problems and meeting systematically defined criteria (Flanagan 1954). It is a research technique that has been used in a range of studies relating to entrepreneurship, management, and marketing (Tjosvold and Weicker 1993; Gremler 2004). This followed the stages of commercialization proposed by Santi et al. (2003) and Duhamel et al. (2014). However, these were modified as follows: (i) initial assessment of the anticipated (potential) rent; (ii) market insertion of the innovation, and feedback from customers and competitors; (iii) assessment of residual (quasi-rent); (iv) review of the firm's resource configuration; (v) assessing the strategies for capturing the value of the appropriable rent (Table 10.1).

10.4 Analysis of the Case

Scanalyse emerged from research work undertaken at Curtin University in Perth, Western Australia (WA). This research was conducted into the application of 3D laser technology to build three-dimensional digital models of objects such as the internal spaces of buildings and structures. The foundation research that led to the creation of the technology was carried out by the Department of Spatial Sciences (DSS) within the university's School of Surveying. The Curtin DSS is a major center within the digital spatial industries cluster in WA. It not only provides bachelor's degree programs for surveying, cartography, and geographic information systems (GIS), but also works closely with industry and other universities to undertake applied research that aims to complement rather than compete with mainstream industry (Mazzarol et al. 2004).

10.4.1 Initial Assessment of the Anticipated (Potential) Innovation Rent

The commercialization pathway for the innovation was not initially clear to the Curtin DSS researchers. The team had access to new 3D laser technology in early

Table 10.1 Coding categories for data analysis

Theoretical categories	Operationalization	Selected references
Innovation rent	Entrepreneurial rent Quasi-rent Ricardian rent Schumpeterian rent	Schoemaker (1990), Makadok (2001), Alvarez and Barney (2004), Alvarez (2007), Lim et al. (2013), Do et al. (2014), Do (2014)
Assessment of the innovation's rent configuration via commercialization	Anticipated (potential) rent Residual (quasi) rent assessment following market insertion Appropriable rent following assessment of firm's resources	Santi et al. (2003), Mazzarol and Reboud (2005, 2006, 2011), Duhamel et al. (2014)
Commercialization process	Lean start-up Lean canvas business model Resource-based view (RBV)	Ries (2011), Osterwalder et al. (2005, 2015), Osterwalder and Pigneur (2010), Trimi and Bergegal-Mirabent (2012), Barney (1991, 2001)
Social capital	Trust Reciprocity Networks	Coleman (1988), Morgan and Hunt (1994), Mayer et al. (1995), Schoorman et al. (1995a, b), Dodgson (1996), Burt (1997), Cooke and Wills (1999), Anderson and Jack (2002), De Carolis and Saparito (2006), Lewicki and Brinsfield (2009), Maisch et al. (2010), Mazzarol (2013)
Strategic networking	Network connectedness Tie strength of network	Granovetter (1973, 1983, 1985), Brown and Reingen (1987), Rothwell (1991), Arias (1995), Valente and Rogers (1995), Senker and Faulkner (1996), Steward and Conway (1996), Echeverri-Carroll (1999), Narayan and Pritchett (1999), Aarikka-Stenroos and Sandberg (2009), Corsaro et al. (2012), Laperche and Liu (2013)
Word of Mouth (WOM) diffusion	Closeness of giver–receiver Organizational advocacy Richness of message Strength of message Valence of message	Brooks (1957), Czepiel (1974), Martilla (1971), Dodson and Muller (1978), Mazzarol et al. (2007), Mazzarol (2011), Kawakami et al. (2013), Sweeney et al. (2008, 2012)

2004 following the acquisition of laser scanner worth over AUD $250,000 that was purchased by the university. Curtin's DSS was investigating applications and uses for the technology that could be applied into local industries in WA. The research team went out and engaged directly with its existing industry networks, and demonstrated the instrumentation to industry people, asking them for ideas on what could be done with the laser technology. The state of WA has substantial mining industries in bauxite, iron ore, nickel, and gold, and is home to major mining companies. It was into this sector that the research team found its initial customers.

One of the first places the researchers went to was Alcoa's alumina refinery based in Wagerup, located in the south west of the state. The staff at Alcoa gave the Curtin DSS research team a range of different applications that they could examine to determine whether they presented an opportunity for future research. One of these was a big grinding mill where Alcoa had a problem with determining when to replace the internal steel liners. The constant use of the grinding mills to process bauxite ore wore away the steel liners within the huge crushing drums. However, it was difficult to predict accurately when the liners needed to be changed due to liner walls wearing too thin. These grinding mills are 10–15 m in diameter and getting access to all parts of the mill in order to measure all points of the steel liner was virtually impossible. In response to this problem, the university researchers developed a process that enabled them to apply the 3D laser technology to the measurement of these grinding mill liners' thickness at all points across the internal surface.

During 2005 the Curtin DSS research team realized that the commercial opportunity for the technology was to develop a method of processing data from the scanner and thereby extracting something that normally could not be extracted from the scanner alone. It was this realization within the university research team that triggered the creation of Scanalyse in 2006. In that year the start-up process commenced, which required the transition from the transfer of the innovation and its prototype technology from the university's laboratory into a spin-out company with its own independent facilities. Offices were acquired in the technology science park adjacent to the university. A company board was formed and a professional CEO, Peter Clarke was recruited to help guide the commercialization process. Peter's background prior to joining Scanalyse involved an extensive career in R&D with roles as a research engineer for the WA State Government's Department of Agriculture, managing director of a small company engaged in robotic and automated systems, and R&D manager for two larger technology companies involving successful commercialization projects. Initial funding for the spin-out company came from the university via seed capital grants supplemented by federal government research grants. The ownership of Scanalyse at this stage vested with Curtin University (Table 10.2).

The creation of the innovation that triggered the birth of Scanalyse was therefore a process of strategic networking and interpersonal communication. The Curtin DSS research team had a strong track record of engaging with local industry partners in undertaking its research work in the field of spatial science (Mazzarol et al. 2004). This helped to position the university's researchers within a strategic network strengthened by the social capital that had been accumulated with partners such as Alcoa. This provided the trust and reciprocal exchange of value to foster the discussions and collaborative problem solving that led to the innovation's creation.

Table 10.2 Assessment of the anticipated (potential) innovation rent—perceptions of Scanalyse 2006 and 2012

With respect to the planned innovation:	2006	2012
• Can the innovation work alone or does it integrate into a system?	Alone	Alone
• Does the innovation substitute an existing product or would it create a new market?	Creates a new market	Creates a new market
• Is the innovation compatible with existing products and processes or does it create a new dominant design or new system?	Creates a new standard or system	Compatible
• How was the innovation generated?	With research centers	Via a wider network
• What would be the potential geographic diffusion of this innovation in your sector?	Very wide	Wide
• What would be the potential annual sales for your innovation on a worldwide basis after 3 years?	\$16–\$20 million	\$5–\$10 million
• What would be the potential diffusion of this innovation within segments of your market?	Very wide (10 or more segments)	Limited (2 to 4 segments)
• What could be the gross profit margin of your innovation?	61–80%	61–80%
• What could be the net profit margin of your innovation?	Over 40%	21–30%
• What best describes your innovation…	Creates new dominant design	Creates a new dominant design
• The technical base of the innovation…	Offers a new technical platform	Offers a new technical platform
• From a technical point of view how difficult is the innovation to copy?	Hard to copy	Easy to copy
• From a legal point of view how difficult is the innovation to copy?	Hard to copy	Hard to copy

During the 2006 interview the Scanalyse CEO assessed the anticipated (potential) rent from the innovation as a "Champion," which is a configuration that has above average volume of sales, rate of profit, and length of lifecycle (Mazzarol and Reboud 2005). As shown in Table 10.2, the innovation was initially viewed as creating a new industry standard or system, with a very wide geographic diffusion across a very wide range of market segments. It was also perceived to offer potential annual sales of between AUD \$16 and \$20 million, with high gross and net profit margins. In terms of its lifecycle the innovation was seen as creating a new dominant design and technical platform that was hard to copy from both a technical and legal perspective.

10.4.2 Market Insertion of the Innovation and Feedback from Customers and Competitors

From early 2007 *Scanalyse* developed its relationship with Alcoa from one based around research into a commercial arrangement whereby Alcoa became the firm's first customer. A new product *MillMapper* was launched with a three-year contract between Scanalyse and Alcoa to service the grinding mills of the big bauxite miner. This initial success with a high profile global corporation such as Alcoa as a lead customer enabled Scanalyse to embark on a pathway to full commercialization of the technology. The presence of many other large mining firms headquartered in the state capital city of Perth was a significant opportunity for Scanalyse. All these companies made use of similar ore grinding mills to that used by Alcoa. Interpersonal networking and WOM communication enabled Scanalyse to secure two additional lead customers for *MillMapper*. With the revenues generated by these three initial customers Scanalyse was able to build up sufficient financial reserves to enable it to purchase its own 3D scanner rather than relying on the device owned by Curtin University. Throughout 2007 the company expanded in size from four employees at the start of the year rising to six by mid-year. Sufficient contracted work had been secured with the three lead customers to underwrite the firm's financial situation for what was to become a sustained growth phase.

Interpersonal communication between the Scanalyse team and the three lead customers provided valuable feedback on the technology and its application. This included some criticism over the value of the initial product offering. From the perspective of the customers the *MillMapper* technology was not viewed as sufficiently usable. For example the CEO of Scanalyse observed that at this time: "*A top issue for us is dealing with the unknowns associated with developing new information and data, and determining how to extract real information and knowledge from that data. No one else has produced this data so we are learning how to extract and read the data*" (CEO Scanalyse).

According to the lead customers the 3D scanner and the data it collected was not by itself of much value. The initial product offered was software that turned the data from the 3D scanner into a thickness map showing the structure of the mill liner walls. While this was of some value, its real potential was in how the thickness maps were interpreted. What the mining companies wanted was a management report based on the data interpretation from the thickness maps that gave them timely recommendations as to when they should change the mill liner walls, and whether all or some of the walls needed to be replaced. The customer value proposition (CVP) in the technology was therefore not a product solution but a service offering. This ability to offer clear evidence of the value of the technology was identified as a challenge by Scanalyse as noted by the firm's CEO: "*Our biggest issue is production of clear case studies demonstrating the value proposition of the technology. We need the ability to demonstrate the value of the technology.*"

This feedback led the Scanalyse team to undertake what Ries (2011) describes as a "*customer needs pivot*" within their business model design, shifting the focus of what was offered from a technology product to a specialist service. This pivot was a major turning point for the company. The work required to create this service package was undertaken extremely quickly. The 3D laser together with the data mapping software wrapped into the trending algorithm could now determine when the liners would need to be swapped out. This provided the breakthrough needed to disruptively change the way this measurement work was done at the crushing mills.

The interaction between the managers from Alcoa and the Curtin DSS research team was of sufficient strength that the company granted them access to a wide range of plant and equipment operations at the Wagerup Refinery. This helped to spawn the development of *MillMapper*. This interpersonal communication between the Curtin DSS researchers and the managers of Alcoa was critical as it enabled them to co-create the product through collaborative problem-solving. This is consistent with earlier research that suggests the level of network connectedness is what facilitates the interpersonal interactions (Valente and Rogers 1995). The closeness between the researchers at Curtin and the senior engineers at Alcoa contributed to the diffusion of the innovation and the development of a broader interest group interested in assessing the innovation. This closeness of the giver and receiver of WOM advocacy along with the strength, richness, and valence of the information provided has been identified as having an important role in enhancing the likelihood of the receiver accepting the message (Sweeney et al. 2008, 2012).

The appointment of Peter Clarke as CEO saw the project move from the university's research laboratories into commercial premises within the Technology Park opposite the university. Negotiations by the CEO with the Curtin University Office of Research and Development resulted in a mutually supportable agreement with all seed funding provided by the university. Strong interpersonal linkages between these two groups saw Scanalyse connect with further mining companies interested in trialing the prototype *MillMapper* product. It was the strength of the connectedness between Scanalyse and their foundation partners that enabled the product's limitations to be identified by the CEO and the subsequent rethink of the business model lead to the final breakthrough innovation. These interpersonal linkages have been recognized as being critical for an innovation to succeed and are often more important than the technical benefits of the innovation itself (Grubler 2000) (Table 10.3).

Table 10.3 shows the responses provided by the CEO of Scanalyse during the 2006 and 2012 interviews in relation to the perceived value of various network actors to the commercialization process. It can be seen that in 2006 the value of customers, in particular lead customers such as Alcoa, was considered to be very high with good value perceived from equity partners, lawyers, and other senior managers in the company. However, by 2012 the view of value in relation to legal advisors, equity partners had changed along with that of suppliers and other business people. In the case of suppliers the initial value of the company providing the 3D scanning technology had reduced. However, the benefit of networking with other business people as the process of commercialization unfolded had increased.

Table 10.3 Value of network actors to strategic decision-making—perceptions of Scanalyse 2006 and 2012

Type of network actor	2006	2012
• Customers, particularly lead customers…	Very valuable	Very valuable
• Suppliers, particularly key suppliers…	Some value	No value
• Directors of your management board…	Some value	Some value
• Equity partners/shareholders…	Valuable	Some value
• Other senior managers in the firm…	Valuable	Valuable
• Family members…	No value	No value
• Friends and social contacts…	No value	No value
• Other business people…	Limited value	Some value
• Accountants…	No value	No value
• Lawyers or legal advisors…	Valuable	No value
• Bankers or providers of debt financing…	No value	No value
• Providers of venture capital financing…	Some value	No value

It should be recognized that the Scanalyse board was comprised of representatives from the university and the venture capital firm who owned the equity in the business. While these actors provided the firm's management with value in the very early start-up phase they were of less value as the company expanded internationally. This highlights the importance of board directors to provide a young, fast growing company with valuable knowledge as well as access to strategic networks that can assist it in its future expansion (Borch and Huse 1993; Pugliese and Zhang 2007). Further, the decision to pivot the business model from a product focus to that of a service delivery mode made the value of legal advice relating to patents of less importance.

10.4.3 Assessment of the Residual (Quasi) Rent

The adoption and acceptance of a new technology depends not only on the customer's perception that the innovation is useful, but also that it is easy to use (Davis et al. 1989). It also needs to be relevant to the work being undertaken by the customer or end user, come from a credible supplier, and be able to demonstrate its ability to offer tangible benefits through pre and post adoption trials (Rogers 1995; Legris et al. 2003). The interaction by the CEO and design engineers at Scanalyse with their counterparts in the lead customer Alcoa played a critical role in assessing the likely adoption and acceptance of the new technology. They helped identify that the value to be created by the innovation was not a stand-alone product comprising a 3D laser scanner modified by sophisticated software to measure wear in large mills and crushers. Although this had some value it was of limited commercial benefit as it required the customer to take on the additional cost and expense of training their own people to use the equipment on top of the up-front cost of the product itself.

The early insertion of *MillMapper* into the market confirmed it needed to be able to trend and anticipate wear and performance. This required the product to pivot and be modified as a continuous service rather than a stand-alone product. The market required a service that would compute and analyze surface damage on the crushers and develop trending algorithms from spot readings. Customers could then utilize the data and undertake planned maintenance which delivered a significant benefit over existing practice. Just in time change out of mill crusher liners also allowed for substantial cost savings in the area of liner wastage. However, there remained a lot more to do with the development of the software if this innovation was to achieve full commercialization. This was noted by the CEO as follows: "*A major challenge for us is highly automating the data processing techniques so as to eliminate the manual data processing currently required, and the aim of automating is to reduce our costs and enable us to scale up very rapidly in response to demand.*"

This "*customer needs pivot*" (Ries 2011) assisted Scanalyse to redesign their business model to deliver best value to the customer. Now with an innovative service available to the market, Scanalyse linked with actors in its social network to put them into direct contact with key decision makers within the mining industry. This phase was critical for the take up of the *MillMapper* service by the mining industry. The social capital available to the company through these actors amplified the diffusion of the innovation which otherwise would have been protracted and difficult (Coleman 1998).

The initial assessment of the anticipated rent from the innovation was fairly realistic although the feedback from the lead customer Alcoa during the early insertion of the technology into the market did result in changes as outlined above. When the innovation rent was examined again during the 2012 interview the configuration of the innovation continued to be that of a "Champion." However, as shown in Table 10.2, the view that the technology would create a new standard or system was modified with the realization that it had to be made compatible with the customer's existing processes and systems. There was also an adjustment of the innovation's development with recognition that it was through the company's strategic network of customers and industry partners that the *MillMapper* was going to be further developed rather than through research in the laboratory.

It was also evident in the 2012 interview that the initial optimism over annual sales of AUD \$16–\$20 million across 10 or more market segments had been moderated to AUD \$5–\$10 million across a more modest 2 to 4 market segments. The net profit margin had also been significantly reduced from over 40% to a more modest 21–30% estimate. The adjustment from a technological product innovation to a technologically supported service innovation was also reflected in the recognition that from a technical point of view the innovation was easy to copy despite the company holding patents and other formal intellectual property (IP) rights protections. This adjustment of the anticipated rent from the innovation to what is often a more modest residual or "quasi-rent" outcome is recognized as logical process through which the commercialization process must move as more information from the market about customer feedback or competitor threats becomes available (Alvarez 2007; Duhamel et al. 2014).

10.4.4 Review of the Firm's Resource Configuration

Following this fundamental adjustment to the firm's business model Scanalyse embarked in 2008 on an expansion into global markets. The first overseas market entry took place in Chile, followed closely by Brazil. Securing access to these markets was facilitated by interpersonal networking and WOM communication between the CEO Peter Clarke and locally based industry group Austmine, a peak body for the Australian mining equipment, technology, and services (METS) sector. This included the engagement of key network actors such as Peter Rossdeutscher, a director of Austmine with an extensive experience in technology and innovation in the mining and resources sector, who assisted Peter Clarke and the Scanalyse team to make the necessary contacts within the South American mining companies.

With the company becoming known through WOM diffusion within the minerals processing sector of the mining industry, a need for further funding became necessary. The company board and management realized they now had an innovation that was ready for full scale commercialization within global markets. Financial investment to fund this expansion was provided by a local venture capital firm Stone Ridge Ventures (SRV). This was facilitated by WOM referral from one of the firm's board of directors to Rob Newman, an experienced venture financier and director of SRV. Thus through informal networking and WOM communication Scanalyse was able to secure sufficient funding to underwrite the business through an investment of venture capital. This brought in a shareholder commercializing partner and the attendant needs of the partner. In this case the need for return of funds was within five years.

WOM diffusion of the benefits of the new technology innovation resulted in the continuation of the company's growth into international markets. In reality the foundation partners that were located in Australia were all part of a very tight global mineral processing network. Distance was not critical for engagement, but what was critical was the trust provided by existing customers to prospective ones in confirming the benefits of the *MillMapper* service. As noted by Narayan and Pritchett (1999) effective social networks consist where there are strong ties between the members. This minerals processing group offered opportunity for Scanalyse that saw the company invited to demonstrate their *MillMapper* service at their first international site in Chile in 2009. It was here the CEO realized that despite the social capital that Scanalyse had on board in terms of references from its foundation partners in Australia more strength was needed to secure its first contract in Chile. As he explained: "*One of our top marketing issues is linking with partners in the international market place, because our service will be delivered by local agents at mine sites.*"

Initially the company was marketing directly from Australia, but soon moved to appoint a Chilean maintenance services company to represent Scanalyse technology. However, it met resistance from the Chilean customer's corporate hierarchy and competitive pressures within the local market. Notwithstanding the visible financial benefits of *MillMapper*, the engagement of Scanalyse (and the introduction of *MillMapper*) was proving to be a difficult decision. The solution came once Scanalyse appointed a Chilean national to take on the role of business development manager. This individual had a strong track record within the local engineering community and was able to significantly increase the level of trust in the Scanalyse product. Trust is an important element in the creation of social capital and can assist in building confidence in the innovation and thereby facilitating its adoption (Mazzarol 2013). This move achieved immediate success and provided the springboard for securing interest from further adjacent markets in Chile and Brazil.

What triggered this move was that Scanalyse was facing difficulties in securing contracts in Chile and it was not until it employed the local Chilean engineer to present for the company that it became successful in this market. The company realized it needed to operate a service-based model requiring strong interpersonal communication and local networking. Contacts had to be trusted and local agents were better able to communicate with and understand the needs of customers in each overseas market. In fact the company found that the best method of securing sales was to physically visit the mine sites where the ore crushing and milling plants were located and demonstrating the system while communicating its benefits. As noted by the firm's CEO: "*A major challenge for us is identifying and partnering with the best industry partners for delivery of onsite services.*"

When interviewed in 2006 before the company entered the Chilean market it was management's view that there was a need to strengthen the firm's marketing and service delivery resources. As stated at the time by the CEO: "*There is a need to look at a possible joint venture with large mining services companies as alliance partners and deal with marketing and site delivery.*" However, once the market entry to Chile had occurred and the reaction of customers experienced, this approach changed.

As explained by CEO Peter Clarke: "*We followed that direction in appointing the Chilean maintenance services company, but it was not successful. It was only when we appointed our own engineer to represent the technology directly that we started to get traction. This experience was repeated when we entered North America in 2009, firstly using a maintenance services company and then going to a directly employed representative with the necessary technical skills to sell the product.*"

Parallel with this international marketing and sales effort Scanalyse embarked on an expansion of its product portfolio. This included the product line extension from *MillMapper* to new offerings *CrusherMapper* and *TransferMapper*. All three products made use of the underlying platform technology of the 3D terrestrial laser scanner to industrial machinery to monitor performance in high cost assets. However, this expansion of the firm's product line also placed pressure on its human resources, as noted by the CEO: *"A challenge for us is developing the in-house capability to provide advisory services in data analysis."*

During 2009 the company achieved further breakthroughs with market diffusion of its products in the Chilean market and market development in Brazil. This expansion required Scanalyse to increase its staffing with more local representatives in its overseas markets to provide the necessary services to international clients. The level of growth had in fact been so strong that the company board decided to resist entry into Europe, despite opportunities to do so, and focus instead on the United States and South Africa so as to make best use of its limited sales and marketing resources.

10.4.5 Assessing the Strategies for Capturing the Value of the Appropriable Rent

The final phase of Scanalyse as an independent firm took place between 2010 and 2012. The company had now secured access to a range of global markets operating a network of agents for the delivery of its technology supported services to major mining companies. The firm's innovation had developed from the initial prototype into a portfolio of service products drawing on the platform 3D scanning technology and the proprietary algorithms underpinning the company's software. It had survived its early years, expanded globally and essentially "crossed the chasm" that confronts technological innovations (Moore 1996).

Having made the decision to build the business model around a service delivery value proposition, Scanalyse built on its initial success in Chile and continued to expand its network of local agents in other countries. Negotiations commenced with further agents in North America and Africa. The model for business development was the same as used in Brazil, employing local agents who possessed the necessary social capital networks within the targeted market segments. Interpersonal WOM communication was the most powerful medium for the promotion of the innovation, as highlighted in the literature (Arndt 1967; Martilla 1971; Czepiel 1974; Dodgson 1996).

Throughout 2010 the company and its innovation gained credibility with initial contracts awarded with commercialization beginning to accelerate. During 2011 the market entry to Europe brought with it a need to build social networks and further challenges were identified with this market. Similarly the company secured its first contracts in Africa. At this point its major shareholders moved to develop an exit strategy that would allow them to on sell their shareholding.

The demand for further investment capital in the business was now recognized by the firm's board as a necessary step to enable Scanalyse to take full advantage of its market opportunity. Annual sales had grown to more than AUD $4 million by the end of 2010 and additional funding was required to undertake future growth within existing markets and expansion into new ones such as Europe. The original investors in the company, such as Curtin University and SRV were keen to see their shareholding returned and were unable to provide the necessary additional funding for this growth. Future funding requirements demanded a shift from the initial foundation investors to new ones willing to provide the capital for market expansion, product upgrades, and new product development.

As explained by the Scanalyse CEO: "*A major issue is the lack of capital for market development and for new product development. Real challenges are getting onto the radar because we are presenting a profitable business, but not with the blue sky potential to give very high returns.*"

The Scanalyse board reviewed the company's business strategy and realized that its only option was to continue its growth path in order to consolidate its existing advantage as a pioneer of the technology. Within the market environment Scanalyse now found itself growth was the only option or competitors would have time to erode its market leadership position with alternatives. However, the raising of additional funding required either an initial public offer (IPO) on the stock market, with significant cost of capital and compliance issues, or look for a trade sale to a larger firm.

In early 2011 the CEO Peter Clarke opened discussions with one of Scanalyse's key competitors Outotec, a Finnish company headquartered in Espoo, Finland that specializes in technologies and services for metal and mineral processing industries. These discussions were aimed at exploring the option of making a trade sale to Outotec. The following year Scanalyse secured its first major contract in Brazil and its first contracts in Africa, which helped to make the firm more attractive to a buyer, but also required additional capital to support the increasing volume of business.

Throughout 2012 the negotiations between Scanalyse and Outotec continued with appropriate due diligence undertaken by the Finnish buyer. Scanalyse's sales were growing strongly over this period and its client base had expanded to include Australia and New Zealand, the United States and Canada, Brazil, Chile, Argentina, Venezuela, Mongolia, Sweden, Vietnam, Papua New Guinea, South Africa, Zambia and other African states. After only eight years Scanalyse had grown from a start-up spin-out from a university research program into a global technology leader. The final trade sale to Outotec was complete by the end of 2012 and saw the company sell for a very substantial premium over the original investment. This allowed the founder investors to exit satisfied and moved the firm's innovation into the marketing channels of a large company with the resources to undertake the required future growth.

10.5 Discussion

The Scanalyse case provides a useful insight into the commercialization process followed by a small company and the important roles played by social capital, strategic networking, and WOM communication in helping to facilitate a successful

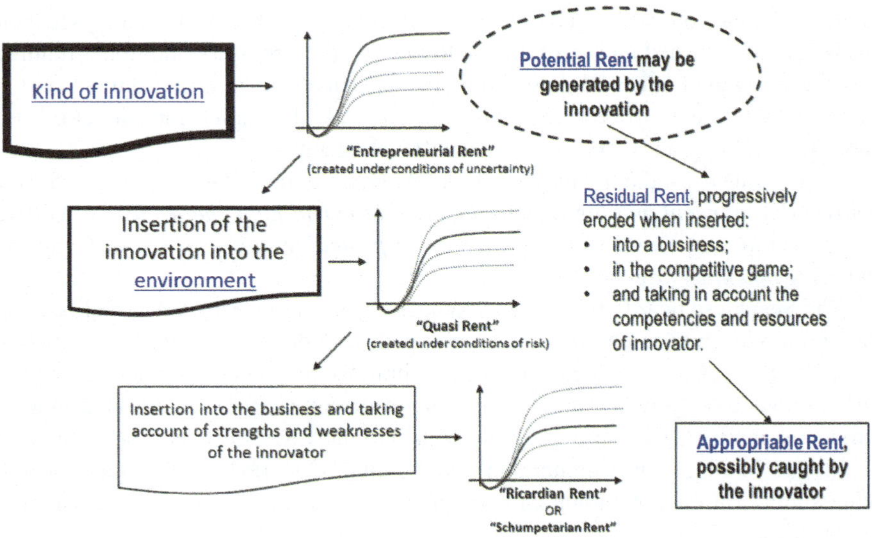

Fig. 10.1 Generation and appropriation of economic rents (adapted from Santi et al. 2003)

outcome. It also highlights the process through which a small, entrepreneurial firm assesses and appropriates the economic rent generated from an innovation. This process is illustrated in Fig. 10.1, where the initial assessment of the potential rent that might be anticipated from the innovation is made. As noted by Alvarez (2007) this is "entrepreneurial" rent, which is created under conditions of uncertainty when entrepreneurial actors combine resources in new ways to generate value. However, it occurs when the real value of these resource bundles is still unknown (Milgrom and Roberts 1990). This was the situation that confronted Scanalyse in its initial years when they were still determining how best to configure their product offer. The early insertion of the innovation into the market with the opportunity to get prompt feedback from the lead customers was made possible due to the social capital and strategic networking that the Curtin DSS research team and subsequently the staff at Scanalyse. Sufficient trust existed for the parties to work collaboratively to co-create the value that eventually emerged from the innovation.

Following the customer feedback to their initial product offer, Scanalyse was forced to undertake a major pivot in order to adapt their innovation from a product to a service offering. This need to get into touch with customers early with a minimum viable product (MVP), receive both positive and negative feedback, and then learn and adapt is a foundation process articulated in the Lean Start-up techniques advocated by Ries (2011). However, it also led the firm to reassess the economic rent that was likely to be extracted from the innovation. The reassessment slightly eroded the initial optimistic forecast, generating a residual or quasi-rent outcome. As noted by Alvarez (2007), quasi-rents are those created by parties under conditions of some certainty or known "probabilistically." While quasi-rents are still being created under conditions of risk, the enhanced knowledge derived from having trialed the technology with lead customers ensures that much of the original uncertainty has been removed.

The final capture of the appropriable rent from the innovation is only achieved once the firm can reconfigure its resource bundles to construct a business model able to sustainably deliver the CVP over time (Kemp and Verhoeven 2002). As with many small firms Scanalyse could not fully exploit the potential of its innovation due to resource constraints. Despite its success in securing initial seed funding from Curtin University and later venture capital funding from SRV, the company still required more financial resources to allow it to build on its early success. Further, the pivot from a product to a service focus shifted the risk from the technical side to the market side. As a service the ability to create legal isolating mechanisms to protect the innovation became more difficult and increased the urgency of getting the innovation diffused into the global market as rapidly as possible.

This need for rapid growth placed significant pressure on the firm's financial resources and led the foundation investors to seek an exit strategy. However, the options for exit were now limited. The high net profit margin forecast in the initial assessment of the firm's potential rent, had been eroded to a more modest level by the time the "reality check" of the market insertion took hold and the residual rent had been estimated. This made the business less attractive to third-party investors and also changed the nature of the appropriable rent from a "Ricardian" to that of a "Schmpetarian" one (Makadok 2001).

"Ricardian" rent is associated with value created through control over assets, while "Schumpetarian" rent is derived from the ability to create value via capabilities (Do et al. 2014). In the transition from the product to the service business model, Scanalyse lost the relative benefits of strong isolating mechanisms such as legal IP rights protections and the sale of tangible products. Instead it was building its competitive edge around a service-based model where its competitive advantage was invested in its ability to develop knowledge and skills within both its in-house staff, and its global network of agents. This presented a more challenging value proposition for future investors and limited the options for the firm's growth to the trade sale to Outotec. Nevertheless, its success in both growth and final exit was enhanced through the application of a board and senior management that leveraged social capital, networks and WOM communication to support its R&D, NPD, and commercialization process, as well as its funding and ultimate trade sale.

This pattern of activity is echoed in an earlier study by Feldman and Klofsten (2000) of a technology spin-out known as IV from Sweden's Linköping University. That company grew rapidly to medium-size only to be subsequently limited in its growth due to financial resource constraints, plus internal organizational and external market challenges. A particular constraint was the ability of the university to continue to provide the necessary support for what became a long-term growth path. While able to supply the necessary R&D and technical experience in the early stage of its life, over time the university was unable to provide the type of support that a firm like IV, with global aspirations and a need for a strong customer-focused culture required. Other problems highlighted in that case were the breakdown to effective communication between the marketing and operations staff, and the need to forge and maintain effective networks with other firms.

While Scanalyse experienced similar financial resource constraints to IV it did not suffer from disconnects between its technical and sales teams in relation to communication, and in fact demonstrated a very close and mutually productive flow of communication, learning, and adaption throughout its lifecycle. Attributes recognized in the literature as being important to successful innovation (Monge et al. 1992; Conway 1995; Kivimaki et al. 2000). It also retained, expanded, and strengthened its networks, a characteristic also found to be related to success in innovation (Arias 1995; Cooke and Wills 1999; Kandampully 2002). Strategic networks of the type developed by Scanalyse have been acknowledged as playing an important role in facilitating the internationalization of high-tech firms. Social networks play a mediating role and help to boost market awareness, while interpersonal relationships and communication provide valuable knowledge about overseas markets (Komulainen 2006).

10.5.1 A Model of Entrepreneurial Innovation Value (EIV)

The Scanalyse case offers some useful insights into the process of value identification and capture within a small entrepreneurial firm. The longitudinal nature of the case over the firm's entire lifecycle provides a unique opportunity to examine what it reveals about the commercialization process of small firms and the relationship this has with some of the theories associated with creation of economic rents from innovation (Alvarez and Barney 2004; Alvarez 2007; Mazzarol and Reboud 2005, 2006, 2008, 2011; Do et al. 2014; Do 2014; Duhamel et al. 2014). Also of interest is the lessons it can provide about the relevance of some of the conceptual frameworks that have been developed to help facilitate better performance in business start-ups and business model design (Ries 2011; Osterwalder et al. 2005, 2015; Osterwalder and Pigneur 2010; Trimi and Berbegal-Mirabent 2012).

Figure 10.2 illustrates a model of Entrepreneurial Innovation Value (EIV) that aims to draw together a range of theories and conceptual frameworks relevant to the recognition, development, and capture of value from the commercialization of an entrepreneurial innovation. The starting point for the process is the identification of the innovation and its initial assessment in relation to the anticipated (entrepreneurial rent) that can be derived. The assessment can be undertaken by assessing the potential volume of sales, rate of profit, and length of lifecycle (Santi et al. 2003; Mazzarol and Reboud 2005, 2006, 2008, 2011; Do et al. 2014; Do 2014; Duhamel et al. 2014).

In the Scanalyse case this process was initially undertaken as the firm was pursuing a product-based CVP and generated a "Champion" configuration. However, once the innovation had been developed into a MVP prototype and

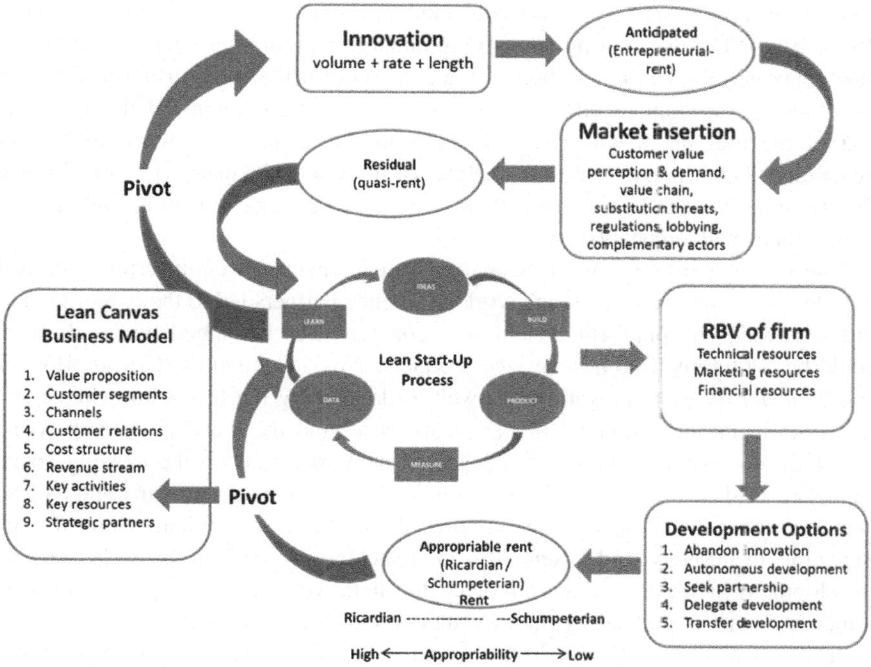

Fig. 10.2 A model of entrepreneurial innovation value (EIV) (Malone et al. 2015)

inserted into the market via lead customers like Alcoa, it became clear that the value perceived by the people at Scanalyse did not accord with that of the customers. The cost and complexity of the 3D laser technology offered as a stand-alone product did not offer customers sufficient ease of use, nor was it compatible with their existing systems. This market feedback required Scanalyse to review their product and undertake a "*customer needs pivot*" (Ries 2011) developing the technology into a service-based CVP. Although this was of more value to the customer, the residual quasi-rent outcome did not capture as much value as originally anticipated.

Although Scanalyse did not specifically use the "Lean Start-up" process (Ries 2011), or "Lean Canvas" business model design principles (Osterwalder and Pigneur 2010) in any formal way, the general process the company followed in its commercialization process mirrors these conceptual frameworks. Further, the role played by social capital and communication within the firm's strategic networks was of prime importance to shaping this process. As suggested by Ostrom (2009), social capital deals with the relationships and shared values that are created by individuals, groups, or organizations to collectively solve problems.

The link with Alcoa provided the genesis for the creation of the product *MillMapper*. However, it was not until the arrival of an entrepreneurial leader such as the CEO Peter Clarke that an innovation value was created. The measuring capability of *MillMapper* for the customer was of some value, but it failed to meet the generic conditions of ease of use and compatibility with existing systems that tend to be the hallmarks of successful innovation adoption and diffusion (Rogers 1976, 1995). The decision by the CEO to undergo a pivot after feedback from lead customers was the key that delivered the creation of *MillMapper* as a service. This breakthrough brought together the process of the static mapping of the wear and tear of the mill liners with the ability to combine a trending algorithm thereby generating a continuous real-time analysis of the liner thickness. This transformed the benefits to customers over their existing processes enabling just-in-time replacement with substantial cost savings.

The interaction between the Scanalyse management and engineering team and their lead customers and wider network of alliance partners led to the exploration of different ways the platform IP and core competencies contained within the firm could be configured into new service products. This generated the *CrusherMapper* and *TransferMapper* innovations, as well as developing the firm's knowledge and networks in the development of new markets around the world. The early years (2004–2006) saw the leveraging of residual networks built on the existing social capital formed by the Curtin DSS researchers. This enable the company and its lead customers to collaborate in order to solve the customer's problems using the 3D Scanning technology and essentially co-create the innovation. In each case the development of these products involved the iterative learning loop between customer and firm as defined by the EIV model shown in Fig. 10.2.

In its growth phase (2007–2009) Scanalyse used networks and WOM communication to strengthen the market acceptance of its innovation. Having developed a radical innovation and being a relatively unknown small firm, Scanalyse lacked brand recognition and significant marketing and sales capacity. As suggested by Ostgaard and Birley (1994) small entrepreneurial firms must rely on WOM and their ability to leverage personal networks to secure competitive strategies. This process provided Scanalyse with the ability to gain its first international customer. Here the company's innovation was diffused through the networking channels of the international mining company maintenance engineers. The positive WOM from the company's foundation customers provided the comfort for the firm's first international customer to sign on.

For innovation to cross over social networks acting as a mechanism for diffusion, there is a need to create 'structural holes' between the networks with the information diffused through influential people providing 'bridges' as 'gatekeepers' (Burt 1992a/b). This reflects the Scanalyse experience in Chile and it became the format for all international contracts going forward. It facilitated the company's subsequent entry into Brazil, and then the USA and South Africa. All four international sites experienced initial problems that related either to the clients' not fully understanding the CVP that Scanalyse was offering, or a lack of identification of the relevant 'gatekeepers' for WOM diffusion.

From the perspective of the conceptual model this understanding reinforces the second step in the assessment of the residual quasi-rent value. Once the innovation was inserted into the market the rent value was adjusted after accounting for the erosion effects on volume of sales, rate of profit, and length of selling of the innovation. Complimentary actors, regulation, and substitutes all contributed to an understanding of the residual value. This provided an opportunity to gain access to expansion capital to drive the company forward. Through utilizing strategic networks the CEO was able to link with a fund and deliver further financial capacity to the company. Over the three-year period, the pivoting of the innovation to adjust for market requirements was undertaken delivering a residual rent.

The final period from 2010 to 2012 saw the company extend its product range and bring out two new products *CrusherMapper* and *TransferMapper*. This extension to the product line increased the target customer base and strengthened the company's capacity to deliver tangible benefits to a broad group of customers. The ability to move quickly was through the diffusion of the innovative product and service through their industry partners. This period saw the innovation go mainstream with sales and revenues rising.

However, as outlined above, the resource constraints and limited exit strategies facing the company led it to opt for the trade sale to Outotec. The CEO Peter Clarke expressed the view that more return to the innovation might have been secured had Scanalyse been able to secure an IPO or form a joint venture alliance. However, the nature of the business model and the board's appetite for further risk in a dynamic global market where speed to market was critical made a trade sale more attractive. Despite this the final exit delivered a very substantial return to the investments made by the original owners. It also ensured that the innovation would continue to provide value to customers around the world through Outotec as an established global service provider.

10.6 Conclusion

The Scanalyse case offers an end-to-end example of a new technology start-up, growth and trade-sale lifecycle. It highlights the stages through which a commercialization process for a small innovator firm can move, and offers evidence to support the EIV model (Malone et al. 2015), and the multi-stage process of how economic rent from innovation is identified, assessed, and finally captured by such firms (Santi et al. 2003; Duhamel et al. 2014). It also shows the importance of social capital, networking, and WOM communication in the diffusion process (Steward and Conway 1996; Partanen et al. 2008). In addition the case demonstrates the argument suggested by Sandberg (2002) that proactivity in the market, particularly at the early stages of a new product launch is important to successful commercialization. It also suggests that a university spin-out can be a success if it has the necessary leadership, industry connections and adaptiveness, as well as an ability to work closely with lead customers and network support actors to co-create the product/service offering. This ability to make best use of the firm's limited resources and remain closely engaged with customers while retaining the full engagement of the original university research team can overcome problems found in other spin-outs (e.g., Feldman and Klofsten 2000).

Scanalyse was able to effectively utilize social capital in gaining access to strategic customers. This provided the company the ability to discuss the concept of *MillMapper* and in the process discover where the real opportunity lay for commercialization. The ability to utilize the social network by the firm's CEO generated the information that was the breakthrough. It was the continuous reporting feature that lifted *MillMapper* from a product of some value to one of immense value. The problem of knowing when to replace the steel liners in the huge crushing mills was a global challenge confronting all mining firms, and until the arrival of Scanalyse it was not being addressed. Historically it had been a function of time with firms typically replacing the liners earlier than required at substantial cost rather than risk a failure before a scheduled replacement occurred. Scanalyse changed maintenance in this area forever. Being able to monitor wear and tear on steel mills allowed real-time control over all maintenance aspects delivering substantial savings to operations. However, the problem would not have been solved without the interpersonal communication between the Curtin DSS/Scanalyse team and their counterparts in lead customer firms such as Alcoa. The strong social capital built on trust and a willingness to collaboratively problem solve was the key to the creation of the *MillMapper* solution.

Bibliography

Aarikka-Stenroos, L., Sanberg, B.: Applying network approach to commercialization of innovations: case study on nets to create markets for innovations. In: IMP2009 Conference, 3–5 Sept 2009. Marseilles, France (2009)

Alvarez, S.A.: Entrepreneurial rents and the theory of the firm. J. Bus. Ventur. **22**(3), 427–442 (2007)

Alvarez, S.A., Busenitz, L.W.: The entrepreneurship of resource-based theory. J. Manag. **27**(12), 755–775 (2001)

Alvarez, S.A., Barney, J.B.: Organizing rent generation and appropriation: toward a theory of the entrepreneurial firm. J. Bus. Ventur. **19**(5), 621–635 (2004)

Amit, R., Schoemaker, P.J.H.: Strategic assets and organisational rent. Strateg. Manag. J. **14**(1), 33–46 (1993)

Anderson, A.R., Jack, S.L.: The articulation of social capital in entrepreneurial networks: a glue or lubricant?. Entrepreneurship. Reg. Dev. **14**(3), 193–210 (2002)

Arias, J.T.G.: Do networks really foster innovation? Manag. Decis. **33**(9), 52–56 (1995)

Arndt, J.: Role of product-related conversations in the diffusion of a new product. J. Mark. Res. **4**(3), 291–298 (1967)

Barney, J.B.: Firm resources and sustained competitive advantage. J. Manag. **17**(1), 99–120 (1991)

Barney, J.B.: Is the resource-based "view" a useful perspective for strategic management research? Yes. Acad. Manag. Rev. **26**(1), 41–56 (2001)

Bazeley, P., Jackson, K.: Qualitative Data Analysis with NVivo (2nd edn.). Sage, London, Thousand Oaks, CA, New Delhi, Singapore (2013)

Borch, O.J., Huse, M.: Informal strategic networks and the board of directors. Entrepreneurship Theor. Pract. **18**(1), 23–37 (1993)

Brooks, R.C.: "Word of mouth" advertising in selling new products. J. Mark. **22**(2), 154–161 (1957)

Brown, J.J., Reingen, P.H.: Social ties and word-of-mouth referral behavior. J. Consum. Res. **14** (3), 350–362 (1987)
Burt, R.S.: Structural holes and good ideas. Am. J. Sociol. **110**(2), 349–399 (1992a)
Burt, R.S.: Structural Holes: The Social Structure of Competition. Harvard University Press, Cambridge, MA (1992b)
Chetty, S.: The case study method for research in small and medium-sized firms. Int. Small Bus. J. **15**(1), 73–85 (1996)
Burt, R.S.: The contingent value of social capital. Adm. Sci. Q. **42**(2), 339–365 (1997)
Christensen, C.M., Johnson, M.W., Rigby, D.K.: Foundations for growth. Mit Sloan Manag. Rev. **43**(3), 22–31 (2002)
Coleman, J.S.: Social capital in the creation of human capital. Am. J. Sociol. **94**(Supplement), S95–S120 (1988)
Collis, D.J., Montgomery, C.A.: Competing on resources: strategy in the 1990's. Harvard Bus. Rev. **73**(4), 118–128 (1995)
Conway, S.: Informal boundary-spanning communication in the innovation process: an empirical study. Technol. Anal. Strateg. Manag. **7**(3), 327–342 (1995)
Cooke, P., Wills, D.: Small firms, social capital and the enhancement of business performance through innovation programs. Small Bus. Econ. **13**(3), 219–234 (1999)
Corsaro, D., Cantù, C., Tunisini, A.: Actors' heterogeneity in innovation networks. Ind. Mark. Manage. **41**(5), 780–789 (2012)
Czepiel, J.A.: Word of mouth processes in the diffusion of a major technological innovation. J. Mark. Res. **11**(2), 172–180 (1974)
Davis, F.D., Bagozzi, R.P., Warshaw, P.R.: User acceptance of computer technology: comparison of two theoretical models. Manage. Sci. **35**(8), 982–1003 (1989)
De Carolis, D.M., Saparito, P.: Social capital, cognition, and entrepreneurial opportunities: a theoretical framework. Entrepreneurship. Theor. Pract. **30**(1), 41–56 (2006)
Do, T.H.: Determinants of innovation commercialization management and anticipated returns: an exploratory typology of SMEs. Int. J. Innov. Technol. Manag. **11**(6), 1–20 (2014)
Do, T.H., Mazzarol, T., Volery, T., Reboud, S.: Predicting anticipated rent from innovation commercialisation in SMEs. Eur. J. Innov. Manag. **17**(2), 183–208 (2014)
Dodgson, M.: Learning, Trust and inter-firm technological linkages: some theoretical associations. In: Coombs, R., Richards, A., Saviotti, P., Walsh, V., Cheltenham, E.E. (eds.) Technological Collaboration: The Dynamics of Cooperation in Industrial Innovation, pp. 54–75 (1996)
Dodson, J.A., Muller, E.: Models of new product diffusion through advertising and word of mouth. Manag. Sci. **24**(15), 1568–1578 (1978)
Duhamel, F., Reboud, S., Santi, M.: Capturing value from innovations: the importance of rent configurations. Manag. Decis. **52**(1), 122–143 (2014)
Echeverri-Carroll, E.L.: Knowledge flows in innovation networks: a comparative analysis of Japanese and US high-technology firms. J. Knowl. Manag. **3**(4), 296–303 (1999)
Eisenhardt, K.: Building theory from case study research. Acad. Manag. Rev. **14**(4), 532–550 (1989)
Fahy, J., Smithee, A.: Strategic marketing and the resource based view of the firm. Acad. Mark. Sci. **99**(10). http://www.amsreview.org/amsrev/theory/fahy10-99.html [online] (1999)
Feldman, J.M., Klofsten, M.: Medium-sized firms and the limits to growth: a case study in the evolution of a spin-off firm. Eur. Plan. Stud. **8**(5), 631–650 (2000)
Flanagan, J.C.: The critical incident technique. Psychol. Bull. **51**(4), 327–358 (1954)
Granovetter, M.: The strength of weak ties. Am. J. Sociol. **78**(6), 1360–1380 (1973)
Granovetter, M.: The strength of weak ties: a network theory revisited. Sociol. Theor. **1**(1983), 203–233 (1983)
Granovetter, M.: Economic action and social structure: the problem of embeddedness. Am. J. Sociol. **91**(3), 481–510 (1985)
Gremler, D.D.: The critical incident technique in service research. J. Serv. Res. **7**(1), 65–89 (2004)
Grubler, A.: Time for a change: on the patterns of diffusion of innovation. J. Reg. Sci. Rev. **125**(3), 19–42 (2000)

Hanna, R., Rohm, A., Crittenden, V.L.: We're all connected: the power of the social media ecosystem. Bus. Horiz. **54**(3), 265–273 (2011)
Hoover, R.S., Koerber, A.L.: Using NVivo to answer the challenges of qualitative research in professional communication: benefits and best practices. IEEE Trans. Prof. Commun. **54**(1), 68–82 (2011)
Kandampully, J.: Innovation as the core competency of a service organisation: the role of technology, knowledge and networks. Eur. J. Innov. Manag. **5**(1), 18–26 (2002)
Kawakami, T., Kishiya, K., Parry, M.E.: Personal word of mouth, virtual word of mouth, and innovation use. J. Prod. Innov. Manage **30**(1), 17–30 (2013)
Kemp, R.G.M., Verhoeven, W.H.J.: Growth Patterns of Medium-Sized, Fast-Growing Firms: The Optimal Resource Bundles for Organisational Growth and Performance. SCALES, Scientific Analysis of Entrepreneurship and SMEs, Zoetermeer (2002)
Kinnunen, J.: Gabriel tarde as a founding father of innovation diffusion research. Acta Sociol. **39**(4), 431–442 (1996)
Kivimaki, M., Lansisalmi, H., Elovainio, M., Heikkila, A.: Communication as a determinant of organizational innovation. R&D Manag. **30**(1), 33–42 (2000)
Komulainen, H.: Social networks in the initiation of a high-tech firm's internationalisation. Int. J. Entrepreneurship Innov. Manag. **6**(6), 3–13 (2006)
Laperche, B., Liu, Z.: SMEs and knowledge-capital formation in innovation networks: a review of literature. J. Innov. Entrepreneurship **2**(1), 1–16 (2013)
Legris, P., Ingham, J., Collerette, P.: Why do people use information technology? A critical review of the technology acceptance model. Inf. Manag. **40**(2), 191–204 (2003)
Lehtimäki, T., Simula, H., Salo, J.: Clarifying the terms launch and commercialization: reflections from theory and practice. In: Proceedings of EMAC-Conference (2008)
Leonard Barton, D.: Experts as negative opinion leaders in the diffusion of a technological innovation. J. Consum. Res. **11**(4), 914–926 (1985)
Lewicki, R.J., Brinsfield, C.T.: In: Bartkus, V.O., Davis, J.H. (eds.) Trust, Distrust and Building Social Capital. Social Capital: Reaching Out, Reaching (pp. 275–303). Edward Elgar Publishing, Cheltenham, UK and Northhampton, MA, USA (2009)
Lim, D.S.K., Celly, N., Morse, E.A., Rowe, W.G.: Rethinking the effectiveness of asset and cost retrenchment: the contingency effects of a firm's rent creation mechanism. Strateg. Manag. J. **34**(1), 42–61 (2013)
Mahajan, V., Muller, E., Kerin, R.A.: Introduction strategy for new products with positive and negative word-of-mouth. Manage. Sci. **30**(12), 1389–1404 (1984)
Mahoney, J.T.: The management of resources and the resource of management. J. Bus. Res. **33**(2), 91–101 (1995)
Maisch, B., Binder, J., Schmid, B., Leifer, L.: The dimensions of trust—building confidence through innovation communication. Innov. J. **8**(1), 1–28 (2010)
Makadok, R.: Toward a synthesis of the resource-based and dynamic-capability views of rent creation. Strateg. Manag. J. **22**(5), 387–401 (2001)
Malone, P., Mazzarol, T., Reboud, S.: Understanding commercialization in entrepreneurial firms: a case example. In: International Council of Small Business (ICSB) 60th Annual Conference 2015, 6–9 June, Dubai UAE (2015)
Martilla, J.A.: Word of mouth communication in the industrial adoption process. J. Mark. Res. **8**(2), 173–178 (1971)
Mayer, R.C., Davis, J.H., Schoorman, F.D.: An integrative model of organizational trust. Acad. Manag. Rev. **20**(3), 709–734 (1995)
Mazzarol, T., Reboud, S.: The role of complementary actors in the development of innovation in small firms. Int. J. Innov. Manage. **12**(2): 223–253 (2008)
Mazzarol, T.: The role of social-capital, strategic networking and word of mouth communication in the commercialisation of innovation. In: Pfeffermann, N., Minshall, T., Mortara, L. (eds.) Strategy and Communication for Innovation (2nd edn., pp. 173–193). Springer, Berlin (2013)

Mazzarol, T.: The role of word of mouth in the diffusion of innovation. In: Hulsman, M.H., Pfeffermann, N. (eds.) Strategies and Communications for Innovations: An Integrated Management View for Companies and Networks (pp. 117–132). Springer, Berlin (2011)
Mazzarol, T., Reboud, S.: Customers as predictors of rent returns to innovation in small firms—an exploratory study. Int. J. Entrepreneurship Innov. Manag. **5**(5/6), 483–494 (2005)
Mazzarol, T., Reboud, S.: The strategic decision making of entrepreneurs within small high innovator firms. Int. Entrepreneurship Manag. J. **2**(2), 261–280 (2006)
Mazzarol, T., Reboud, S. (eds.): Strategic Innovation in Small Firms: An International Analysis of Innovation and Strategic Decision Making in Small to Medium Sized Enterprises. Edward Elgar Publishing, Cheltenham, UK, Northampton, MA, USA (2011)
Mazzarol, T., Patmore, R., van Heemst, N., Wong, J., Adam, D.: Digital Communities: A Study to Determine the Feasibility of Creating a Digital Content Industry Cluster in Western Australia. Perth, Centre for Entrepreneurial Management and Innovation, Graduate School of Management, UWA (2004)
Mazzarol, T., Sweeney, J.C., Soutar, G.N.: Conceptualizing word-of-mouth activity, triggers and conditions: an exploratory study. Eur. J. Mark. **41**(11/12), 1475–1494 (2007)
McGuirk, H., Lenihan, H., Hart, M.: Measuring the impact of innovative human capital on small firms' propensity to innovate. Res. Policy **44**(4), 965–976 (2015)
Miles, M.B., Huberman, A.M., Saldaña, J.: Qualitative Data Analysis: A Methods Sourcebook (3rd edn.). Sage Publications Inc, Los Angeles, London, New Delhi and Singapore (2014)
Milgrom, P., Roberts, J.: Bargaining costs, influence costs, and the organization of economic activity. In: Alt, J., Shepsle, K. (eds.) Perspectives on Positive Political Economy (pp. 57–69). Cambridge University Press, Cambridge, UK (1990)
Mintzberg, H., Waters, J.A.: Tracking strategy in an entrepreneurial firm. Acad. Manag. J. **25**(3), 465–499 (1982)
Monge, P.R., Cozzens, M.D., Contractor, N.S.: Communication and motivational predictors of the dynamics of organizational innovation. Organ. Sci. **3**(2), 250–274 (1992)
Moore, G.: Crossing the Chasm. Harper Business, New York (1996)
Morgan, R., Hunt, S.: The commitment-trust theory of marketing relationships. J. Mark. **58**(3), 20–38 (1994)
Narayan, D., Pritchett, L.: Cents and sociability. Econ. Dev. Cult. Change **47**(4), 871–889 (1999)
Osterwalder, A., Pigneur, Y.: Business Model Generation: A Handbook for Visionaries, Game Changes, and Challengers. Wiley, Hoboken, NJ (2010)
Osterwalder, A., Pigneur, Y., Tucci, C.L.: Clarifying business models: origins, present, and future of the concept. Commun. Assoc. Inf. Syst. **2005**(16), 1–25 (2005)
Osterwalder, A., Pigneur, Y., Bernarda, G., Papadakos, T.: Value Proposition Design: How to Create Products and Services Customers Want. Wiley, Somerset (2015)
Ostgaard, T., Birley, S.: Personal networks and firm competitive strategy—a strategic or coincidental match? J. Bus. Ventur. **9**(4), 281–306 (1994)
Ostrom, E.: What is social capital? In: Bartkus, V.O., Davis, J.H. (eds.) Social Capital: Reaching Out, Reaching (pp. 17–38). Edward Elgar Publishing, Cheltenham, UK, Northampton, MA, USA (2009)
Partanen, J., Möller, K., Westerlund, M., Rajala, R., Rajala, A.: Social capital in the growth of science-and-technology-based SMEs. Ind. Mark. Manage. **37**(5), 513–522 (2008)
Pugliese, A., Zhang, P.W.: Board members' contribution to strategic decision-making in small firms. J. Manage. Governance **11**(4), 383–405 (2007)
Ries, E.: The Lean Startup: How Constant Innovation Creates Radically Successful Businesses. Portfolio Penguin Books, London (2011)
Rogers, E.M.: New product adoption and diffusion. J. Consum. Res. **2**(4), 290–301 (1976)
Rogers, E.M.: Diffusion of Innovations. The Free Press, New York (1995)
Rogers, E.M., Shoemaker, F.: Communication of Innovations: A Cross-Cultural Approach. Free Press, New York (1971)
Rothwell, R.: External networking and innovation in small and medium-sized manufacturing firms in Europe. Technovation **11**(2), 93–111 (1991)

Rovai, A.P., Baker, J.D., Ponton, M.K.: Social Science Research Design and Statistics: A Practioner's Guide to Research Methods and SPSS Analysis, 1st edn. Watertree Press, Chesapeake (2013)

Sandberg, B.: Creating the market for disruptive innovation: market proactiveness at the launch stage. J. Targ. Meas. Anal. Mark. **11**(2), 184–197 (2002)

Santi, M., Reboud, S., Gasiglia, H., Sabouret, A.: Modèle de valorisation et de protection intellectuelle des innovations des PEI (63p). July, HEC/INPI (2003)

Schoemaker, P.J.H.: Strategy, complexity and economic rent. Manag. Sci. **36**(10), 1178–1192 (1990)

Schoorman, F.D., Mayer, R.C., Davis, J.H.: Including versus excluding ability from the definition of trust. Acad. Manag. Rev. **21**(2), 339–340 (1996a)

Schoorman, F.D., Mayer, R.C., Davis, J.H.: Organizational trust: philosophical perspectives and conceptual definitions. Acad. Manag. Rev. **21**(2), 337–340 (1996b)

Senker, J., Faulkner, W.: Networks, tacit knowledge and innovation. In: Coombs, R., Richards, A., Saviotti, P., Walsh, V. (eds.) Technological Collaboration: The Dynamics of Cooperation in Industrial Innovation (pp. 76–97). Edward Elgar, Cheltnam (1996)

Steward, F., Conway, S.: Informal networks in the origination of successful innovations. In: Coombs, R., Richards, A., Saviotti, P., Walsh, V. (eds.) Technological Collaboration: The Dynamics of Cooperation in Industrial Innovation (pp. 201–222). Edward Elgar, Cheltenham, UK (1996)

Sweeney, J.C., Soutar, G.N., Mazzarol, T.: Factors influencing word of mouth effectiveness: receiver perspectives. Eur. J. Mark. **42**(3/4), 344–364 (2008)

Sweeney, J.C., Soutar, G.N., Mazzarol, T.: Word of mouth: measuring the power of individual messages. Eur. J. Mark. **46**(1/2), 237–257 (2012)

Tarde, G.: The Laws of Imitation. Henry Holt and Company, New York (1903)

Tjosvold, D., Weicker, D.: Cooperative and competitive networking by entrepreneurs: a critical incident study. J. Small Bus. Manage. **31**(1), 11–21 (1993)

Trimi, S., Berbegal-Mirabent, J.: Business model innovation in entrepreneurship. Int. Entrepreneurship Manag. J. **8**(4), 449–465 (2012)

Valente, T.W., Davis, R.L.: Accelerating the diffusion of innovations using opinion leaders. Ann. Am. Acad. Polit. Soc. Sci. **566**, 55–67 (1999)

Van de Ven, A.H., Poole, M.S.: Methods for studying innovation development in the Minnesota innovation research program. Organ. Sci. **1**(3), 13–35 (1990)

Van Hemert, P., Nijkamp, P., Masurel, E.: From innovation to commercialization through networks and agglomerations: analysis of sources of innovation, innovation capabilities and performance of Dutch SMEs. Ann. Reg. Sci **50**(2), 425–452 (2013)

Yin, R.K.: Case Study Research: Design and Methods, 4th edn. SAGE Inc, Thousand Oaks, CA, London, UK, and Singapore (2009)

Chapter 11
Managing Emotions Matters—A Balanced Framework for Communicating Innovations in Companies

Claudia Mast

Abstract Nowadays, innovations are crucial for nations, the progress of societies and the success of companies. But only if they are noticed and appreciated by people in their daily lives, employees, customers, experts, academics, journalists, and other opinion leaders, they can fully be effective. Therefore, communication management has to take into consideration, that the advantages of innovations can be re-evaluated from different points of view—employees worrying about their jobs, companies struggling for sustaining their competitiveness, or customers who mistrust products with the label "innovative". It is the balance between facts and figures on the one hand, and emotions on the other, which is often neglected in internal and external corporate communication. The article presents a balanced framework for creating a communication strategy for innovations and discusses results of surveys from communication directors in the top 500 companies in Germany.

11.1 Introduction

So what else is new? Never before we have been asking this question as often as we do today. And that is because information and products are more and more short-lived. The news of today is outdated tomorrow. New goods and services are constantly being announced, especially in the technical and digital sector. Thus, companies need to hurry if they want to hold the line. Above all, companies are still busy recovering from the major economic and confidence crisis, which have caused a major rift for communication experts. Innovations can—if properly staged—pave the way back to the stakeholders, but this requires a functioning, strategic communication. The conditions for such a communication management have changed dramatically though. More and more companies have to cope with negative media coverage and a skeptical public atmosphere. Employees, customers, journalists, and politicians have begun to question the social and ecological responsibility of the

C. Mast (✉)
Universität Hohenheim, Fruwirthstraße 49, 70599 Stuttgart, Germany
e-mail: sekrkowi@uni-hohenheim.de

N. Pfeffermann and J. Gould (eds.), *Strategy and Communication for Innovation*,
DOI 10.1007/978-3-319-49542-2_11

companies and they care about whether the technologies, products, and services offered by companies are truly an advantage to them. In addition, editorial offices have emphasized the arising uncertainties of innovations and the consequences more than the opportunities.

That means innovation communication poses more and more particular challenges. In this article innovation communication is defined as two-way interactions between organizations and their stakeholders, dealing with new products, services, and technologies. These innovations are often novel and complex. Their positive and negative implications may be hidden for some time and their concrete applications may be explained later on. Therefore, the environment for the emerging field of innovation communication has changed and the most important stakeholders in this field (employees, customers, and journalists) have to deal with growing uncertainty, mistrust, and sometimes fear. In their eyes, innovations and developments of change often go together. Many innovations initiate a change process or are an essential part of it. So far, some experiences made in the wide field of change communication can be transferred to communicating innovations (Mast 2011). This is essentially the management of emotions, time, and the way of reporting news to stakeholders who are emotionally affected and involved.

11.2 Creating a Communication Strategy for Innovations

The accelerating process of innovations causes the fear of individual, organizational, and social risks, which may dominate the hope, confidence, and chances in parts of societies and companies as well. In this case, the communication system is critical to business success as long as it enables a constant process of change and manages the balance between cognitions and emotions. On the opposite side, communication can destroy the company's values, including support, involvement, and employee commitment.

Strong companies are those which manage innovations and change quickly, without public trouble or damaging the motivation of stakeholders. But first of all, they have to win the hearts, minds, and participation of their employees who are more and more worried about the individual and social impact of new products, services, and technologies (Mast and Simtion 2016). Some examples are job cuts, increasing stress in the office, media coverage of dangerous products, environmental pollution, and unfair trade. People tend to re-evaluate who is taking advantage of an innovation, which kind of innovations are acceptable and the social costs of new products or procedures.

Changes and innovations (from the employee's point of view) have become nearly one and the same. Here are some reasons why it is so difficult to communicate innovations:

- *Lack of insight*: If people do not understand why a certain innovation is necessary for their company, they are not motivated. They will resist the change process: Why change?
- *Lack of acceptance*: People emotionally desire a break in the never ending process of change. Most of them feel that they are captured by anonymous forces and cannot really participate: Who benefits from this innovation?
- *Lack of trust*: The trust of many employees in their managers has become fragile. In some companies the trust and the credibility have been decreasing dramatically for several years. Many people do not believe any longer that the public announcements and reasons for the flood of changing projects are really true: Why should we trust the managers this time?

The crucial point for a successful communication strategy is: How can communication support the ability of companies and employees to innovate? How can communication contribute to the "innovation readiness" (Zerfaß 2005, S. 7)? Certainly, one important influential factor is corporate communication. But there is no doubt, that the topicality, comprehensibility, and meaning of the messages are just as important as the basic attitude of the communicators toward the stakeholders' interests and needs. And one of the main interests of the stakeholders is to be well informed in a trustful way with meaningful concrete statements by their executives.

Therefore, using buzz words is not a solution. But "innovation" and "change" are buzz words of our time. Nobody knows exactly what they mean. Nevertheless, these terms are constantly used in our daily business lives. Many authors emphasize the importance and impact of innovations for the economic and the social system of countries. Others stress the aspect, that an innovation must be something really new and successful in the market (Zerfaß and Möslein 2009). This is—from their point of view—the difference between an idea and an innovation. The survey INNOVATE (Mast et al. 2005, 2006) however highlights: It is the inflation of buzz words like innovation which causes a general mistrust amongst stakeholders. The misuse of these terms has increased the stakeholders' skepticism about technologies and products which are described as "innovative".

Similar effects can be seen when analyzing change processes. What comes to mind when employees hear the word "change"? Is it "Yes, we can" or rather "Please, not again a change project"? Employees and executives all have their own perspectives on changes affecting the daily routine in their companies (Deekeling and Barghop 2009). Literature is full of advice about what can go wrong in change management and how people should act (Pfannenberg, 2013, Klewes and Langen 2008). There is, however, very little empirical evidence about how the change processes actually proceed and how they affect people's emotions (Mast 2011, 2010).

But consistently over the past years roughly two-thirds of all change projects have had little or no success (Houben et al. 2007). Of course, communication is only one of many drivers for organizational change within companies. But even well-planned communication management fails almost half of the time. The most

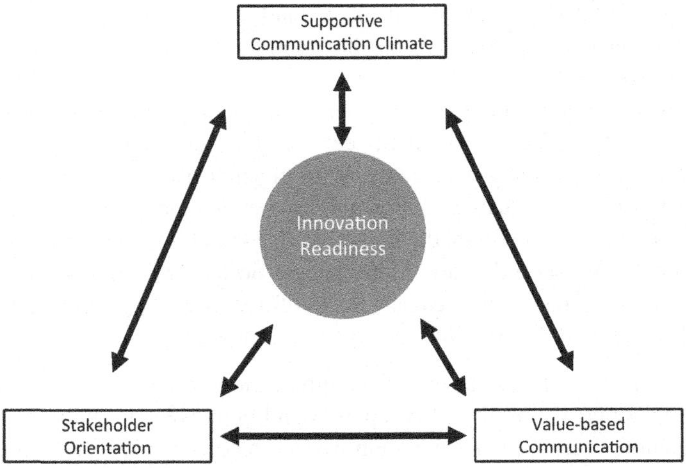

Fig. 11.1 Framework for creating a communication strategy for innovations

important success factors for change communication are (these are the results of the survey among communication directors in the top 250 companies in Germany) a strategic concept with a clear focus on stakeholders, organizational conditions which foster cooperation, and a quick flow of information (Mast 2011, 2009; Mast and Zerfaß 2005).

To get better results in communicating innovations in our time there is some work to be done on creating new and adequate strategies. Taking the current results of studies on change communication into account innovation communication management from now on should stress the following three factors (see Fig. 11.1).

Value-based communication: Are tangible and especially intangible corporate values points of reference for the communication practice (Mast 2016)? Do employees really understand the corporate vision, mission and current objectives? And do they know the written and unwritten rules of how to behave in case of struggling corporate values? Could they produce their own picture of how the value innovation applies to their job and can they contribute to the creation of new products or procedures? And the most important point: What about "Making the Connections" (Quirke 2008) and turning strategy into action? Does communication really paint pictures of a "strategic alignment" (Van Riel and Fombrun 2007, S. 209) so that employees understand and are able to enact their company's objectives?

Stakeholder orientation: What kind of image crosses the mind of communication managers when they are planning communication activities and talking to employees, customers, and journalists (Mast and Spachmann 2015)? Are they just "target" groups for them which are easy to handle if the right "tool" has been chosen? Are they purely considered as "objects" for certain applications of "tools" or furthermore as "subjects" who are balancing the rational and emotional side of communication? Above all, stakeholder orientation means that innovation communication reflects two sides of the coin—the rational and emotional needs of stakeholders—especially when communicating complex innovations.

Supportive climate: Companies are often unaware of the fact that how they communicate is as important as what they say. Stakeholders who are unsure or even mistrusting are very sensitive to the tonality of corporate communication, to the nuances of statements and the choice of which person says something in which channel and especially when. Stakeholders, including employees, feel like a seismograph whether they are respected by a communicator and told the truth. Managers in their role as communicators can change the climate in a company more quickly and sustainably than media ever could (Kinter et al. 2009). This kind of microclimate has a great effect on business nowadays and is often neglected. Reasons why managers have problems in reaching their stakeholders are quite different. One reason can be seen in prejudices against the company. In addition, clichés and misinformation play a decisive role when stakeholders do not feel up paying attention to the company's messages. These results, gathered by a survey[1] among communication directors in the top 500 companies in Germany, once more emphasize the ability of communicators to identify negative emotions like mistrust and their capability of dealing with them.

Recent surveys (Mast 2011, 2009; Houben et al. 2007) have revealed that most of the change projects have failed because the communication managers misjudged the emotional state of mind of the stakeholders who they wanted to convince. Major reasons were: The messages did not aim for the needs and questions of employees, relevant information was transmitted lately, and the employees could not paint a picture of how they should behave, whom they could trust and rely on, and whether they could feel as a part of the company in the future. There was no dialogue with stakeholders, too much media communication and too few conversations, insufficient management communication or cultural characteristics were ignored (Mast and Stehle 2016). All together—the strategic communication of innovations neglected the most important point: the management of emotions.

11.3 Success Factors of Strategic Communication of Innovations

Stakeholders are weighing up manifest and latent messages, time lags, symbolic functions of communication channels and especially, meaningful statements being addressed without using buzz words. Their daily balanced decisions are based on the following factors:

[1]Sample: TOP 500 enterprises (turnover) and DAX enterprises of Germany; $n = 121$, survey period: December 2012/January 2013.

- *Consistency*: Messages should signalize an alignment to corporate values like innovations and current objectives alike. Otherwise, stakeholders perceive mixed signals and watch more carefully to see how these inconsistencies will be resolved.
- *Credibility*: Leaders are credible when their communication behavior is seen as open, honest, and reliable and when they are telling the truth and the "whole story" of an innovation, not just the good news.
- *Orientation*: People want to see a clear direction, where a company is going and how it is doing. Stakeholders want to understand how they fit in, how they can contribute or how they are affected by innovations.
- *General attitude and esteem toward stakeholders*: Do stakeholders play an active or passive role in the practice of innovation communication? Are they estimated as partners, sources of ideas and experiences or even handled as "target" groups which are only expected to show desired "reactions"?
- *Participation*: Innovation communication has to avoid producing feelings of being captured or subjected to anonymous processes or procedures. Communication is challenged to make the connections between the concerns, preoccupations and agendas of stakeholders and those of the company and also to keep its finger on the pulse of stakeholder's sentiment. And finally it should explain apparent and hidden contradictions between the desires of people and the realities of business processes.

11.4 Transferring Experiences from Change Communication

These, and more, are key success factors for innovation communication, which can be transferred from change communication. These factors play a profound role when the top 250 companies in Germany think about their experience as to why communication was successful (Mast 2011) (see Fig. 11.2).

Sample: DAX enterprises and Top 250 enterprises (turnover) of Germany; $n = 83$, survey period: June/July 2008; percentage values. Question: "Thinking about your experience with change communication: What's your opinion, why was communication successful?" (multiple answers permitted).

11.4.1 Success Factor 1

Almost two-thirds of the companies point out the need for a long-term strategy including precise goals and consistent implementation. This includes thorough preparation of all measures, the right timing ("as early as possible") as well as detailed implementation plans and—in the case of campaigns—an elaborate

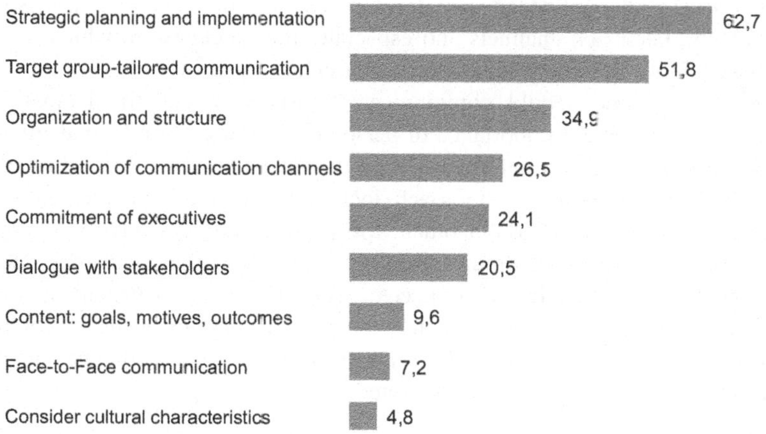

Fig. 11.2 Success factors of change communication (Mast 2011)

dramaturgy of all the elements. Formulating a consistent, intelligible, and vivid "change" or "innovation" story is considered to be as important as issues management which reduces complex issues to core statements and clear and simple messages. It is also vital to truly involve all stakeholders in the concept and not to forget anybody.

According to the communication directors' self-critical judgments clear strategies are rarely established in the daily business. This, communications experts claim, is often due to inconsistent decisions the management surprises them with. Others admit they simply underestimated the challenges of change communication or communicating innovations. Thus, a strategic approach, including a concept, is missing, resulting in erratic issues planning without "one-voice policy" or "key speaker". When the top management is quick in laying the blame for bad news on others the effects on the communication climate can be disastrous. One out of four companies admits that bad timing resulted in the creation of rumors. Some companies start too late, others plan communication programs which do not last long enough. Stakeholders are still interested in news, but can no longer learn about them through official media channels.

11.4.2 Success Factor 2

Target group aligned communication—a matter of course—is identified as the reason why about half of the companies succeeded realizing change projects or innovations. This implies a continuous information process including all employees and their respective cognitive and—even more importantly—emotional needs. Even the news that there is no news is important information in change processes.

Whether trust and confidence can develop depends on openness, honesty, and transparency. Feedback channels and especially the quickness in which questions are answered, are important measures which create trust.

Even though one should communicate consistently via all channels many companies underline the importance of personal communication by executives and the top management's commitment as a decisive stimulus. According to one company the "direct, simple approach including an emotional element" is the pivotal factor of success, or—in other words—the communicative integration of employees into effective communication networks, as well as managers operating as communicators often leads to success. The culture of management is under suspicion when a company self-critically admits: "The employees have been treated as objects and not as subjects". It is no surprise, then, that change encounters resistance and innovations are not accepted.

11.4.3 Success Factor 3

One out of three companies polled pointed out that organizational processes, responsibilities, and structures of cooperation between human resources, public relations, marketing, and investor relations are important. This coordination and the quick transfer of information between departments are just as critical for success as the early integration of communication experts in the top management's planning process. The adoption of project teams is generally judged positively.

11.5 Innovation Communication in a Changing Media Landscape

A changing media landscape holds opportunities as well as risks. While at the beginning of the year 2013, nearly 85% of the communication directors in the surveyed top 500 companies (see Footnote 1) in Germany announced that it is easy to communicate innovations, this appraisal could be totally different in the years to come. In spite of this quite significant number, companies should not rest on their laurels. The increasing importance of the Internet, especially with regard to the rising relevance of social media, poses a challenge for many companies. Today already 44% of the surveyed top 500 companies have problems in reaching opinion leaders via the World Wide Web. At the same time 59% have noticed that it is important to invest in new media and to communicate, e.g., via Facebook or Twitter.

Besides the appropriate choice of the right medium, it is important to meet the needs of the stakeholders when composing the message. According to the survey INNOVATE (Mast et al. 2005, 2006) emotional news value and dramaturgy are crucial in media relations when innovations are hard to explain. Storytelling, personalization, and visualization are preconditions for a positive media image of innovations. Because of this, media relations for new products, services and technologies need to work—much more than other kinds of communication—with illustrations and examples, stories, personalization, and concrete benefits for the individual. Innovations have to be prepared for journalists and other stakeholders in such a way that they can be experienced and felt. This is especially important since the Internet asks for information which is easy to understand and gather as well as eye-catching and attention-grabbing. 18% of the surveyed communication directors mention that one reason why stakeholders are difficult to reach is the complexity of a topic. In addition, 16% announce that it is a success factor if you meet the needs of the stakeholders concerning the way you are communicating.

To date the emerging field of management communication as an essential part in communicating innovations has been underestimated. Zerfaß and Huck (2007, S. 112) identified four different types of promoters: "Expert promoters" who have an intensive knowledge about the topic concerned, "authority promoters" who have power and resources as a result of their hierarchical position, "process promoters" who have an excellent organizational knowledge to foster the collaboration of people, and the "relationship promoters" who have a large personal network and know the right people. These are only some roles of managers communicating innovations.

In summary: Managers as communicators should be analyzed in further research. How do they see themselves as communicators? What is their view of stakeholders' roles? What about their ability to tell stories? Do they appreciate that their communicator role is a central duty in order to create a supportive climate?

Strange as it may sound, developing stakeholder aligned change or innovation communication continues to be the most important challenge for the near future—for managers in their role as communicators as well as for communication experts. It is a paradox, however, that on the one hand executives believe that innovation communication programs can be planned and need to be managed, while on the other hand there is a lack of intuition and knowledge about stakeholders' interests, desires and emotional needs. The orientation toward stakeholders often called for in science and business practices is implemented rather slowly in innovation communication.

Bibliography

Deekeling, E., Barghop, D. (eds.): Kommunikation im Corporate Change: Maßstäbe für eine neue Managementpraxis, 2nd edn. Gabler, Wiesbaden (2009)

Houben, A., Frigge, C., Trinczek, R., & Pongratz, H.J.: Veränderungen erfolgreich gestalten: Repräsentative Untersuchung über Erfolg und Misserfolg im Veränderungsmanagement. C4 Consulting, Düsseldorf (2007)

Kinter, A., Ott, U., & Manolagas, E.: Führungskräftekommunikation: Grundlagen, Instrumente, Erfolgsfaktoren: Das Umsetzungsbuch. F.A.Z.-Institut, Frankfurt a. M. (2009)
Klewes, J., Langen, R. (eds.): Change 2.0: Beyond Organisational Transformation. Springer, Berlin (2008)
Mast, C.: Sizing up change challenges. Communication Director, **02**(2009), 60–63 (2009)
Mast, C.: Gang in die Offensive. Pressesprecher **01**(2010), 44–47 (2010)
Mast, C.: Innovationen in der Unternehmenskommunikation. Ergebnisse von Umfragen bei DAX-Unternehmen, Analysen und Meinungen. LIT, Berlin (2011)
Mast, C.: Unternehmenskommunikation: Ein Leitfaden, 6th edn. UVK, Konstanz, München (2016)
Mast, C., Simtion, A.: Employer Relations. Arbeitgeberkommunikation—neues Handlungsfeld für Unternehmen. UVK, Konstanz, München (2016)
Mast, C., Stehle, H.: Energieprojekte im öffentlichen Diskurs. Erwartungen und Themeninteressen der Bevölkerung. Springer VS, Wiesbaden (2016)
Mast, C., Spachmann, K.: Thesenjournalismus statt ergebnisoffener Recherche? Erfahrungen von Pressesprechern mit Rechercheanfragen von Journalisten. Bundesverband deutscher Pressesprecher e. V, Berlin (2015)
Mast, C., Zerfaß, A. (eds.): Neue Ideen erfolgreich durchsetzen: Das Handbuch der Innovationskommunikation. F.A.Z.-Institut, Frankfurt a. M. (2005)
Mast, C., Huck, S., Zerfaß, A.: Innovation communication: Outline of the concept and empirical findings from Germany. Innov. Journalism **2**(7), 1–14 (2005)
Mast, C., Huck, S., Zerfaß, A.: Innovationskommunikation in dynamischen Märkten: Empirische Ergebnisse und Fallstudien. LIT, Berlin (2006)
Pfannenberg, J. (ed.): Veränderungskommunikation: So unterstützen Sie den Change-Prozess wirkungsvoll, 3rd edn. F.A.Z.-Institut, Frankfurt a. M. (2013)
Quirke, B.: Making the Connections: Using Internal Communication to Turn Strategy into Action, 2nd edn. Gower, Aldershot (2008)
Van Riel, C.B.M., Fombrun, C.J.: Essentials of Corporate Communication: Implementing Practices for Effective Reputation Management. Routledge, London (2007)
Zerfaß, A.: Innovation readiness: a framework for enhancing corporations and regions by innovation communication. Innov. Journalism **2**(8), 1–27 (2005)
Zerfaß, A., Huck, S.: Innovation, communication, and leadership: new developments in strategic communication. Int. J. Strateg. Commun. **1**(2), 107–122 (2007)
Zerfaß, A., Möslein, K. (eds.): Kommunikation als Erfolgsfaktor im Innovationsmanagement: Strategien im Zeitalter der Open Innovation. Gabler, Wiesbaden (2009)

Chapter 12
The Role of Communicators in Innovation Clusters

Bettina Blasini, Rani J. Dang, Tim Minshall and Letizia Mortara

Abstract Innovation clusters continue to be an important focus of economic development policies in many nations. Leading innovation clusters demonstrate that regional concentration strengthens the innovative capability and can lead to successful competitiveness on a global level, as demonstrated by regions such as Silicon Valley (US), Cambridge (UK) and Sophia Antipolis (France). However the successful creation of clusters still presents a challenge to policy makers as efforts to do so regularly fail. The development of innovation clusters has therefore received much academic and policy maker attention. While past research has examined a variety of factors as drivers for clustering effects, the role of communication within the cluster—and, specifically, the role of key individual communicators—in underpinning successful cluster development has received almost no academic attention. In this chapter, we will draw upon the relevant literature to develop a conceptual framework that will underpin research on this important topic by investigating the role of communicators in innovation clusters. Building on communication theories, the framework suggests that there are four influence levels that shape and impact the role of communications in innovation clusters: the Individual Level, the Organizational Level, the Cluster Level and the Context. The interdisciplinary view on clustering effects contributes valuable insight to both communication studies and cluster theories. The framework developed within this chapter provides a structure to aid future research on the role of communicators within innovation clusters.

B. Blasini · T. Minshall
University of Cambridge, Cambridge, UK

R.J. Dang (✉)
University of Nice Sophia-Antipolis, Nice, France

L. Mortara
Centre for Technology Management, Institute of Manufacturing, University of Cambridge, Cambridge, UK

N. Pfeffermann and J. Gould (eds.), *Strategy and Communication for Innovation*
DOI 10.1007/978-3-319-49542-2_12

12.1 Introduction

Innovation clusters continue to be an important focus of economic development policies in many nations (Uyarra and Ramlogan 2012). Leading innovation clusters demonstrate that regional concentration strengthens the innovative capability and can lead to successful competitiveness on a global level, as demonstrated by regions such as Silicon Valley (US), Cambridge (UK) and Sophia Antipolis (France). However the successful creation of clusters still presents "[..] a unique challenge to policy makers" as efforts to do so regularly fail (Clark 2013: 6). The development of innovation clusters has therefore received much academic (e.g. Porter 1998; Karlsson 2008) and policymaker (e.g. European Commission 2008; OECD 2012) attention. While past research has examined specialized supply and demand, collaboration and competition, the infrastructure and mobility of goods as crucial drivers for clustering effects, the role of communication within the cluster—and, specifically, the role of key individual communicators—in underpinning successful cluster development has received almost no academic attention.

Perceptions gathered from interviews[1] in one of Europe's leading clusters indicates that this may be an important omission: "Public Communication is critical [..]. If it wasn't for the communication, the cluster wouldn't exist. It is very important that people within the cluster talk to each other and that the cluster communicates with the outside world" (PR Consultant, Cambridge, UK). For complex clusters, which are characterized by diverse internal and external stakeholders, communication was seen by the same interviewee as the "segment that holds it all together". Gathering and sharing information, communicators build interrelations between the actors and create a communication network. "We joined up and connected the companies in the Cambridge area and we also connected those companies with trade collaborators in other parts of the world. We are the catalyst for growth," explained an editor in chief of a newspaper in Cambridge. Connecting and informing relevant stakeholders, communicators describe their work as, in the words of one interviewee, "breaking down boundaries. That's the core of communication".

Such statements point to an important yet under-researched issue, which merits attention in order to enrich our understanding of how innovation clusters develop. In this chapter, we will draw upon the relevant literature to develop a conceptual framework that will underpin research on this important topic by investigating the role of communicators in innovation clusters. Building on communication theories, the framework suggests that there are four influence levels that shape and impact the role of communications in innovation clusters: the Individual Level, the Organizational Level, the Cluster Level and the Context. The interdisciplinary view on clustering effects will contribute valuable insight to both communication studies and cluster theories. On a theoretical level this study provides a

[1]Eight professional communicators working in the Cambridge (UK) cluster were interviewed as part of an on-going research project in April–May 2013.

strong foundation for further research in the field of innovation communication and on a practical level it identifies aims and strategies of communicators in innovation clusters.

12.2 Economic and Sociological Cluster Theories

The geography of innovation shows a clustered economic landscape, characterized by a regional concentration of innovative capability. Established cluster theories discussed this phenomenon from different perspectives and identified factors, which support the local agglomeration. Marshall (1890, 1920) introduces the notion of external economies, an environment characterized by skilled labour, specialized goods, face-to-face contact and trust, which enables spillovers and sparks innovative activity. Porter (1990, 1998, 2000) highlights the importance of external value and identifies supply and demand conditions, competition and collaboration and the context of the firm as most important for innovative capability. Krugman (1994, 1998) focuses on the dynamics of resource allocation across activities and location by identifying tangible and intangible goods that shape a complex economic environment. The research by Marshall, Porter and Krugman established a strong foundation to understand the dynamics of clustering effects and influenced the academic perspective on how innovation takes places. While these papers have mostly been discussed in terms of their economic contribution, they also imply the importance of sociological aspects. Discussing externalities, Marshall, Porter and Krugman refer to information gathering, knowledge sharing and the resulting relations between the actors (as summarized in Table 12.1).

The sociological factors, as shown in Table 12.2, are kept implicit within in the economic theories and are not explained in depth. This leads to a blurred understanding of sociological externalities based on interrelations, common knowledge and information spill-overs. Focusing on the characterisation of interrelations within a network and the resulting information gathering and sharing processes, sociological theories complement the economic perspective on cluster dynamics.

Table 12.1 The role of communication in economic cluster theories

Economic cluster theories	Role of communication
Marshall (1890, 1920)	Marshall introduces the idea of knowledge-spillover, which is based on the face-to-face contact of the people. According to Marshall, individuals move from firm to firm and exchange knowledge and ideas
Porter (1990, 1998, 2000)	In his Diamond-Model, Porter discusses knowledge about new innovations and early perceptions of new possibilities due to the close relationships between the actors
Krugman (1994, 1998)	Krugman's notion of centripetal forces can be understood as external economies, which are based on relationships and shared information

Table 12.2 The role of communication in sociological cluster theories

Sociological cluster theories	Author	Role of communication
Network theory	Granovetter (1937)	Close relationships that are characterized by strong ties are more likely to share knowledge than those who communicate infrequently or who are not emotionally attached. Weak ties, i.e. acquaintances, support the diffusion of knowledge, the creation of new opportunities and the emergence of new collaborations
	Burt (1992, 2001)	Structural holes define potential connections between units that are not connected and lead to non-redundant information, as the sources are more additive than overlapping
Knowledge-based view	Polanyi (1958)	Tacit knowledge can be understood as "not codified" knowledge. As it defies easy articulation or codification, it is difficult to exchange over long distances
	Asheim and Gertler (2006)	Tacit knowledge depends on shared conventions and norms that have been fostered by a common institutional environment and relies on a mutual language and communication codes. Thus it is transferred through face-to-face interactions between individuals
Gatekeeper Studies	Dang et al. (2011)	To access tacit knowledge and context-laden information, "listening posts" are created in order to build channels inside and outside the cluster
	Lezaric et al. (2008)	Gatekeepers are characterized by three functions: To search information from external sources To transcode and translate the meaning of information To transfer information and to disseminate accumulated and local knowledge

Network Theory, Knowledge-Based View and Gatekeeper Studies provide a valuable insight to the research field as shown in Table 12.2.

Both economic and sociological cluster theories highlight the importance of networking, knowledge gathering and information sharing. While economic theories imply such action as given processes, sociological theories acknowledge a deliberate and organized way of connecting and communicating. Gatekeeper studies point to the importance of specific actors managing information in order to build relationships inside and outside the cluster. Communication studies have a long tradition in analyzing actors who pursue these aims on a professional basis, but have never been applied to the study of clusters. This discipline opens new perspectives of analyzing and understanding cluster processes. Thus the next section will introduce the different role of communicators as discussed in communication studies.

12.3 Communication Studies

Originally communication studies focused on journalism, which has been defined as public mass communication that targets a broad and heterogeneous audience in order to inform (Pürer 2003: 75). Over the last century, corporate communication, especially public relations, gained a lot of academic attention. Public relations can be also understood as public communication but addresses defined stakeholders to pursue specific messages. In their interplay, journalism and public relations target a broad audience and thus shape and influence the public opinion. The following two sections introduce the actors in these professions:

12.3.1 Journalists

The understanding of the journalists' role has been changing over time in academic research. Based on the News-Bias studies, early research regarded journalists as powerful "gatekeepers" who decide what is newsworthy according to their own principles. Further studies took social aspects into account: gatekeeping as part of a profession, influenced by social norms and values, political and social standards and criteria of production such as time pressure or constrained wordcount. Research on news factors and news values strengthened the perspective that journalists are not isolated actors, but part of a social process that is influenced by journalism routines.

Targeting the public, journalism has always been attributed a strong impact on society and has been discussed in terms of its framing, priming and story-telling potential. Recent research assumes that mass media structures knowledge and opinions of the recipients and define what they perceive as important (Rössler 1997). Thus journalists do not influence what recipients think, but what they think about.

12.3.2 Public Relations Consultants

Discussing the impact of journalists on the public opinion, public relations has become prominent in communication studies. Public relations is persuasive communication following a certain strategy to evoke publicity by both functioning as a source for journalism and targeting stakeholders directly. Addressing journalism, public relations consultants have been understood as influential actors in the dynamics of public communication. Ever since Bearns (1985) stressed that public relations consultants determine journalism through timing and content, the relationship of influence has been a recurring theme in communication research. More recent studies show a rather balanced relation: the intereffication model by Bentele et al. (1997) analyzed the daily collaborations between public relations consultants and journalists and showed a bilateral give and take interrelation (Bentele et al. 1997). Based on these results, they created the model of intereffication. "Intereffication" stems from the Latin terms "inter" and "efficare" and means "to enable each other". While journalists needs the basic information of public relations, public relations benefits by the broad and heterogeneous audience of journalism and its significance. Therefore analyzing public communicators both journalists and public relations consultants should be taken into account.

12.4 Innovation Communication

Understanding the traditional roles of communicators in communication studies offers valuable insight to the role of communicators in innovation clusters. Going back to Porter, clusters can be understood as "geographic concentrations of interconnected companies, specialized suppliers and service providers, firms in related industries and associated institutions (e.g. universities, standard agencies, and trade association) in particular fields, that compete but also cooperate" (Porter 2000: 253). Consisting of heterogeneous members, clusters are characterized by many communicators taking part in public communication, pursuing different aims and strategies. To understand the role of communicators in innovation clusters, the young and developing research field innovation communication provides significant contribution.

Introducing the concept of "Innovation Journalism" in 2004, Nordfors sparked the academic debate about how communication might affect innovation innovations or innovation clusters (Nordfors 2004a, b; Nordfors and Ventresca 2006; Nordfors and Uskali, forthcoming; Nordfors 2009). His research highlights the leverage of communication in two ways: Journalists start a public discussion and thereby create a public agenda. Furthermore journalists may explain complex innovations and create meanings.

Building on Nordfors' research, Zerfass (2005) introduces a broad view on communication, which includes not only journalism but also public relations and interpersonal communication to meet the challenges of innovation clusters. According to Zerfass, the complexity of innovation clusters calls for manifold public and bilateral relations driven by communication. Innovation journalists as defined by Nordfors play an important role in facilitating information flow, which allows collaborations and the identification of entrepreneurial opportunity. Innovation public relations consultants aim to systematically plan, implement, and evaluate communication strategies in order to create an understanding of and trust in innovations. Finally innovation-related leadership communicators seek to influence attitudes towards innovations by mediating meaning in asymmetrical, social relations (Zerfass 2005: 11).

Pfeffermann (2011) highlights the strategic perspective on communication by discussing innovation communication as a cross-functional dynamic capability of an innovative company or cluster. As defined by Teece et al. (1997) dynamic capabilities are the firm's capacity to integrate, build and reconfigure internal and external resources and competences to address and shape rapidly changing business environments (Teece et al. 1997: 516). Pfeffermann shows that communicators can achieve this aim by introducing ideas and concepts, generating and highlighting context-issues, presenting the organization's innovative capability, building up new stakeholder schemata or modifying existing ones (Pfeffermann 2011: 263). Though Pfeffermann's research is only valid for public relations consultants—journalists aim for neutral information—it gives new and valuable insight to the research field of innovation communication.

Nordfors, Zerfass and Pfeffermann provide important contributions from different perspectives to the research field innovation communication. By highlighting the potential of communicators in innovation clusters their studies complement cluster theories on different levels.

In summary, according to Nordfors, Zerfass and Pfeffermann communicators may:

- Name and explain innovations (Nordfors 2004a)
- Create a public news agenda (Nordfors 2004b)
- Formulate a shared vision of the cluster (Zerfass 2005)
- Connect stakeholders inside and outside the cluster (Zerfass 2005)
- Create collaborations and entrepreneurial opportunities (Zerfass 2005)
- Introduce new ideas and concepts (Pfeffermann 2011)
- Build new knowledge schemata and or modifies existing ones (Pfeffermann 2011)
- And thus creates and maintains a cluster's innovative capability (Pfeffermann 2011)

These highlighted issues point to the need for more research on the role of communicators in innovation clusters. The analysis of communicators requests an interdisciplinary approach to take both business and communication studies into account. In the next section, the conceptual approach to analyze communicators will be introduced.

12.5 Conceptual Approach

Communication studies have a broad tradition in analyzing communicators in various contexts. Its origins go back to sociological studies by Weber, Durkheim and Bourdieu who discussed action theory and identified the possibilities and limits of the individual's action. This can be seen as response to system theory, which dominated the field over a long period of time and drew attention away from the individual and towards field structures. Communication studies benefited from both research perspectives and developed complex perspectives on communicators, its personal potential of action and the influences by its environment. Based on the influential sociological studies, communication studies show different approaches to analyze the role of communicators by identifying the elements and composition of the "role". The interest of communication studies can be explained by the potential influence of the communicator's role on the content of news. Thus some approaches focus on the media content, yet offer a valuable basis to understand and identify influencing factors on the communicator and of the communicator. Table 12.3 summarizes the influential communicator studies by Shoemaker and Reese (1991), Weischenberg (1992), Esser (1997) and Preston (2009).

Table 12.3 Conceptual approach on the role of communicators

Conceptual approach	Influence levels and categories
Shoemaker and Reese (1991)	Individual Level: Personal aspects such as professional backgrounds and experiences, professional roles and ethics, personal attitudes, values and beliefs and the power within the organization Routines Level: Routinized and repeated practises, which can be viewed as both enabling and constraining Organizational Level: Organizational roles, organizational structure, organizational policies Extra-Media Level: Institutions in society, government, advertisers, public relations, influential news sources, interest groups, and other media organizations Ideological Level: Social interest and the construction of meaning
Weischenberg (1992)	Media-Actor: Demographic Data, social and political opinions, perception of the role, image of Recipient, professionalism and socialization Media-Message: Origin of information, reference groups, patterns of presentation and news, construction of reality effects and retroactive effects Media-Institution: Economical imperatives, political imperatives, organizational imperatives and technological imperatives Media-System: Societal conditions, historical and legal foundations, communication policy, professional and ethical standards
Esser (1997)	Individual Level: Subjective values, political attitudes, work motivation, self-perception, professionalism and demographic data Organizational Level: Job profiles and practices, organizational structure, distribution of competences, work processes, control and technology within the media organization Legal-Normative and Economic Level: Economic conditions of the media market, press law, self-control in media, ethic foundation, trade unions and associations, education of journalists Historic-Cultural Level: Freedom of the press media history, perception of the press, journalistic tradition, understanding of objectivity, political culture and socio-political conditions
Preston (2009)	Individual Level: Personal characteristics, background, values of the communicator, definitions and perception of their professional roles Media Routines: Taken-for granted institutional practices and norms, that frame and shape how individuals work and function within complex settings Organizational Influences: Organizational values, strategic goals, policies and power structures of the company Political-Economic Factors: Political and economic culture, distribution of power in society Cultural and Ideological Power: Norms, values and cultural background

12.5.1 Theoretical Foundation

The Hierarchy of Influences approach by Shoemaker and Reese (1991), the Zwiebelmodell by Weischenberg (1992), the Integrative Multi-Level Model by Esser (1998) and the Clusters of Influences approach by Preston (2009) contribute valuable insight to the influences on the role of communicators. Though they follow different research interests, they show similar patterns and recurring sets of structures as summarized in Table 12.4.

As shown in Table 12.4, influence approaches usually draw on four or five influence levels to cover the complex power structure in the communication field. Whether four or five levels are identified, depends on the structuration of influence levels: While Shoemaker and Reese (1991) and Preston (2009) differentiate between economic and cultural background, Weischenberg (1992) combines these influences in a Media System level. Furthermore, the level Media Routines appears in most of the approaches as a level on its own, only Esser mentions routines in terms of patterns and structures in the context of the Organizational Level. Besides these minor distinctions, the approaches show a common structure:

- Individual Influences
- Organizational Influences
- Communication System Influences
- Cultural, political, economical background Influences

The hierarchical order of the levels also shows the same composition: The Individual Level is in the centre of influence structure, embedded in the Organizational Influence Level, covered by the System Influences, surrounded by the Macro Influence Level, such as cultural, political and economical conditions. While the early studies by Shoemaker and Reese (1991) and Weischenberg (1992) do not explicitly take interrelations between the levels into account, the more recent approaches by Esser (1998) and Preston (2009) highlight the reciprocal influences between the levels. According to Esser, the levels must be understood as "open" and not as "closed" systems (Esser 1998: 33).

To express in which context each of the levels works, graphic models provide important insight to the dynamic complexity of the communication field: By visualizing, models simplify and structure the influences. Furthermore, they create a profound foundation for an empirical analysis as they define the relevant factors and

Table 12.4 Structure of influence levels

Levels of influence	Shoemaker and Reese	Weischenberg	Esser	Preston
Individual level	✓	✓	✓	✓
Media routines level	✓	✓		✓
Organizational level	✓	✓	✓	✓
Political and economic level	✓	✓	✓	✓
Culture and ideology level	✓		✓	✓

point out to interrelations. In the next section, a new model will be created in reference to the established models in order to build a theoretical framework for the analysis of communicators in innovation clusters.

12.5.2 *Creation of a Conceptual Framework*

The research on communicators in innovation clusters requires a new and specific framework. Unlike the frameworks discussed earlier, which focus on journalism only, this research includes different kinds of communicators to meet the challenges of innovation clusters. This points to the need for an interdisciplinary understanding of influence levels and categories. The framework proposed in Fig. 12.1 classifies four influence levels:

1. Individual Level
2. Organizational Level
3. Cluster Level
4. Context

The centre of the model shown in Fig. 12.1 positions the research interest of the study, the role of communicators in innovation clusters. This role is influenced by the levels evolving around it: The individual influences, the organizational influences and the cluster influences. Graphically these influence levels are structured in a tetrahedron, which visualizes the reciprocal interaction between them. The openness of the levels is highlighted by the dotted lines that separate the levels. This visualization overcomes the problems of hierarchy as the influence levels take place on the same levels and therefore all interact. This interaction takes place in front of a societal background, which surrounds the influence levels and the role of the communicator. It indicates that everything takes place in a certain context that has been established by various factors, such as historical, cultural, political and economical conditions, which have to be taken into account. In the following section, each level will be explained in detail.

12.5.2.1 Individual Level

The Individual Level deals with the personal and unique attributes of the communicators. This level appeared in all models discussed earlier, as it covers important information about the personal position of the interviewees and thus builds the foundation to understand their ideas throughout the whole study. As the level implies various attributes, the model suggests three categories: Disposition, Socialization and Self-Perception. Disposition covers the demographic data, such as gender and age. Socialization also refers to individual attributes, covering the communicator's education and professional career. This category provides

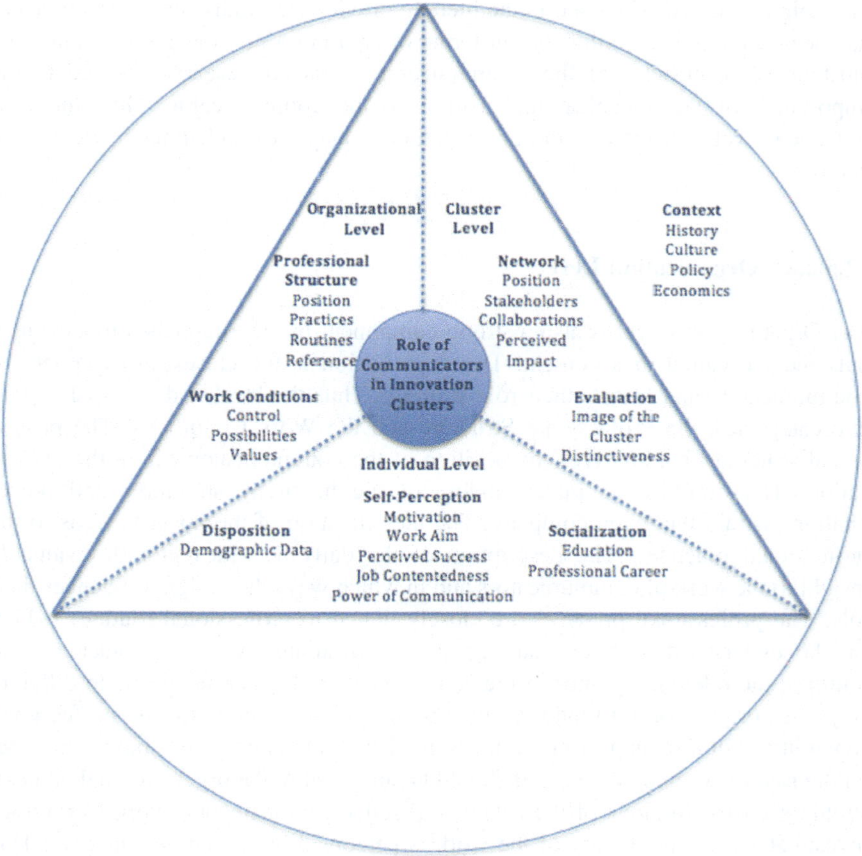

Fig. 12.1 The role of communicators in innovation clusters

information on the interviewee's background and thus creates a deeper understanding for the communicator's position nowadays. Besides individual facts as dispositions and socialization, the Individual Level also covers the category Self-Perception. This category structures complex aspects of individual perceptions regarding the interviewee's role. Work motivation will provide valuable insight to the attitudes and expectations of the communicator. Closely linked is the definition of the work's aim, which identifies beliefs and ideas of the profession in general. Subsequently it is very interesting to analyze the perceived success of these aims to check the broad aspirations with the reality. This leads to the next aspect, the job contentedness, which deals specifically with the satisfaction in the profession and links back to realization of aims and motivations. Finally, the category Self-Perception points out to the importance of the perception of the power of communication. This aspect completes the idea about their profession, what they want to do, what they plan to do, if and how they achieve it and what greater power they might hold. As discussed in the literature review, the individual communicator has been seen as very influential in terms of creating a news agenda. Though the attributes and attitudes of the communicator is crucial, research pointed to the importance of the organizational setting of the communicator. Thus the next influence level will refer to the organizational categories and influence factors in detail.

12.5.2.2 Organization Level

The Organization Level is also a strong component in the established models as it puts the individual in a context. The organization is the closest environment of communicators and frames their role strongly. Thus the level is distinguished into two categories, the Professional Structure and the Work Conditions. The professional structure first analyzes the position of the communicator within the organization. That implies the understanding of the hierarchy structures and power relations within the given company. The identification of the position leads to the professional practices. The description of the daily activities provides valuable insight to the work of communicators and in which ways they might influences their role. The professional practices are closely linked to professional routines, which can be understood as both enabling and constraining. Analyzing practices and routines, the reference groups of the communicators play a crucial role in different ways: as origins of information sources, as guidance in terms of professional capabilities but also as peer group pressure. Thus, reference groups have an impact on the role of communicators and should be analyzed in the organizational context. Building on the findings of the category Practices, the category Work Conditions investigates how the quality of the work is perceived by the communicators. This category implies internal control, which could be developed trough hierarchal structures, peer pressure or organizational tendencies. However the possibilities and perceived freedom will also be analyzed. The organizational atmosphere is much influenced by constraints and tolerance, and leads to the professional values that characterize a company. Values are a broad field and could be based on a journalistic education, organizational codes or general ethic norms that influence the organizational context. In sum, the Organizational Level acknowledges that the communicators work in an organizational context that shapes and conditions their professional role. The research interest requests to understand the role of communicators also in a broader context, which impacts both individual and organization: the cluster.

12.5.2.3 Cluster Level

The Cluster Level is highly specific to this research interest, as it has not been discussed in communication studies before. While the models summarized earlier in Sect. 5 focus on communication structures, such as media-institutions or the journalism field, this study takes place in a very distinctive environment: innovation clusters. This requests distinctive categories, which characterize—and thus analyse—the specifics of clusters. Applying a journalistic model to a business research field, previous models only benefit in terms of level analysis. To identify relevant categories for the Cluster Level, the cluster theories provide valuable contributions. As discussed in the literature review, the benefit of clusters lies in the network structure, which makes information flow and knowledge spillovers possible. Thus the Network is the first categories of the Cluster Level. This category refers to the interrelations and connections in terms of communication. First of all, the position in a network is influential for the role of communicators. Whether they are located centrally or rather distanced might impact the way they communicate. The position is connected with the important role of stakeholders: it is important to analyze who the stakeholders of the relevant communicators are and how they work with them. Also, if there are different communication strategies for each stakeholder could be very interesting. Bearing in mind the insights from the literature review, which identifies internal and external stakeholders of clusters, communicators can be seen as central actors who gather and spread knowledge. At this point it is very interesting to analyze the perceived success of their work: the perceived impact. How the communicators perceive their communication strategies are influencing their stakeholders, provides interesting insight to their professional role and evaluates their work experience. This leads to the second category of the Cluster Level, the Evaluation of the cluster. In this category, the perceived image of the clusters will be analyzed. The image communicators have about a cluster will influence their behaviour, their opinions—their role. While the image can be created by other people's views, the second aspect only measures the individual opinion about the cluster. Analyzing the distinctiveness from the communicator's point of view completes the evaluation as this aspect summarizes the perceived most important characteristics about the cluster and thereby reveal the personal "image".

Defined as regional agglomeration, clusters are influenced by their specific environment. This environment must be understood as broad background which influences not only the cluster but also the organizations and actors of the cluster. Thus it is important to set the so far identified levels of influence in a certain context.

12.5.2.4 Context

The three influence levels Individual Level, Organizational Level and Cluster Level are embedded in a specific Context. As explained earlier, the context impacts the influence levels and the role of communicators in innovation clusters. The Context is identified as history, culture, policy and economics. These conditions are highly specific to geographic areas and frame the cluster, organizations and actors. The history factor examines the historical evolution of the cluster and takes important milestones into account. History is closely linked to further factors, namely culture, policy and economics. While history focuses on the development of a cluster, the culture factor looks at cultural environment of the cluster, for example the relation to universities and research institutions. The policy factor investigates how political decisions and support influences the cluster and shapes its situation. Political initiatives might lead to financial support, which will be examined by the factor economics. This latter factor concentrates on the financial situation and external support. The Context is important for the cluster, the organization and the actors and conditions the role of communicators in innovation clusters.

Together, the ***Individual Level***, the ***Organizational Level***, the ***Cluster Level*** and the ***Context*** create a valuable framework to analyze and understand the role of communicators in innovation clusters. Taking various levels and perspectives into account, the framework offers a strong foundation for further studies, which will be discussed in the following section.

12.6 Conclusions

The preceding sections brought together a diverse body of academic literature on cluster theories and communication studies in order to find interrelations between the two research fields. The classic economic cluster theories by Marshall (1890, 1920), Porter (1990, 1998, 2000) and Krugman (1994, 1998) indicate the importance of communication, which informs but also connects actors inside and outside the cluster. While economic cluster theories refer to communication as given spillovers, sociological cluster theories acknowledge communication as an organized, deliberate process and thus complement the economic theories. The insights of Network Theory (Granovetter 1937; Burt 1992), Knowledge-Based View (Polanyi 1958; Asheim and Gertler 2006) and Gatekeeper Studies (Lezaric et al. 2008) highlight the importance of communicators who build connections to gather and share information. While there is no research on communicators in innovation clusters, the young and developing research field innovation communication discussed by Nordfors (2004a, b, 2009), Zerfass (2005) and Pfeffermann (2011) offers valuable insight and strengthens the demand for further research on the role of communicators in innovation clusters.

To achieve this aim, a conceptual framework has been created based on established conceptual approaches by Shoemaker and Reese (1991), Weischenberg (1992), Esser (1997) and Preston (2009). Though these studies focus on different research interests, they share a common quest into the interrelation between a communicator and the context that influences in both constraining and enabling ways. As discussed in the literature review communicators are not isolated individuals but part of an organizational and social context. To analyze the role of communicators, the relevant variables are heterogeneous and complex. The discussed models identify different influence levels and allocate influence factors appropriate to the specific research question. Also this study's research interest requested a unique conceptual framework to take the specification of innovation clusters into account. The tetrahedron-shaped framework identifies four influence levels: the individual level, the organizational level, the cluster level and the context which cover influence factors that shape and condition the communicator. The framework offers a strong foundation to investigate the role of communicators in innovation clusters.

This study addresses an unexplored interdisciplinary research area at the interface of communication and business studies and thus makes a series of academic and practical contributions.

Innovation Communication

Innovation communication is a new research field, which developed over the last decade. So far, research concentrated on innovation communication in terms of either innovation journalism or innovation communication on a corporate level. This research focuses on communication at a cluster level and therefore views innovation communication from a complex perspective: a cluster's communication is published by many different actors in different positions, pursuing different aims with different strategies. While previous communicator studies focused on a specific profession in public communication, this study includes different professions, which create the public message of a cluster. This approach allows the comparison of different professional positions and will make patterns visible. The inclusion of heterogeneous professions also allows the analysis of interaction. As discussed in the literature review, the interrelations between journalists and public relations consultants are intense and influential. Including everybody who engages in public communication about the cluster provides a holistic and comprehensive perspective on innovation communication.

Cluster Theories

Addressing innovation communication at a cluster level, this research contributes also to cluster theories. While both economic and sociological cluster theories refer to communication, the role of communicators stays implicit. Though sociological theories highlight connections and the management of information, the range of actors remains unclear. Investigating the people who make the message sheds light on a research angle, which has not been explored. The conceptual framework allows an understanding of the position of the communicators—who they are, what they do, what aim they pursue and which strategies they apply. This insight addresses the research gap between the assumed importance of communication in clusters and the missing data on it. The conceptual framework offers a valuable foundation for empiric research on the role of communicators in innovation clusters. The formulated categories, influence factors and their interrelation lead to a systematic structure for a qualitative interview guideline with relevant communicators. The results will establish how communicators connect in clusters, how they gather knowledge and share information and thereby create certain roles of communicators. Understanding the meaning of communicators will contribute valuable insight to classic cluster theories in terms of clustering dynamics. In addition to established cluster factors such as specialized supply and demand, collaboration and competition and the infrastructure and mobility of goods, communicators and their information networks should also be seen as crucial drivers for a cluster's development. Based on this understanding, further future research will be able to focus on the impact of communicators on the cluster's success and investigate causal effects on a quantitative basis. It will be interesting to compare different clusters in different developmental stages and in different locations to make similarities and differences visible.

Practical Contributions

This understanding provides valuable practical implications. Innovation clusters are based on the premise of interrelations and common knowledge and demand the creation and maintaining of a network of information inside and to the outside of the cluster. This points to three major aims for communicators:

1. Create connections and dialogue with **internal stakeholders**
2. Create interrelations and communication with **external stakeholders** and the outside world
3. Create a **common language**

Internal communication is crucial to connect the members within a cluster, to create a shared environment and a common vision. Only if members are connected and talk to each other will collaborations and entrepreneurial opportunities be possible. Knowledge spillovers are rarely coincidence but based on a frequent dialogue between actors. As one journalist of one of the Cambridge High Tech Clusters explains "We actually joined up companies in the business community. We helped them collaborate with one another. Before you would have had companies operating in isolation. But you have got to have networks and connections. And we formed that role by publicizing companies and their role and what they were doing. We acted as catalyst for collaborations." It shows the important aim of communicators to connect and relate members of the cluster and thereby create a vibrant community. This practical insight refers back to cluster theory and provides valuable contribution to the studies of Marshall, Porter and Krugman who implied the importance of collaboration and shared knowledge by indicating the role of communication. Only if the cluster's members are aware of their environment, collaborations and entrepreneurial opportunities arise.

Furthermore **external communication** plays an important role as it connects the cluster to the outside world and allows collaborations with other clusters. External communication also has a representative role, aiming to attract potential members or financial supporters. A public relations consultant in Cambridge explained: "For Cambridge and the Cluster, to get this information out there is really key because you have people from around the world that are coming to Cambridge to look at it, and some are coming here to invest, that's because they are hearing about it, they are hearing the positive news. This is one of the key places in the world. If we didn't communicate that would they think to come to Cambridge? It's really important to get that message across". The strategic communication to the outside attracts attention from all over the world and thus positions the clusters in a global competition. This result can be drawn back to theoretical position of gatekeeper studies as discussed by Dang et al. (2011) and Lezaric et al. (2008), who highlight the importance of creating channels not only inside but also outside the cluster. By representing the cluster to the outside world, communicators attract potential members and financial supporters.

To communicate successfully to the inside and the outside of the cluster, communicators have to create a **common language** by explaining complex innovations in order to make them accessible to a broader audience. A public relations consultant in the Cambridge cluster describes this process as "translation": "This is translation. The raw material compared to what is then written, you have to translate what can be something quite complex and put it into language. That is part of the communication, getting the language right that you are not dumbing down the technology or the invention but in that you are still caring up consistent messaging so that people will pick up and start to understand." Only by means of public communication, internal and external stakeholders will be able to comprehend unknown innovations. Communicators have to make sure that the message is consistent and understandable so that it reaches the recipients successfully. This has been discussed by Nordfors (2004a, b) in terms of journalism. Anecdotal evidence shows that also public relations consultants are aware of the importance of "translation" and pursue this aim and thus contributes to the research on innovation communication.

The identification of communicators' aims in innovation clusters make strengths and weaknesses of clusters visible and point to specific improvements in terms of internal and external communication and the creation of a common language. Understanding the role of communicators in innovation clusters provides first insights to the interrelations of communication and a cluster's development. This understanding allows future research on the dynamics of communication and creates a new perspective on the evolution of innovation clusters.

Bibliography

Asheim, B.T., Gertler, M.S.: The geography of innovation. In: Fagerberg, J., Mowery, D.C., Nelson, R.R. (eds.). The Oxford Handbook of Innovation, pp. 291–317. Oxford University Press, Oxford (2006)

Baerns, B.: Öffentlichkeitsarbeit oder Journalismus. Zum Einfluß im Mediensystem. Verlag Wissenschaft und Politik, Köln (1985)

Bentele, G., Liebert, T., Seeling, S.: Von der Determination zur Intereffikation. Ein Integriertes Modell zum Verhältnis von Public Relations und Journalismus. In: Bentele, G., Haller, M. (eds.), Aktuelle Entstehung von Öffentlichkeit. Akteure, Strukturen, Veränderungen, pp. 225–250. UVK, Konstanz (1997)

Burt, R.S.: Structural Holes. Harvard University Press, Cambridge, MA (1992)

Burt, R.S.: The social capital of structural holes. In: Guillén, M.F., Collins, R., England, P., Meyer, M. (eds.) New Direction in Economic Sociology, pp. 201–247. Russell Sage Foundation, New York (2001)

Clark, J.: Tech country report. Retrieved from: http://admin.bvca.co.uk/library/documents/Tech_Country.pdf (2013). [13.05.2013]
Dang, R.J., Mortara, L., Thomson, R., Minshall, T.: Developing a technology intelligence strategy to access knowledge of innovation clusters. In: Hülsmann, M., Pfeffermann, N. (eds.) Strategies and Communications for Innovations. An Integrative Management View for Companies and Networks. Springer, Berlin (2011)
Esser, F.: Die Kräfte hinter den Schlagzeilen. Englischer und deutscher Journalismus im Vergleich. Verlag Karl Alber, München (1998)
European Commission: Towards world-class clusters in the European Union: implementing the broad-based innovation strategy. *European Commission*. SEC(2008) 263 (2008)
Granovetter, M.S.: The strength of weak ties. Am. J. Sociol. **78**(6), 1360–1380 (1973)
Karlsson, C.: Handbook of Research on Cluster Theory. Edward Elgar Publishing, Cheltenham (2008)
Krugman, P.: Increasing returns and economic geography. J. Polit. Econ. Geogr. **99**(3), 483–499 (1991)
Krugman, P.: Complex landscapes in economic geography. Am. Econ. Rev. **84**(2), 412–416 (1994)
Lezaric, N., Longhi, C., Thomas, C.: Gatekeepers of knowledge versus platforms of knowledge: from potential to realized absorptive capacity. Reg. Stud. **42**(6), 837–852 (2008)
Marshall, A.: Principles of Economics. Macmillan, London (1890)
Marshall, A.: Industry and Trade. Macmillan, London (1920)
Nordfors, D.: The role of journalism in innovation systems. Innov. Journal. **1**(7/8), 1–18 (2004a)
Nordfors, D.: The concept of innovation journalism. And a program for developing it. Innov. Journal. **1**(1), 1–14 (2004b)
Nordfors, D., Ventresca, M.: Innovation journalism: towards research on the interplay of journalism in innovation ecosystems. Innov. Journal. **3**(2), 1–18 (2006)
Nordfors, D., Uskali. T. (forthcoming). The role of journalism in creating the metaphor of Silicon Valley, 1–20
Nordfors, D.: Innovation journalism, attention work and the innovation economy. A review of the innovation journalism initiative 2003–2009. Innov. Journal. **6**(1), 1–46 (2009)
OECD: Cluster policy and smart specialisation. In: OECD Science, Technology and Industry Outlook 2012. OECD Publishing (2012)
Pfeffermann, N.: Innovation communication as a cross-functional dynamic capability: strategies for organizations and networks. In: Hülsmann, M., Pfeffermann, N. (eds.) Strategies and Communications for Innovations. An Integrative Management View for Companies and Networks, pp. 257–292. Springer, Berlin (2011)
Polanyi, M.: Personal knowledge: towards a post-critical philosophy. Routledge & Kegan Paul, London (1958)
Porter, M.: The competitive advantage of nations. Harvard Bus. Rev. 73–93 (1990)
Porter, M.: Clusters and the economics of competition. Harvard Bus. Rev. 77–90 (1998)
Porter, M.: Location, clusters and company strategy. In: Clark, G., Gertler, M., Feldman, M. (eds.) The Oxford Handbook of Economic Geography, pp. 253–274 (2000)
Pürer, H.: Publizistik- und Kommunikationswissenschaft. Ein Handbuch. UTB, Konstanz (2003)
Preston, P.: Making the News: Journalism and News Cultures in Europe. Routledge, London & New York (2009)
Rössler, P.: Agenda Setting. Theoretische Annahmen und Empirische Evidenzen einer Medienwirkungshypothese. Westdeutscher Verlag, Opladen (1997)
Shoemaker, P.J., Reese, S.D.: Mediating the Message: Theories of Influence on Mass Media Content, 2nd edn. Longman, White Plains, NY (1996)
Teece, D.J., Pisano, G., Shuen, A.: Dynamic capabilities and strategic management. Strateg. Manag. J. **18**(7), 509–533 (1997)

Uyarra, E., Ramlogan, R.: The effects of cluster policy on innovation. UK National Endowment for Science, Technology and the Arts (NESTA). Working Paper 12/05 (2012)

Weischenberg, S.: Journalistik. Medienkommunikation. Theorie und Praxis. Opladen, Medienetik, Medieninstitutionen (1992)

Zerfass, A.: Innovation readiness. Innov. Journal. **2**(8), 1–27 (2005)

Chapter 13
Integrated Communication in the Innovation Process—An Approach to Integrated Innovation Communication

Manfred Bruhn and Grit Mareike Ahlers

Abstract As innovation processes become accessible to consumers and other interested public parties in the sense of open innovation, innovation communication faces new challenges. The interface existing between internal and external commercial interests must be systematically coordinated in order to ensure that the development process is efficient and effective and that the developed innovation is successfully implemented on the market. Innovation communication plays a key role here in securing that the points liaising internal and external interests are integrated over the length of the innovation process. This is a complex task and involves coordinating communication objectives and publicity, integrating communication instruments and, not least of all, aligning the numerous target groups—from the research specialists in R&D to the Internet bloggers. This chapter uses impulses stemming from integrated communication to develop a phase-oriented concept for integrated innovation communication that is capable of guaranteeing a systematic coordination of the interfaces involved and providing a central support in promoting a satisfactory outcome for the innovation process.

13.1 New Challenges for Communicating Innovations

13.1.1 From Closed Innovation to Open Innovation

Corporate innovation management has always faced multiple challenges. Progressive internationalization, shortened product lifecycles as well as a dramatic growth in information relating to products and services pose substantial risks for product development.

The sobering failure rates for new products (up to 70%) in the consumption goods markets clearly indicate that innovation processes, more often than not, fail

M. Bruhn
University of Basel, Basel, Switzerland

G.M. Ahlers (✉)
Prof. Bruhn & Partner AG, Basel, Switzerland

N. Pfeffermann and J. Gould (eds.), *Strategy and Communication for Innovation*,
DOI 10.1007/978-3-319-49542-2_13

to meet the needs of the target groups (N.U. 2006). In this context, the quest for strategies promising sustainability and promoting innovation has become critically important, especially with regard to the organization of the innovation's development proposal.

Up until a few years ago, the innovation process was primarily aimed at the company's internal processes, being organized as a so-called closed innovation. According to this concept, the company's own staff, in particular its researchers and developers, are seen as the central contributors to innovation. The ideas produced are handled confidentially and the new products are developed on the drawing board. Collaboration with clients and research institutes is not prohibited by the process, but usually only takes place at the beginning of the innovation process. The problem-solving phase, and thus the innovation process itself, is conducted extensively within the closed confines of the company (Chesbrough 2006; Trott 2011; Herzog 2011).

While a "closed" innovation policy can be seen as offering the advantages of minimal complexity and maximal controllability, the disadvantages and risks of such a strategy become more apparent in a global, dynamic competitive environment. On the one hand, internal innovation demands that a company invests heavily in accumulating internal know-how and establishing internal technical facilities. On the other hand, an absence of an adequate information transfer between product developers and product users presents the danger that innovations will not correspond to client needs and that development processes will be protracted (Lindman 2002; Ernst 2004; Reichwald and Piller 2009; Herzog 2011).

The concept of open innovation was introduced in 2003, extending the discussion beyond the confines of closed innovation (Chesbrough 2006; Prahalad and Krishnan 2009; Chesbrough 2012). The fundamental idea here is to establish a corporate innovation policy that is collaborative in its outlook and that aims to integrate internal and external stakeholders within the innovation process, thereby closing gaps between technology and the market. Innovations, ready for market launch, are no longer developed solely within the confines of the corporation, but are the product of processes that dovetail internal and external processes. New external parties such as suppliers, competitors, end clients, and online communities (Ayuso et al. 2006; Mahr and Lievens 2012) are increasingly joining the ranks of the traditional innovation partners (research institutes, market research consultancies, etc.). In the United States, 30% of a basketball community will be engaged in innovation activities, for example. Involvement in the innovation process is, in fact, the main stimulus driving the majority of members to join the club (Jawecki et al. 2009). In particular, the new use of consumer-based knowledge and creativity is the result of developments in the media environment, in the applications of social networking, and the "you-too"-Internet, now such important aspects of daily experience. Modern forms of innovation research such as "crowdsourcing" (e.g., Howe 2008; Poetz and Schreier 2012), Consumer-Co-Creation (e.g., Payne et al. 2008; Filieri 2013) and Nethnography (e.g., Kozinets 2002; Kozinets 2012) are subsequent developments of this trend and an expression of a change in attitude to innovation policy.

13.1.2 Consequences for Innovation Communication

In addition to reflecting an open and modern innovation culture, open innovation policy is charged with the task of identifying all the relevant internal and external stakeholders and drawing them into the innovation process (Trott 2011). While within the closed innovation concept most interactions take place within the company, the open innovation process is characterized by numerous real and virtual interfaces liaising internal and external activities. This requires a process of control and alignment that spans the length of the innovation process in order to avoid friction losses that would jeopardize the opportunities that the open innovation policy has to offer (Reichwald and Piller 2009; Vesshoff and Freiling 2009). Effective and efficient communication management is therefore essential to the success of an open innovation policy.

From this perspective, innovation communication assumes a new importance in the context of the innovation process. Its function is no longer to simply communicate innovations to the external environment, but is now also responsible for aligning all internal and external interfaces throughout the innovation process. Academic and business communities are increasingly discussing the consequence that this has on the organization and management of innovation communication (see Mast and Zerfaß 2005; Zerfaß and Möslein 2009 as well as the study by Zerfaß and Ernst 2008). So far, no consistent approach has yet established itself definitively.

Many valuable impulses for innovation communication can be drawn from integrated communication. Integrated Communication Management is primarily aimed at aligning all internal and external communication instruments and messages—with the goal of conveying a consistent image of the communication's reference object (see in detail Bruhn 2009). The referenced object of the communication might be the company itself, a specific trade name, a product or, as in this case, an innovation. The alignment decisions for integrated communication are primarily based on conceptual criteria. Later, when it comes to the innovation's implementation phase, organizational and company-specific measures also have to be considered. The next section discusses the approaches for innovation communication derived from this in more detail.

13.2 The Concept of Integrated Communication as the Basis for Integrated Communication in the Innovation Process

Starting with Communication Deficiencies

With the advent of the twenty-first century, integrated communication is no longer a novelty: It is a well-recognized and long accepted necessity. As far back as the mid-1970s, the need to integrate communication instruments and measures has been defined as the most important challenge facing companies. This imperative has not lost any of its force; instead, it has gained more and more momentum. Many developments have contributed to this: in particular, increasing competition in communications, the atomization of the media, the flood of information and stimuli inundating the consumer, as well as the dynamic increase in new channels of communication such as the social media. Developments such as these present new tasks and challenges for the integration of communication measures (Peltier et al. 2003).

A central task of integrated communication is to eradicate *communication deficiencies* in companies. Communication deficiencies occur where various implemented measures are not harmonized with regard to content, form, or time. Certain communication interfaces present a basis for systematizing communication deficits: the relationship between locations where communication takes place (internal and external) and the relationships between levels where communication takes place (horizontal and vertical).

Classic deficiencies in company communications occur where there is a mismatch between *internal and external communications,* where a company's staff is not informed about proposed communication measures. *Internal communication deficiencies* may also occur, either horizontally in the alignment of inter- or intra-departmental communications or vertically in the hierarchical communication of messages between functions where the messages between staff and management do not share consistency in form or content. Communication deficiencies in *external communications* on a horizontal level will occur because a company employs different market-related tools for communication, which are nonaligned with regard to content, form, or time. On the other hand, communication deficiencies may occur on multiple market levels because the marketing intermediaries that are engaged to provide services to supply end users have not had the required message content communicated to them (Johnson and Chang 2000; Sieg et al. 2010).

The deficiencies sketched here serve to highlight a corporation's need for integration and thus the necessity of instituting integrated communication. In view of the increasing importance of open innovation, the need for integration can be extended to the innovation process. The more acute communication deficiencies are between internal and external, and horizontal and vertical interfaces, the greater the danger is that important information will fail to be exchanged or will be late, that ideas will get lost, innovations will miss their targets, or costly time will be wasted (Sleeswijk Visser et al. 2007; Vesshoff and Freiling 2009).

Conceptual, Organizational, and Employment Approaches to Integrated Communication

The concept of integrated communication provides a framework for making conceptual, organizational, and employment decisions to eradicate communication deficiencies and achieve the optimum level of effectiveness and efficiency in communications.

On a *conceptual level,* three types of *communication integration* take place: content, form, and time integration. While the first type is principally concerned with guaranteeing consistency by using thematic associations, formal integration aims to strengthen recognition by employing uniform design principals. Time integration deals with aligning communications instruments during and between planning periods. In the framework of innovation communication, content and time integration are particularly important, where content integration presents the greatest challenges. Here, clearly defined guiding principles have to align communication goals, messages, and measures across all points of contact between internal and external groups. This process employs a so-called *conceptual framework of integrated communication*: On the one hand, this gives general rulings on the composition of communication content, and on the other hand, it provides explicit instructions on how the communication content should be implemented in daily work.

Beside the planning measures, the *organizational implementation* of integrated communication is extremely important. This entails setting up a management process that is capable of involving all the relevant departments and employees in the communication effort. This has implications for organizational implementation: It means that all overly rigid forms of structural organization have to be dropped and replaced by flexible methods of process organization, in particular process management and forms of team organization (Ahlers 2006).

Employment measures are closely associated with organizational structures and processes. These have to be specified within the context of the individual company, whereby integrated communication basically supports the case for institutionalizing the roles of a superordinate communications manager or team of specialists (Sonnenwald 1999; Johnson and Chang 2000).

This position functions as a kind of coordination office and is responsible for the supra-disciplinary planning, implementation, and monitoring of communication (Sonnenwald 1999; Johnson and Chang 2000). Basic conceptual as well and organizational and employment information can be adopted from integrated communication for innovation communication. The integrated innovation communication process will illustrate this more clearly.

Development of a Process for Integrated Innovation Communication

Identifying the relevant integration needs is a precondition for planning and implementing innovation communication successfully. The innovation process maps the way for this and has to be specified more clearly for this purpose; it has to be structured within an innovation process for development and implementation (see Fig. 13.1).

The *innovation development process*, in this context, refers to the cycle of innovation during which the integration requirements are considered for the generation of the ideas, for the selection of a specific idea, and for its realization. The innovation *implementation process* covers the product or service lifecycle and thus

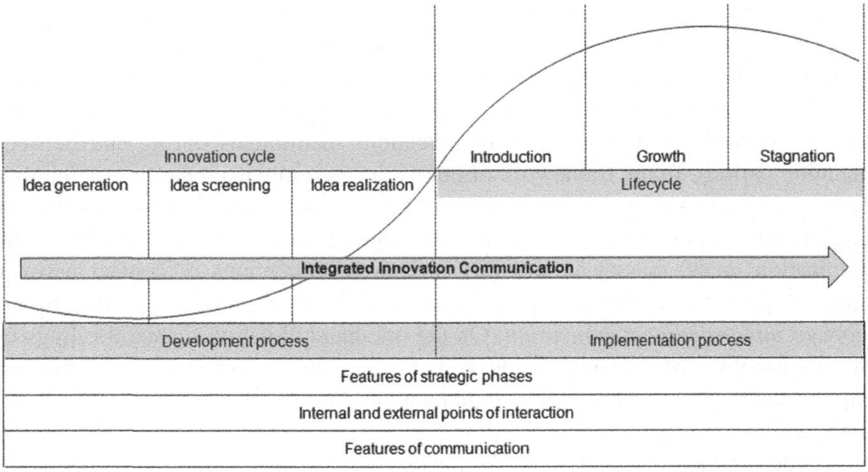

Fig. 13.1 Integrated innovation communication process

also the transfer of the innovation to the market. The requirements for integration are assessed along the course of the innovation's market life from market introduction, to market growth, and on to stagnation. In accordance with this systematization, the concept of integrated innovation communication (IIC) can be defined as follows:

> *Integrated innovation communication* is a process that aims to identify internal and external contact points in the development and implementation process of an innovation, as well as to develop and implement communicative measures that guarantee the alignment of these interaction points in order to achieve a maximum level of development efficiency and effectiveness internally as well as optimal market saturation externally.

Integrated innovation communication covers a multilevel process, the phases of which are marked by different characteristics and points of interaction. Consequently, communication plays a particular role in the individual phases, being defined by different goals and target groups as well as by specific communication instruments and contents. The interplay of these factors, again, has an influence on the alignment and integration needs between departments and employees as well as between internal and external stakeholders. Table 13.1 summarizes the features of innovation communication in the innovation development and implementation process. The following sections will discuss the individual phases more closely.

Table 13.1 Features of integrated innovation communication in the development and implementation process

Features	Idea generation	Idea screening	Idea realization	Introduction	Growth	Stagnation
Points of interaction						
Internal points of interaction	Organization of the innovation process, idea acquisition procedure	Appraisal and/or broad and narrow selection of the product ideas	Product test procedure	Product training	Interaction with product users	Interaction with product users
External points of interaction	Information about the innovation project, activation of the innovation community, acquisition of ideas	Broad and narrow selection of product ideas, feasibility studies	Testing procedure for products	Distribution of the new product on the market, communication and marketing mix	Sales process, communication and marketing mix, Interaction with product uscrs	Sales process, interaction with product users
Features of communication						
Internal communication target groups	R&D. market research. Customer Service, product marketing, etc.	Compare target groups for acquisition of ideas as well as production and finance areas	Person responsible for pro duct testing	Product marketing, Product-PR, Customer Sen/ice, Distribution, etc.	Product Marketing. Product-PR, Customer Service, distribution, etc.	R&D. market research, customer service, product marketing, etc.
External communication target groups	Research institutes, consultants, lead-users, clients, opinion makers, online community, etc.	Innovation community, selected target groups for special test processes	"True" customers in e.g.. laboratory and field experiments	Innovators, early adopters, first customers	First and subsequent buyers, broad mass, product users, online community	Latecomers, second- time purchasers, broad mass, product users, online community
Communication objectives	Identification and motivation of the innovators, stimulating a fast rate of ideas,	Steer and control idea testing, emotional bonding of innovation community,	Acquiring suitable test individuals, information exchange between test procedures, and efficient feed back	Announcements, stimulating initial sales. activating networks, positive word-of-mouth. contagious effects	Emotionalization, Raising/stabilizing purchase frequency, observing	Cushioning emotionalization, identifying the reasons for the fall in sales, and

(continued)

Table 13.1 (continued)

Features	Idea generation	Idea screening	Idea realization	Introduction	Growth	Stagnation
	steering the production of ideas	generating positive word-of-mouth			user-buzz on the Internet	improvement opportunities
Internal communication instruments	Personal communication, internal blogs	Personal communication	Personal and written communication/documentation	Training, workshops, personal communication, Intranet, internal blogs	Internal blogs	Personal communication
External communication instruments	Relevant homepages, blogs, communities, social networks, special online platforms	Blogs, communities, social networks, special online platforms, personal communication	Blogs, communities, social networks, special online-platforms, personal communication	Media advertising, sales promotions, direct marketing, online-communication, social media, etc.	Supportive media advertising, sales promotions, online monitoring, social media	Online monitoring, Social Media, personal communication
Communication content	Innovation framework, stimuli	Realistic assessment of innovation idea, emotional bonding of the innovation community.	Data and facts of the test results	Sales arguments, product merits, price advantages, product image	Consolidating sales arguments, reaction to user responses	Reaction to user-responses
Integration requirements	In particular, alignment between internal and external acquisition of ideas	Coordination of internal and external appraisals of the idea	Information exchange between product development and product testing	Coordination or market communication, alignment of internal and external communication	Coordination of market communication, harmonization of user responses and internal communications	Harmonization of user responses and product improvements

13.3 Integrated Innovation Communication in the Development Process

Within the framework of innovation planning, multilevel planning processes are employed to provide a systematic base for decision-making in product innovations. Classically, the process follows an established sequence of phases: searching for product ideas, gathering a broad selection of ideas, testing product concepts, making a narrow selection, and finally introducing the new product onto the market (e.g., Homburg 2012; Meffert et al. 2012). For simplicity, a three-stage innovation process can be introduced here, consisting of three phases: "idea generation", "idea screening", and "idea realization".

The Idea Generation Phase

The phase of generating new product ideas marks the beginning of the development process. Product ideas are both systematically gathered and generated here.

The objective here is to accumulate as many ideas as possible to compensate for subsequent rejections.

The company can decide to generate its ideas for the product or service either internally or externally. The varying number of internal and external *interaction points* that have to be managed by communications depends of the degree of "openness" that the innovation process has. Information communicated here relates to initiating the innovation process, giving instructions on procedure, to actually recording product ideas.

The *communication target groups* are closely associated with the specified interaction points. The internal staff engaged here is primarily drawn from R&D, Market Research, Customer Service and Product Marketing.

Ideally, a cross-functional idea pool should be established, enabling a constant exchange of ideas and information. In this phase, companies typically access external providers such research institutes or market researchers and collaborate with information brokers or work together with lead-users (Lilien et al. 2002; Ernst et al. 2004; Lichtenthaler 2011). In line with the concept of open innovation, an ever-growing body of customers, opinion makers, and interested parties are also involved in generating ideas. This open procedure offers the company substantial advantages; for example, an increased wealth of ideas, sharpened concentration on target groups, and speed in producing ideas. On the one hand, the "right" external agents have to be drawn into the development process and, on the other hand, the "right" ideas have to be generated for these advantages to be exploited. In this phase, communication goals aim at identifying suitable external innovators as well as motivating them to participate in the innovation process. A major goal of communication is to elicit the best ideas in the shortest time. In addition to motivating the "innovation community", communications have a lead function in issuing timely and detailed information regarding the innovation's requirements and objectives, be these "hard" factors such as technical framework conditions or "softer" factors such as "the cultural fit" of the product ideas (Ernst 2004).

At the onset of the innovation process, *communications instruments* are also employed to enlist the help of external parties in generating ideas, and in organizing and directing the process. To accomplish this, the development proposal has to be publicized to the relevant target groups through technology or design institutes and blogs or social networks such as Facebook and StudiVZ in order to mobilize "creative masses" by "crowdsourcing" (Füller and Mühlbacher 2004; Gardlo et al. 2012). OSRAM, for example, brought their development of a new "emotional lighting" concept to the attention of more than 200 websites, communities, and blogs and was able to attract 910 participants from nearly 100 countries in under 11 weeks to participate in the development of their product. While OSRAM brought its own online platform to life, in the mean time numerous publically accessible "idea marketplaces" have sprung up (e.g., InnoCentive, Fellowforce, or Openinnovators), where companies can call upon their target groups to submit ideas for product development. The possibilities offered by Web 2.0 extend even further by offering internal company support in the innovation process as well (McAfee 2006; Bertoni and Chirumalla 2011). In addition to the classic forms of personal communication, special blogs can also be set up within innovation teams for the purpose of exchanging and collecting product ideas. Whether one chooses to set up a blog for specialists or a blog inviting staff from other departments will depend on the degree of openness that innovation process offers.

With regard to communication contents, during the phase of idea generation, communication focuses on directing the quantity and quality of the product ideas. In addition to stimulating participation in the innovation process, information relating to the central framework conditions for the development process should be fed into the "fuzzy front end" of the innovation process as early as possible. This applies internally, but at the same time pays particular attention to the external "innovation community".

In connection with the communication deficiencies discussed earlier at the beginning of this paper, the greatest *need for integration* arises in the idea generation phase in order to align internal and external communication. Here, especially, the product ideas that have been generated internally have to be aligned with those generated externally so that they can be jointly evaluated later on in the process, and, if necessary, developed further. If communication deficiencies arise at this stage, there is the danger that promising ideas (in particular, those acquired externally) may go unnoticed and "founder". Apart from aligning internal and external communication processes, the requirements for innovation can also be positioned internally: This may be done horizontally within the individual development teams, or vertically by aligning operational units with management. This insures that the innovation framework conditions are clearly communicated (Rothwell and Robertson 2002). Improvements to external communications should also be considered, where the use of different communication facilities causes deficiencies (e.g., idea generation via blogs, communities, etc.). The more comprehensively that a company can satisfy its coordination needs in the idea generation phase, the better prepared it will be in the idea screening phase.

Idea Screening Phase

In the framework of the idea screening phase, the main task is to sort out less promising ideas effectively and efficiently, reducing the risk of failure so that available resources can be concentrated on the ideas that have the best chance of success (Trott 2011).

Points of interaction occur throughout the idea screening process from the broad and narrow selection of ideas through to the feasibility check. While the phases of the selection processes have both internal and external interaction points, the points of interaction in the feasibility analysis phase are mainly internal.

During the idea screening phase, the *principle communication target groups* consist of employees who have been commissioned with individual ideas. In addition to the R&D departments, these areas also have direct contact with customers and have first-hand experience of the product's use. At this stage, the production area and the finance area should also be viewed as communication target groups and thus guarantee the smooth running of the economic feasibility study. With regard to the external online community, communication at this stage concentrates on the active innovation community: Starbucks, for example, at www.mystarbucksidea.de asks its customers and readers not only to suggest product ideas, but also to discuss and appraise the ideas of other participants. Moreover, select customers may even be personally contacted and enlisted for special phases of the idea screening process, such as (Web-based) conjoint analyses (Gustafsson et al. 2007).

The *communication goals* of this stage are concerned principally with directing and controlling monitoring activities. It is not only important for a company to gather information: The information has to subsequently be actively fed back into the innovation process so that a real exchange of ideas is guaranteed. This is important as it demonstrates the company's recognition of the innovation community's contribution and cements its commitment to the on-going innovation process. In this phase, generating positive word-of-mouth is an important communication goal. Members of the innovation community should be inspired to publicize the innovation process in their context and, in so doing, either bring more innovators on board or arouse curiosity in the new product (Horbel and Woratschek 2009).

Personal communication has an important role internally as one of the *communication instruments* in the monitoring phase. Communication can be conducted externally via online platforms. Where select external target groups are to be drawn more tightly into the process, a transition to more personal forms of communication is recommended.

The content of communication can be subdivided into rational and emotional messages during the idea screening phase. While, on a rational level, products are primarily appraised on a technical basis, companies will employ emotional communication to strengthen and extend the bonds they have with innovation communities, not simply allowing the associations to fade out once the idea has been acquired (Kunz and Mangold 2004; Teichert et al. 2004). It is not unusual for communication managers to assume the role of moderator during this phase and to mediate between parties when critical idea appraisals take place.

The *requirements for integration* during the monitoring phase primarily concern coordinating the evaluation of the idea. A large number of employees and departments may be involved in the innovation process, which spans from product idea selection to the feasibility analysis, and only an on-going and comprehensive information policy can ensure that ideas are monitored and justified as being suitable and consistent. This is all the more important when external target groups are also drawn into the process. Basically, the requirements for integration are satisfied in this phase, once all the relevant ideas have been thoroughly checked, all the internal and external innovators have been informed about the monitoring process and the selected ideas have gained a high level of acceptance.

Idea Realization Phase

The idea realization phase focuses on building and testing prototypes as inexpensively and as quickly as possible, terminating the development process: The aim is to secure the product's market success with appropriate budgeting and to plan its market introduction.

Interaction points between internal and external target groups arise principally during the individual product tests of the idea's realization. A wide range of procedures is available here, from the concept test to mini test markets, each of these occasioning different interactions (Mahajan and Wind 1992).

When following a closed innovation approach, the innovation process is opened for the first time to technically unqualified people during the product tests. Acceptance tests are conducted to determine specific design features such as color matches and packaging sizes. The importance of external interaction points is well established with open innovation. In this phase, the communication target group is no longer the Internet innovation community, but is now made up of "real" customers who actually test the products in the field or in laboratory experiments.

During the idea realization phase, communication measures concentrate on promoting the efficiency of the realization process. *Communication goals* are particularly concerned with sustaining the exchange of information between staff involved in the realization process. Feedback from the test procedure has to be obtained, evaluated, and fed back into product development process. With regard to external target groups, communication primarily focuses on securing suitable test individuals as well as informing the innovation community about main developments and sustaining their commitment.

Communication instruments that transmit information on the results of the relevant product tests are especially important during the realization phase. This may be communicated personally and also impersonally in the form of protocols and experiment documentation. Established blogs and online forums can be used for external communications. For the recruitment of test subjects, individuals must be contacted personally or at least in writing. The larger the company's data repository of customer contacts is, the smoother this process usually is.

The idea realization phase is all about exchanging information on the test results and forwarding suggestions for improvements. *Communication contents* consist of rational data and facts. In spite of the large number of external interfaces, the requirements for integration in the realization phase should, primarily, take place internally. Here, a continual exchange of information between the development team and the managers in charge of the test procedure must be guaranteed. The external requirement for integration consists in keeping the innovation community regularly informed about the realization process.

Once the test phase has been successfully completed, commercialization begins when the new product is introduced to the market. Innovation communication as part of the development process is followed by innovation communication as part of the implementation process.

13.4 Integrated Innovation Communication in the Implementation Process

The implementation process for innovation communication consists of the communication phases that involve internal and external target groups that are involved in scheduling the market development of a product or service. This process usually consists of the phases: introduction, growth, maturity, saturation, and decline (i.e., Meffert et al. 2012, p. 849f.; Bruhn 2011, p. 63f.). To simplify this somewhat, a three-stage lifecycle is also introduced here, consisting of the phases: introduction, growth, and stagnation (see, Table 1).

While innovation communication still focuses on a limited target group in the development phase—even if this is relatively broad where open innovation is concerned, in the implementation phase the communication process is open to as wide a public as possible in order to achieve the greatest possible market saturation.

Introduction Phase

The introduction phase is often the most important phase for a new product. Here one decides whether the original product ideas can be feasibly turned into a financially successful product. Often the introduction phase carries the greatest marketing and communication costs, which often means calculating with losses. It is all the more important, therefore, that communication measures are specified by precise directives and synergy effects are exploited.

Interaction points are established in the introduction phase through the sales and distribution of the new product on the market as well as through the launch of supportive, sales-promoting communication activities.

Often these measures are aimed at external *communication target groups* during the product's introduction. In particular, innovators and early adopters should be identified as potential customers and measures aimed at stimulating purchases should be used to provide them with information about the new product. In addition to potential customers, media representatives and other opinion makers should be considered as an important target groups during the introduction phase. Ideally, these target groups will have already participated in the innovation process, producing synergy effects across the different phases. In addition to customers and opinion makers, internal target groups should be purposefully involved in innovation communication during product introduction. Where the communication of innovation is directed towards the outside arena, (product-) marketing and (product-) PR assume a central role. These departments can only fulfill this function however, if they have received adequate and timely information about the new product. The same applies to staff working in Sales and Distribution or in Customer Service or all staff in general who deal with customers and make use of the new product themselves or issue information on its use.

A central *communication goal* during the introduction phase is the goal of advertising the new product, its features and advantages in order to stimulate trial samples and first purchases. In addition to this, communication measures aim at activating opinion makers and generating positive word-of-mouth and contagion effects. Opinion makers are not necessarily or solely taken from the ranks of press representatives, but are to be found more and more in the Internet, and often have a substantial negative or positive influence on product appraisals (Helm et al. 2010). Only recently the biggest diaper innovation in 25 years by Procter & Gamble almost failed due to negative word-of-mouth on the Internet claiming the new pampers would cause itchiness (N.U. 2010).

When selecting *communication instruments*, the company has more or less the whole spectrum of marketing communication at its disposal in the introduction phase. This spans from traditional media advertising, to direct marketing and sales promotion at the point of sale, and on to the interactive possibilities of online and social media communication. The possibilities offered by social media can also be employed for addressing internal target groups, promoting the new product through blogs, for example, and by simultaneously setting up a platform for exchanging experiences (McAfee 2006; Bertoni and Chirumalla 2011); however, workshops and user training courses are necessary for conveying more detailed product information relating to particularly demanding business areas.

Communication content in the introduction phase features both factual and emotional messages. While factual *communication content* deals primarily with sales arguments relating to product features, price advantages, and application possibilities, emotional communication content deals with building up a particular image of the new product.

The *requirements for integration* during the introduction phase correspond largely to the classic alignment requirements for integrated communication: The coordination of the content and scheduling of all the market communication measures is of particular importance in order to create a uniform and convincing market appearance. Here, it is equally necessary to integrate internal and external communication measures in order to guarantee that employees are informed about the product features that are advertized on the market and respond adequately to them.

Growth Phase

The use of marketing and communication measures usually increases awareness of the product, leading to above-average rates of growth.

The *points of interaction* in this phase correspond substantially to those in introduction phase, whereby, having gained initial experience in using the new product, the focus, here, is on the external interaction points.

Communication activities thus focus on external *target groups* in the growth phase. Here, early adopters are no longer of prime interest, but rather first-buyers and recurrent purchasers who contribute to the increasing market saturation of the product. Media representatives usually become less important in this phase, while the (critical) online community is more important. In view of the fact that product descriptions gain a high level of credence in online-user forums and other communities, they can both positively and negatively impact the success of a product.

On the one hand, *communication goals* in the growth phase are associated with increasing emotionalization, and an increase in purchasing frequency. On the other hand, the observation of Internet target groups once again becomes more important in this phase, for the purpose of identifying evidence of product faults and clues to product improvements.

Communication investments are generally recouped in the growth phase, whereby, depending on the level of communication pressure one wishes to apply, media advertising and sales promotion are still employed as *communication instruments*. A professional monitoring system should be set up for observing target groups in the Internet and be capable of not only analyzing product commentaries but also of entering into direct dialogs with the target groups (Berkman 2008).

In the growth phase, companies do not usually channel new *communication content* into the communication process. Here, it is more a matter of ensuring that the messages already communicated in the introduction phase are consistently repeated and reinforced. Beyond this, communication content consists of consumer comments, complaints about insufficient information ascribed directly or indirectly to the company, and how the company responds.

In the growth phase, *integration requirements* are concerned with aligning the different forms of market communication. At the same time, the need for alignment between external and internal communication increases as commentaries on products are picked up externally and have to be referred to internal desks.

Stagnation Phase

In the stagnation phase product turnover becomes regressive for the first time. Market potential has been exhausted and the market is saturated. Consequently, the intensity of communication activities also declines.

This does not mean, conversely, that the *interaction points* between internal and external target groups become less important: On the one hand, the sales process has to be precisely aimed at target groups just as before. On the other hand, it is now important to engage more intensively with product users in order to discover the reasons why sales have fallen as well as to identify new ways of improving or adapting the product.

External *target groups* therefore are of central importance during this phase. Current product users should be observed on online platforms and "listened to", to gain indications of the product's weak points (Berkman 2008). Beyond this, one can seek direct contact with select customer groups in order to collect concrete details for improving specific product features. At the same time, internal target groups can also be drawn into the communication process in order to pursue possible (further) product development from within the firm. Most often, the core team of developers is permanently occupied with product improvements; however, in the phase, they should be more involved in confronting requests for improvement from external sources.

Communication objectives are met by following two tracks in the stagnation phase. On the one hand, they aim to stabilize sales at the highest possible level by using emotionalization to limit the trend. On the other hand, the company must prepare a transition from the implementation phase to a fresh phase of development by identifying the reasons for the drop in sales and seizing strategies to renew stimulus. This does not necessarily mean that a renewed innovation process will develop out of the stagnation: It is more likely that products will simply be improved or new varieties of the product devised.

In the stagnation phase, costly *communication instruments* are usually dispensed with; however, it is all the more important in this phase, to exploit forms of "passive social media communication" by monitoring, and communicating with members via user forums and blogs (Berkman 2008).

The communication content used in the stagnation phase corresponds to the content used in the growth phase. No new communication content is created; however, it is the way in which a company responds to its customers' enquiries—be these from the Internet, the company's Service Center or via sales field services—that is extremely important here. The more open and personal a company's approach to its customers is here, the more able it will be to win over customer support in a renewed process of innovation and improvement.

Where the *requirements for integration* had essentially concerned the alignment of external communications during the introduction and growth phases, in the stagnation phase the coordination between internal and external communications become much more important. This involves systematically gathering opinions expressed outside the company and presenting these to internal development teams. The achievement of this opens a path to recovery, from product stagnation to a new product idea.

13.5 Elements of Strategic Communication as Joining Brackets

The description of the individual phases of the innovation process highlights the complexity of the process, from idea generation through to market implementation. Coordination requirements arise in response to communication goals, communication instruments and contents, and, not least, in response to the communication target groups.

In order to guarantee the integration of these components, a strategic concept is required that can stipulate and coordinate communication programs that are consistent over the long term, plausible and synergistically aligned. The development of a comprehensive *integrated communication strategy* is core to such a concept. This strategy must apply throughout, irrespective of the actual phase of innovation communication, in specifying the common framework for all communication measures. There are fundamentally three core components for guaranteeing integration in this context:

1. *Strategic positioning of the innovation*: Strategic positioning is the desired image that a company aims to achieve for its innovation: what it wishes to communicate about its product. Here the company decides which of its new product's features should primarily be promoted, where the main customer benefit is, what advantages the innovation has over competitive products. Independent of how definitive and concrete the innovation's image is within the company and of how "open" the company's innovation process is, the innovation's strategic positioning can be decided early at the beginning of the development process or concretized during the idea generation phase. Strategic positioning provides a map for all of the communication measures and presents communications' overriding aim. To achieve this, it has to be formulated in general terms on a composite level so that it is not compromised by the component interests of specific target groups. The relevant features of the innovation have to be reduced to a "common denominator" (Fig. 13.2).

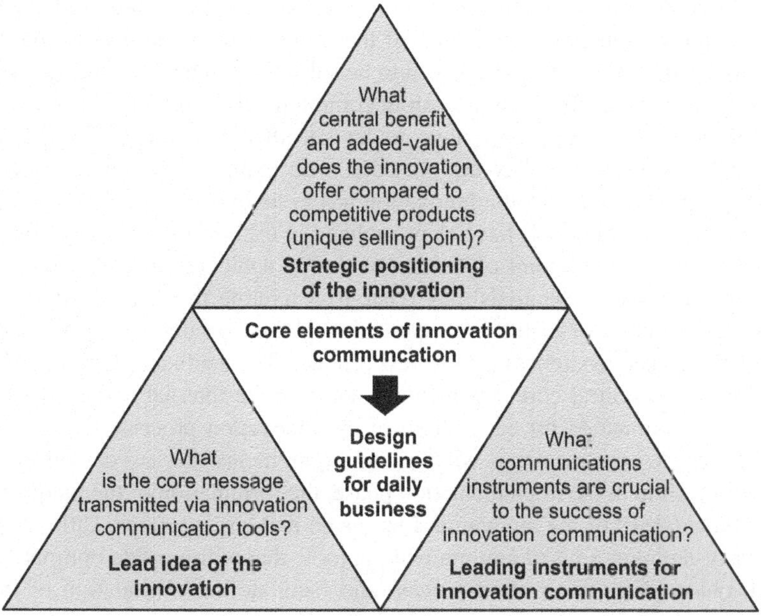

Fig. 13.2 Core components of an integrated innovation communication strategy

2. *Lead idea of the innovation*: The strategic positioning of the innovation has to recur in all communication messages. A lead idea is developed for the innovation for this purpose; i.e., a fundamental statement containing the most predominant features of the innovation. This lead idea has to be considered for all phases of the innovation process and lays the guidelines governing communication with both internal and external groups. The characteristic "optimized user friendliness" thus provides the innovation community with bearings for its activities in the development process as well as a core message for advertising the innovation in the implementation process.
3. *Specifying the leading communication instruments for innovation*: The palette of communication measures available to innovation communication is almost limitless today. However, these communication measures, which are so essential to the success of an innovation and which provide the crash barriers for implementing other communication instruments, have to be defined. Here, it is recommended that a maximum of two leading instruments are employed for the development and the implementation phase together. During the development phase, Internet forums and communities are suitable in the framework of open innovation processes, in addition to personal internal and external communications. Media advertising, as ever, holds a position of major importance in the implementation phase for many branches of industry, whereby forms of social media communication are becoming ever-increasingly important, in particular for stimulating contagion effects.

The three core elements outlined above, the definition of strategic positioning for an innovation, the formulation of a lead idea and the specification of leading communication instruments, constitute the essential guidelines for establishing consistent innovation communication during the development and implementation phase of innovation. The formulation of these guidelines must necessarily remain relatively abstract at this level and need to be substantiated and elaborated explicitly for the purpose of practical application in communication and innovation activities. These three strategic elements have to be specifically defined for each of the individual phases so that they are provided with explicit procedural instructions. The strategic positioning of the innovation has to be explicitly laid out, stipulating which communication goals have to be realized in each of the individual phases so that the strategic positioning of the innovation is achieved on the market. In the development phase, these goals may relate to activating the "right" external innovators, for example, or in the implementation phase, to building up a sufficiently high level of brand awareness for the new product. The lead idea should be "broken down" until the central core statements addressing the internal and external target groups are documented for each phase of the innovation process. In the development phase, the core messages contain the requirements that are demanded of the innovation, while in the implementation phase, the formulation of the unique selling propositions (USP) of the innovation focuses on real target groups. Ultimately, it is a matter of defining a kit of communication tools that will provide optimal support at each phase of the innovation process and facilitate the exploitation of synergy effects. This serves, particularly in the implementation phase of innovation communication, to substantially increase the effect of media advertising through the use of promotions and targeted PR-sales promotions.

The strategic components of innovation communication constitute the content of the mission statement for coordinating all communication activities within the framework of the innovation process. They establish the conceptual foundation for efficient and effective innovation communication and are of central importance to the innovation process. Apart from providing this conceptual basis, successful innovation communication also places demands on cultural, employment, and organizational measures within the company. The major success factors arising from a company's culture of open innovation are its willingness to collaborate across departmental boundaries as well as its cooperation with external innovators and the establishment of cross-functional forms of collaboration.

Communication thus presents companies, researchers, developers, and communication managers with numerous new challenges in the open innovation process. The opportunities and enormous innovation potential that are made available in this way can only boost impetus to embrace these challenges.

Bibliography

Ahlers, G.M.: Organisation der Integrierten Kommunikation. Gabler Verlag, Wiesbaden (2006)

Ayuso, S., Rodriguez, M.A., Ricart, J.E.: Using stakeholder dialogue as a source for new ideas: a dynamic capability underlying sustainable innovation. Corp. Gov. **6**, 475–490 (2006)

Bartl, M.: Netnography. Einblicke in die Welt der Kunden. Planung & Analyse **5**, 83–89 (2007)

Berkman, R.: The Art of Strategic Listening—Finding Market Intelligence Through Blogs and Other Social Media. Paramount Books, Ithaca, NY (2008)

Bertoni, M., Chirumalla, K.: Leveraging Web 2.0 in new product development: lessons learned from a cross-company study. J. Univ. Comput. Sci. **17**, 548–564 (2011)

Bruhn, M.: Integrierte Unternehmens- und Markenkommunikation, 5th edn. Schäffer-Poeschel, Stuttgart (2009)

Bruhn, M.: Marketing. Grundlagen für Studium und Praxis, 11th edn. Gabler Verlag, Wiesbaden (2011)

Chesbrough, H.: Open Innovation: The New Imperative for Creating and Profiting from Technology. Harvard Business School Press, Boston (2006)

Chesbrough, H.: Open innovation. Where we've been and where we're going. Res. Technol. Manag. **55**, 20–27 (2012)

Ernst, H.: Virtual customer integration—Maximizing the impact of customer integration on new product performance. In: Albers, S. (ed.) Cross-Functional Innovation Management—Perspectives from Different Disciplines. Gabler Verlag, Wiesbaden (2004)

Ernst, H., Soll, J.H., Spann, M.: Möglichkeiten der Lead-User-Identifikation in Online-Medien. In: Herstatt, C., Sander, J.G. (eds.) Produktentwicklung mit virtuellen Communities – Kundenwünsche erfahren und Innovationen realisieren. Gabler Verlag, Wiesbaden (2004)

Filieri, R.: Consumer co-creation and new product development: a case study in the food industry. Mark. Intell. Plan. **31**, 40–53 (2013)

Füller, J., Mühlbacher, H.: Community-based Innovation – Ein Ansatz zur Einbindung von Online Communities in die Entwicklung neuer Dienstleistungen. In: Bruhn, M., Stauss, B. (eds.) Dienstleistungsinnovationen. Gabler Verlag, Wiesbaden (2004)

Gardlo, B., Ries, M., Hoßfeld, T., Schatz, R.: Microworkers vs. facebook: The impact of crowdsourcing platform choice on experimental results. In: Quality of Multimedia Experience (QoMEX), 35–36 (2012)

Gustafsson, A., Herrmann, A., Huber, F.: Conjoint analysis as an instrument of market research practice. In: Gustafsson, A., Herrmann, A., Huber, F. (eds.) Conjoint Measurement, 4th edn. Springer, Heidelberg (2007)

Helm, R., Möller, M., Rosenbusch, J.: Offline- versus Online Multiplikator: Eine kritische Betrachtung des klassischen Meinungsführerkonzepts, Jena Research Papers in Business and Economics—Working and Discussion Papers, 03/2010 (2010)

Herstatt, C., Sander, J.G.: Online-Kundeneinbindung in den frühen Innovationsphasen. In: Herstatt, C., Sander, J.G. (eds.) Produktentwicklung mit virtuellen Communities – Kundenwünsche erfahren und Innovationen realisieren. Gabler Verlag, Wiesbaden (2004)

Herzog, P.: Open and Closed Innovation: Different Cultures for Different Strategies, 2nd edn. Gabler Verlag, Wiesbaden (2011)

Homburg, C.: Marketingmanagement. Strategie – Instrumente – Umsetzung – Unternehmensführung, 4th edn. Gabler Verlag, Wiesbaden (2012)

Horbel, C., Woratschek, H.: Integration von Weiterempfehlungsgebern als Erfolgsfaktor für die Vermarktung komplexer Dienstleistungen. In: Bruhn, M., Stauss, B. (eds.) Kundenintegration. Gabler Verlag, Wiesbaden (2009)

Howe, J.: Crowdsourcing: Why the Power of the Crowd is Driving the Future of Business. Crown Business, New York (2008)

Jawecki, G., Füller, J., Matzler, K.: Innovation creation in online consumer groups. Der Markt **48**, 117–123 (2009)

Johnson, J.D., Chang, H.-J.: Internal and external communication, boundary spanning, and innovation adoption: An over-time comparison of three explanations of internal and external innovation communication in a new organizational form. J. Bus. Commun. **37**, 238–263 (2000)

Kozinets, R.: The field behind the screen. Using netnography for marketing research in online communications. J. Mark. Res. **39**, 61–72 (2002)

Kozinets, R.V.: Marketing netnography: prom/ot(ulgat)ing a new research method. Methodol. Innov. Online **7**, 37–45 (2012)

Kunz, W.H., Mangold, M.: Hybride Communities als Treiber des Kundenwertes. In: Herstatt, C., Sander, J.G. (eds.) Produktentwicklung mit virtuellen Communities – Kundenwünsche erfahren und Innovationen realisieren. Gabler Verlag, Wiesbaden (2004)

Lichtenthaler, U.: Open innovation: past research, current debates, and future directions. Acad. Manag. Perspect. **25**, 75–93 (2011)

Lindman, M.T.: Open or closed strategy in developing new products? A case study of industrial NPD in SMEs. Eur. J. Innov. Manag. **5**, 224–236 (2002)

Mast, C., Zerfaß, A. (eds.): Neue Ideen erfolgreich durchsetzen. Das Handbuch der Innovationskommunikation. Frankfurter Allgemeine Buch, Frankfurt am Main (2005)

Mahajan, V., Wind, J.: New product models: Practice, shortcomings and desired improvements. J. Product Innov. Manag. **9**, 128–139 (1992)

Mahr, D., Lievens, A.: Virtual lead user communities: Drivers of knowledge creation for innovation. Res. Policy **41**, 167–177 (2012)

McAfee, A.P.: Enterprise 2.0: the dawn of emergent collaboration. MIT Sloan Manag. Rev. (2006)

Meckel, M.: Innovationskommunikation 2.0 – Das Beispiel Apple iPhone. Mark. Rev. St. Gallen **26**, 42–46 (2007)

Meffert, H., Burmann, Ch., Kirchgeorg, M.: Marketing: Grundlagen marktorientierter Unternehmensführung. Konzepte - Instrumente – Praxisbeispiele, 11th edn. Gabler, Wiesbaden (2012)

N.U.: Diaper battle being waged over new Pampers product. Retrieved from: http://www.topnews.com.sg/content/21702-diaper-battle-being-waged-over-new-pampers-product (2010). [Accessed 20 May 2010]

N.U.: Innovationsflops kosten zehn Milliarden Euro. *absatzwirtschaft*, 6/2006, 35 (2006)

Payne, A.F., Storbacka, K., Frow, P.: Managing the co-creation of value. J. Acad. Mark. Sci. **36**, 83–96 (2008)

Peltier, J.W., Schibrowsky, J.A., Schultz, D.E.: Interactive integrated marketing communication: combining the power of IMC, the new media and database marketing. Int. J. Advert. **22**, 93–115 (2003)

Poetz, M.K., Schreier, M.: The value of crowdsourcing: Can users really compete with professionals in generating new product ideas? J. Product Innov. Manag. **29**, 245–256 (2012)

Prahalad, C.K., Krishnan, M.S.: Die Revolution der Innovation. Wertschöpfung durch neue Formen in der globalen Zusammenarbeit. Redline Verlag, München (2009)

Reichwald, R., Piller, F.: Interaktive Wertschöpfung – Open Innovation, Individualisierung und neue Formen der Arbeitsteilung, 2nd edn. Gabler Verlag, Wiesbaden (2009)

Rothwell, R., Robertson, A.B.: The role of communications in technological innovation. Res. Policy **2**, 204–225 (2002)

Sleeswijk Visser, F., van der Lugt, R., Stappers, P.J.: Sharing user experiences in the product innovation process: participatory design needs participatory communication. Creat. Innov. Manag. **16**, 35–45 (2007)

Sieg, J.H., Wallin, M.W., von Krogh, G.: Managerial challenges in open innovation: a study of innovation intermediation in the chemical industy. R&D Manag. **40**, 281–291 (2010)

Sonnenwald, D.H.: Communication roles that support collaboration during the design process. Design Stud. **17**, 277–301 (1999)

Teichert, Th., Rost, K., von Wartburg, I.: Aufbau von Sozial- und intellektuellem Kapital in virtuellen Communities of Practice. In: Herstatt, C., Sander, J.G. (eds.) *Produktentwicklung mit virtuellen Communities – Kundenwünsche erfahren und Innovationen realisieren*. Gabler Verlag, Wiesbaden (2004)

Trott, P.: Innovation Management and New Product Development, 5th edn. FT Prentice Hall, Essex (2011)

Vesshoff, J., Freiling, J.: Kundenintegration im Innovationsprozess – Eine kompetenztheoretische Analyse. In: Bruhn, M., Stauss, B. (eds.) Kundenintegration. Gabler Verlag, Wiesbaden (2009)

Zerfaß, A., Möslein, K.M.: Kommunikation als Erfolgsfaktor im Innovationsmanagement – Strategien im Zeitalter der Open Innovation. Gabler Verlag, Wiesbaden (2009)

Zerfaß, A., Ernst, N.: Kommunikation als Erfolgsfaktor im Innovationsmanagement. Ergebnisse einer Studie in deutschen Zukunftstechnologie-Branchen, Leipzig (2008)

Part III
Integrated Perspectives on Innovation

Chapter 14
Innovation Implementation: Leading from the Middle Out

Alan Belasen and Elliot B. Luber

Abstract The Competing Values Framework, a model for understanding paradoxical tension in an organization, can explain how managers help translate new ideas into actionable improvements to ultimately fit an organization's culture and operations by managing tasks and culture. Middle managers' central roles at the crossroads of defining culture, strategy, process and markets allow them to act as a fulcrum for leaders to pry people and systems toward delivering meaningful change, yet also serve as a lynch pin to hold organizations together amid stress. These managers play alternating roles, first as agents of change and then buffers to temper the same, internalizing competing values stress then forging its resolution. An ambidextrous ability to shift focus then de-focus through skillful communication can yield significant results.

14.1 Introduction

In today's fast-changing and disruptive global business environment, continuous redesign of organizational systems, processes, and structures is critical for sustaining business success and competitive advantage. As continual redesign becomes a core competency, effective organizations learn to develop creative and flexible design options to cope with the challenges of rapidly increasing global competition, shrinking product life cycles, and changing customer needs. The key to effective management of these rapid and disruptive changes is building and sustaining agility and resilience through adaptive capacity. Adaptive capacity refers to an organization's ability to implement innovation in response to changes in the marketplace (Belasen and Lus 2013). Adaptive organizations have a great competitive advantage

A. Belasen (✉)
SUNY Empire State College, 113 West Avenue, Saratoga Springs, NY 12866, USA
e-mail: alan.belasen@esc.edu

E.B. Luber
SUNY Empire State College, 3589 Wyanet St., Seaford, NY 11783, USA
e-mail: elliot.luber@esc.edu

N. Pfeffermann and J. Gould (eds.), *Strategy and Communication for Innovation*,
DOI 10.1007/978-3-319-49542-2_14

in times of instability and uncertainty. The key to success of many of today's businesses such as Microsoft, Intel, and IBM is their ability to change continuously to keep pace with rapidly changing market environments and technology. In their study of 471 North American companies, McCann et al. (2009) show that adaptive organizations that build agility and resiliency are more competitive and profitable, even with higher levels of turbulence.

This chapter focuses on examining how successful corporations leverage competitive advantages through the management of innovation. It points out the competing commitments associated with the challenge of combining high-volume operations with elements of complex systems. On the one hand, striving for a higher level of efficiency helps boost organizational reputation as a stable, high reliability organization; On the other hand, maintaining a higher level of stability creates an organizational mindset that resists change or that limits the ability of the organization to adapt to customer needs or respond quickly to crises. Within this context, we explore the critical roles of middle managers in managing the tension between these competing commitments and their significant contributions to innovation implementation in complex adaptive organizations.

14.2 Adaptive Versus Unadaptive Organizations

Most organizations are variants or a mix of two archetypes: volume operations and complex systems (Moore 2005). Volume operations are based on standardization and scalability or mass production. Firms in this category tend to be more formalized, centralized, mechanistic, and with a larger span of control and less-educated workers. Complex systems, on the other hand, are based on unique and specialized operations, and build-to-order production processes. Firms in this category tend to be decentralized, organic, and with a narrower span of control. Volume operations such as Procter and Gamble, Nike, eBay, and Amazon build their operations around thousands of transactions with an average cost per transaction with multiple bases of businesses and consumers generating high revenue streams for the companies. Complex systems such as IBM and Boeing organize their transactions, on the other hand, around fewer dependable businesses with a higher than average cost per transaction.

Complex systems use marketing to align partners and allied firms around unique products and high-touch service quality. High-volume operations compete on cost leadership while complex systems use focused differentiation to command premium prices for their customized products. Competition between these two business models occur when high-volume companies offer substitutes at a lower price, although they cannot compete well on variety and service quality. In contrast to high-volume operations which standardized their operations around low cost components, source at the mean, and employ control systems to minimize inventory costs and maximize on-time deliveries, complex systems implement lean manufacturing to reduce production costs and increase operating efficiencies.

Complex systems source at the margin and zealously monitor scheduling to avoid or reduce cost overruns. The supply chain in high-volume operations is fully connected and geared towards the end result where end users are expecting deliverables reliably and predictably. Branding, therefore, is an important marketing strategy for complex organizations as evident in Apple Inc., which remains at the forefront of the Smartphone market by launching its iPhone 6.

High-volume operations draw on standardization of work processes to achieve consistency and scalability. The sequentially interdependent units generate end-to-end products that compete on volume and price rather than quality and innovation. The risk of setting the assembly line wrong or producing unwanted products, or going through rework and complexity is too high. Disruptive change (e.g., regulation, technology, competition) creates inefficiencies as the subsystems are not designed to adapt quickly. High-volume operations that are stuck in their own inertia and fail to introduce innovation struggle and go through decline. In *Escape Velocity: Free your Company's Future from the Pull of the Past* (2011), Geoffrey Moore includes Sperry Univac, Honeywell, Digital Equipment, Prime, Kodak, Polaroid, Lucent, Nortel, Compaq, Gateway, Lotus, Ashton Tate, Borland, Novell, Nokia, Tower Records, Borders, Barnes and Noble, and Blockbuster among companies that declined because they failed to craft an appropriate response to disruptive technology. For example, the disruptive change of Netflix's method to delivering movies eventually led to the downfall of Blockbuster.

The common denominator across these unadaptive organizations was their inward focus and extensive reliance on rules and codified decisions, separation of planning and doing, and suppression of creativity and innovative thinking. Their product life cycle model was based on linear technology with basic science in order to manage the entire lifecycle of a product from inception through design, engineering, manufacturing, marketing and selling the finished product (Giannopoulou et al. 2010). Essentially, they managed to perform every step of the process in-house in a "closed", linear and highly controlled process: they discover new breakthroughs, develop them into products, create high-volume operation, distribute, finance, and service those products—all within the four walls of the company (Chesbrough 2003).

The failed, unadaptive organizations manifest Christensen's (2011) innovator's dilemma. Disruptive innovations are the changes that toppled the industry leaders mentioned above. These innovations create low cost, high value products in a new market or in a niche that the industry leaders avoid because of small scale. However, often the performance of the disruptive technology grows faster than users' needs, with time catching up to, and surpassing the more high-end or mainstream technologies that are the domain of industry leaders, those that cannot act quickly due to structural and systemic constraints.

Contrast the failed organizations with adaptive organizations that mix operating efficiency and scale with elements of complex systems and innovation compatible with applications of the ambidextrous model discussed later. They include Nestlé, Procter and Gamble, Nike, Dell, Apple, Sony, Hertz, Hilton, United Airlines, Microsoft, Adobe, Electronic Arts, eBay, Google, IBM, Cisco, SAP, Goldman

Sachs, Boeing, Tektronix, Honeywell, Bechtel, Accenture, Apache, Halliburton, Burlington Northern, and Amazon. These organizations are driven by R&D and B2B marketing research to meet the expectations of customers for highly differentiated products or services. They use tested experiments, investments in human capital, business development groups that interact with externals (e.g., customers, researchers, competitors), and inter-functional collaboration to implement innovation.

14.3 Innovation from Everywhere

Advances in technology and broadband wireless access have made incremental innovations easily imitated by competitors and consequently put pressure on many large scale companies to revert from long term to short term innovations. With knowledge so widely distributed and easily accessible, companies move to acquire inventions or intellectual property from other companies to offset R&D costs, reduce risk, accelerate time to market, and increase their competitiveness (Huang et al. 2010). Others, especially in high-technology industries, have adopted the strategy of Open Innovation (OI) (De Wit et al. 2007). For example, Magneti Marelli is a subsidiary of the Fiat Group that maintains 12 R&D centers. After using OI to stimulate creativity and innovation, which promotes the development of ideas, products and services, only 15% of the total ideas came from the internal R&D team—the other 85% came from users around the world (Day 2015).

OI helps companies reduce the cost of product development and process improvement, accelerate time to market for new products, improve product quality, and access customer and supplier expertise outside the organization (Wallin and Krogh 2010). For OI to be successful, it has to be distributed both internally and externally through collaborative networks or ecosystems (Chesbrough and Appleyard 2007; De Wit et al. 2007; Dooley 2007; Giannopoulou et al. 2010). For example, AstraZeneca, the big pharma company, is a group that embraces OI. Astra has spent the past few years forging partnerships with groups including health charities, academic researchers and even its industry rivals, such as GlaxoSmith-Kline. The collaborations work by openly sharing early-stage research (sometimes, even, further along research), and allowing all the parties to use the knowledge to solve the particular disease problem. Often they involve Astra scientists that share the lab with researchers from other organizations. The company noted: “It enables us to get more out of our research dollars than by just doing things on our own. You’re creating an ecosystem where you can collaborate and get more out of it than if you were just doing something on your own” (Roland 2014).

Crowdsourced innovation involves soliciting ideas or proposals, ranking or prioritizing ideas, evaluating, or even prototyping ideas (Phillips 2010). For example, Siemens Automation’s user forum involves quality. Contributors are encouraged to participate with useful information. Other users rate the validity and usefulness of the information and contributions are ranked. Another example is

Cisco's "Entrepreneurs in Residence," an incubation program that supports early-stage entrepreneurs who are working on projects that may benefit the organization (https://eir.cisco.com/about/). A few of the areas that Cisco focuses on are big data, cloud computing, and enterprise security and the entrepreneurs selected to partake in the event receive financial support, industry expertise, and access to business groups. The program events include opportunities to present the ideas to investors, and work side by side with the experts at Cisco. This is advantageous for both parties, as the entrepreneurs enjoy the benefits aforementioned, and Cisco is able to bring in fresh minds and revolutionary ideas from outside of the company.

Crowdsourcing is great as well as it does not overwhelm the company or cause a diversion of the company's energy due to information overload that can actually lessen productivity (Hemp 2009). Crowdsourcing also tends to restrict the amount of managerial oversight and control over the company's direction as reporting becomes a key time allocation in this model. One example is Frito Lay and its "Do Us a Flavor" campaign for flavors for their Doritos brand. The four flavors in the final were chosen from over 14.4 million public submissions. These ideas do not come free as Lay's needs to pay a team of market researchers and personnel to sift through these submissions, conduct demand analysis for such flavor ideas, manufacture, and conduct quality tests to eliminate unappealing and impractical proposals. LEGO is another example of an overwhelmed company that mismanaged its crowdsourcing strategy or failed to develop an effective implementation plan. Still, an unintended bi-product of crowdsourcing and OI is the appreciation and excitement of users who feel that the company actually cares about their opinions and how they perceive the brand. In the case of Lays, the company is determining the winner through America's vote at "DoUsAFlavor.com". This is beneficial for the company because it significantly helped increase sales, as customers try the new flavors, while also creating publicity and enhancing Lay's social media presence.

14.4 Outside-In and Inside-Out Perspectives

There are "inside out" and "outside in" approaches to OI, where underdeveloped ideas are incorporated into others' innovation processes. Complex adaptive systems (CAS), for the most part, draw on an outside-in perspective and evolve within a network of highly linked organizations. Within such a context, change is seen in terms of co-evolution with all other linked systems, rather than as adaptation to a separate and distinct environments. This implies that a decision or action by one part within a system tends to influence all other related parts but not necessarily in any uniform manner. CAS are characterized by autonomous engagement, emergence, exploitation, disequilibrium, reintegration, and disintegration of structures and processes. These systems rely on nonlinear dynamics of punctuated equilibrium (switching between temporary structures of stability and chaos) to transform and self-organize into a structure more compatible with changes in the marketplace. The results are companies, like Honda and Canon, with ambidextrous structures and hybrid technologies and production lines that are geared simultaneously towards high-end and low-end markets. HP printing technology for the high-end office market in addition to its consumer products exemplifies the ambidextrous approach discussed below. AMD in the semiconductor industry and Pfizer in the pharmaceutical industry are examples of companies that combine the strengths of the two models to achieve the strategy of joint optimization. Others sought partnerships with academic institutions. Procter and Gamble works with Virginia Tech to develop improved analytics and engineering approaches for manufacturing.

For companies concerned with intellectual property or losing their distinctive competencies to their competitors, one application of combining the strength of high-volume operations with elements of adaptive complex systems is the "inside-out" perspective. GE's *FirstBuild* program is designed to foster groundbreaking ideas from line employees who can submit their proposals, and if selected, even win monetary awards. FirstBuild allows GE to act like a startup and crowd-source breakthrough ideas. The "inside-out" perspective is consistent with the principles of TRIZ (Dager 2014). Instead of relying on open source to solve a problem, TRIZ relies on internal (functional) knowledge that people outside the company may not have, allowing the organization to work on confidential or sensitive data and receive input from "the crowd" without having to expose its details to the crowd. TRIZ also allows companies without much success with OI methods to still become innovative.

Since the main premise of OI is to extend the pursuit for and commercialization of new products beyond the existing boundaries of the organization, the first thing that an organization interested in this strategy should do is open a channel of communication that allows for other firms, suppliers, and customers to connect with the organization. Of course, there can and should be multiple channels, rather than only one. Today, with the benefit of instant communication, Twitter has become a large platform for such innovation. However, before any of this can be feasible, the organization must first be sure that they are willing to, at least partially, give up the control that a "closed" strategy embodies. There is a plethora of ways in which a company can reap the benefits of OI.

14.5 Leading Ambidextrously: The Role of Middle Managers

The ambidextrous structure follows what Miles and Snow (1978) called "Analyzer" strategy. This strategy is aimed at combining the operating efficiencies and cost leadership of a defender strategy and breakthrough innovation and quality through customization offered by the prospector strategy (Belasen and Rufer 2013) and flexible manufacturing systems (Belasen and Lus 2013) while at the same time making imitation costly for competitors (Arend 2009). Ambidextrous organizations rely on organic and mechanistic systems of management to initiate and implement change (O'Reilly and Tushman 2004). This implies that both organic and mechanistic processes are needed for innovation implementation. Organic processes are needed in the initiation stage of innovation. Mechanistic processes are needed in the implementation stage to ensure successful execution.

Middle managers make substantial qualitative and quantitative contributions to their organizations and have the know-how to move work forward with tact and velocity. They understand the complexity of their work environment and feel energized by being the conduit between corporate thought, action and outcomes (Haneberg 2010). As a result, they are able to provide much richer views to problem solving communication with diverse insights. Middle managers are centrally located within the chain of command and typically perform or directly influence three important tasks depicted in Fig. 14.1: (a) Technical tasks linked to the operating core, routines, and compliance; (b) People tasks involving leading, motivating, and developing direct reports; and, (c) Strategic tasks associated with financial reports, analysis and strategic communication including direct involvement in CQI projects and boardroom discussions.

Middle managers handle multiple roles and deal with diverse sets of internal and external stakeholders. They engage in dual lines of reporting with top down implementation roles and bottom up championing roles. Functioning as intermediaries across hierarchical levels, middle managers must strike a delicate balance between creativity and efficiency, transformational and transactional roles. Many scholars would liken the middle manager's responsibilities to those of a strategist or change architect, facilitating the communication about corporate thought, action, and results (Haneberg 2010; Wooldrige et al. 2008). Middle managers accomplish this through synthesizing information as well as championing and executing plans (Belasen and Rufer 2013). Synthesizing information consists of gathering intelligence on the feasibility of new programs, communicating the activities of

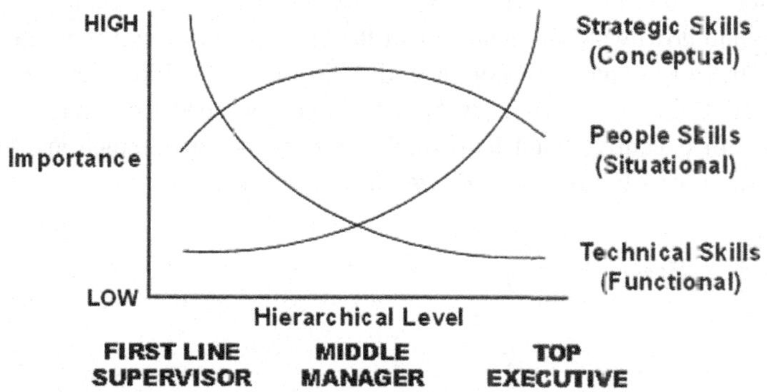

Fig. 14.1 The scope of middle managers' tasks

competitors and suppliers as well as assessing changes in the external environment. Championing involves justifying and defining new programs, evaluating the merits of new proposals, searching for new opportunities and proposing programs or projects to higher level managers (Mantere 2008). Middle managers' duality of operating experience and proximity to decision authority centers make their championing roles critical for sustaining competitive advantage (Morrison et al. 2000; Mumford 2002).

According to Brache (2004) if the organization's strategy is clearly established and communicated, middle managers can use it responsibly to guide their operational decisions and linked them with strategy. In fact, Mollick (2011) found that it was middle managers, rather than innovators or organizational strategy, who best explains variation in firm performance. Managers accounted for 22.3% of the variation in revenue among projects, as opposed to just over 7% explained by innovators and 21.3% explained by the organization itself—including firm strategy, leadership and organizational practices. Additionally, even in a young industry that rewards creative and innovative products, innovative roles explain far less variation in firm performance than do managers. Or, as Mollick (2011) explained: "High-performing innovators alone are not enough to generate performance variation; rather, it is the role of individual managers to integrate and coordinate the innovative work of others." So while innovators may come up with new games and new concepts, managers play the more crucial role of deciding which ideas are actually given resources. The role of the middle manager in the strategic process is broad and complex where both formulation and implementation are well connected (Belasen and Rufer 2013).

Others argue that allowing middle managers to play a more active role in decision-making and in the execution of strategies contributes to a dynamic enrichment of organizational roles (Beck and Plowman 2009). Indeed, middle managers with active involvement in the strategy process have been found to demonstrate higher levels of commitment to organizational goals and contribute to the success of strategic initiatives (Belasen and Frank 2004; Huy 2001; Vila and Canales 2008). They support the organization through their know-how, experience, networking and ability to strengthen the social fabric and core values of the organization (Valentino 2004). Evidence shows that middle managers' upward leadership and downward influence affect the alignment of organizational activities with the strategic context (Balogun and Johnson 2004; Floyd and Lane 2000; Rouleau and Balogun 2011). They help facilitate the need for change in communications with executives and help implement change in interactions with lower levels (Belasen and Frank 2010; Kuyvenhoven and Buss 2011).

14.6 High-Impact Middle Managers

Implementing innovation requires effective balance of strategic awareness with operating experience and an optimal mix of leadership and management roles. This is important as crowdsourcing often creates different identities for the organization that also require effective strategic communication (Belasen 2008). A broader identity gives work units permission to engage in opposing strategies—to exploit existing products and services while simultaneously explore new offerings and business models. Middle managers more so than senior managers or operating-level managers have the communication competence to handle the complexity of information and number of potential interactions (Beck and Plowman 2009).

Another mix between leadership and management roles involves interpersonal communication and strategy implementation. Middle managers have to deal with the process of managing the collaboration process with external sources (Slowinski et al. (2009). According to Wallin and Krogh (2010), when firms invite users to contribute their knowledge to innovation, they cannot apply traditional rules or centralized authority to directing, incentivizing, and monitoring their efforts. Instead, positive and supportive communication as well as a climate that embraces informal climate should be used.

High-impact middle managers optimize their functions by facilitating adaptation while acting in the capacity of innovator and broker roles and by implementing deliberate strategy using the director and producer roles depicted by the Competing Values Framework (Quinn 1988). Figure 14.2 demonstrates that middle management resides at the point of friction between leadership and management and of internal versus external focus. They must thusly internalize the stress not only between leadership and management goals and methods but also between loyalty to the company's goals and objectives and interests and needs of client bases. This is where leadership strengths, the fine communication skills high-impact middle managers develop through socialization, come to play as they look to avoid direct confrontations by seeking out "both/and" strategies (Belasen 2008).

Middle managers are not just the fulcrum on which an organization pivots, but serving as the lynch pin that at times holds it together when pulled in opposite directions between marketing and operations, manufacturing and sales, or strategy versus employees. In the case of normalizing innovative change, middle management must build a system with people where it does not exist, and enable sound practices to be put in place without losing the "special sauce" of the innovation—which will deliver competitive advantage to the market. It is here where the payback is manifested despite studies (Belasen and Frank 2004) that found that these employees are the first to go during mergers and acquisition because they are seen as redundant cost, a shortsighted strategy according to that study.

Facilitating adaptability requires that the middle managers relax control to get new projects started, secure time for experimental programs, locate, and provide resources for trial projects, and encourage informal discussion and information sharing. They can implement intended strategy by monitoring activities to support

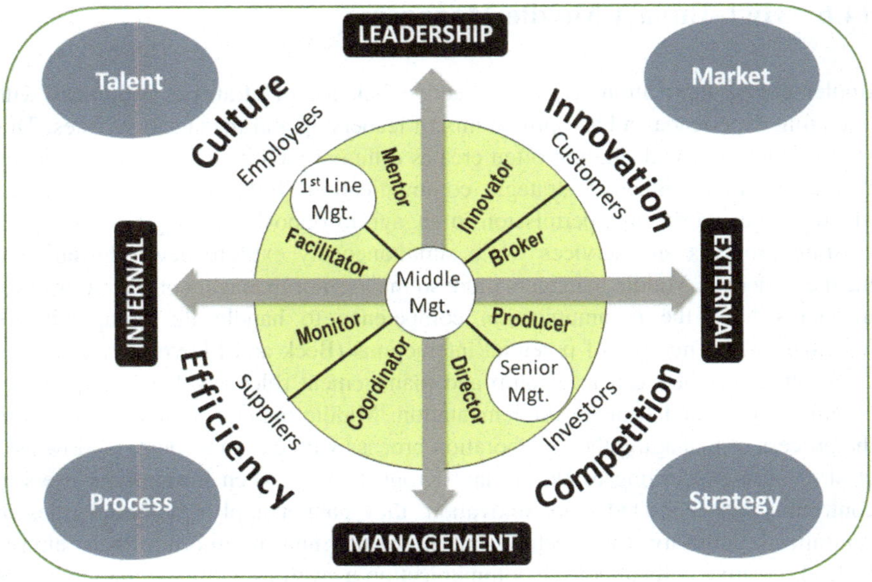

Fig. 14.2 The context for high-impact middle managers

top management objectives, translate goals into action plans and work unit objectives, and present top management initiatives to subordinates. Middle managers, enacting the mentor and facilitator roles, have the knowledge and skills to function effectively with both groups and the flexibility to shift from the role of a manager to the role of *team builder and team leader* as well as implement innovation.

Team leadership evolves through two different phases: the transition phase and the action phase (Morgeson et al. 2010). Throughout each of these phases, the team responds to different needs. For example, during the transition phase the team needs a team charter outlined with goals and objectives, developing positive team norms, clarifying task performance strategy, sense making and providing feedback. In the action phase, monitoring output inside and outside the teams, coordinating team actions, engaging in effective communication and maintaining boundaries are important aspects of positive relationships and good team performance.

In order to fulfill all these needs, a driving force must be present—a high-impact middle manager who is both task-and-relationship oriented with situational skills. Wei et al. (2010) found that most leaders at the R&D collaborative group use a “steering” rather than “directing” approach, motivating innovators, communicating, and clarifying goals and expectations. Within the R&D collaboration, cross-functional teams are often the most prominent and a leader of such collaboration requires technical skills, personal skills, cognitive ability, political, and project management skills.

As shown in Fig. 14.3, leadership is exercised through social (e.g., broker, facilitator) and task (e.g., coordinator, monitor) behaviors (Madlock 2008).

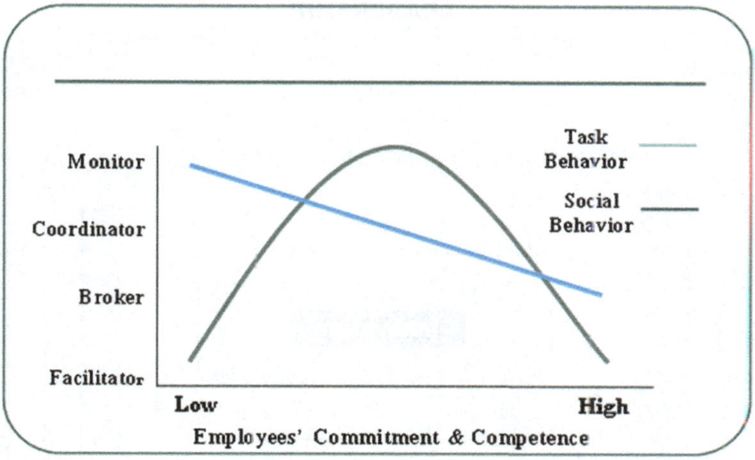

Fig. 14.3 Situational leadership

In addition to the persuasion aspect of managerial communication and the ability to control language, gestures, and tone of voice, for managers to be perceived as competent communicators by their subordinates, Shaw (2005) suggested that they must also share and respond to information in a timely manner. Moreover, they should integrate the activities of inter-functional groups, monitor progress towards implementation, evaluate and control the process.

Effective leadership in the external environment means first developing a situational understanding of key stakeholder motivations, and communicating to them within the prism of their own ecosystemic needs (Belasen 2008), at least acknowledging and demonstrating empathy for their positions—even when moving against their stated interests. A middle manager's understanding and exploitation of such nuance can make a considerable difference during times of significant change, impacting the organization's overall reputation or credibility.

While upper management does most market communication, message alignment falls to middle managers. The same way that middle management balances the needs of talent versus corporate objectives; they must balance the corporate identity against the brand promises and the company's sense of responsibility against its earnings consistency (Belasen et al. 2015). Are the employees delivering the brand experience promised in the company's advertising? This is a matter of communicating culture and instructions (Belasen 2008).

Thus not only do middle managers play a significant role in normalizing innovation, but they align company messages so that the market credits the company with receiving and delivering the innovation. Figure 14.4 illustrates the interconnected circles of innovation implementation that affect both the business success and the company's external image as it encompasses both market and investor perspectives to ultimately drive stock price or other perceived equity measurements

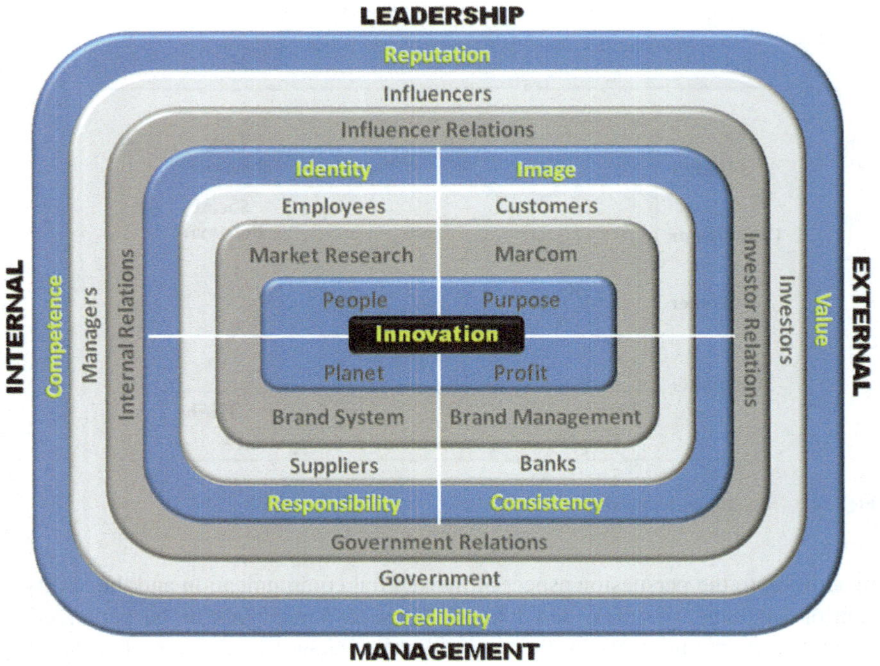

Fig. 14.4 A framework for innovation communication

(Belasen 2008). Middle management helps make innovation happen for the customer and enables the company to communicate its success through OI initiatives.

The long road from innovation to value travels a tangible path of operations and transactions, and is likely measured by an ERP system. The intangible path of being recognized as an innovator travels a road less measured and less directly controlled, yet this plays a significant role in value recognition as it drives the stock price of publicly traded companies. It should be noted that the level of intangibility is changing in an age of big data, Web analytics and social media.

We define value as the public perception—all-in—of the company's market position and owner's pending equity, the anticipated direction of a stock price. Competence is the company's ability to deliver on its stated plan from both a cultural and process perspective, thus it measures the tangible executional capability and intangible identity of the company. Reputation is a comparison of the brand promise versus identity—does the company set out to be the kind of company promised in its advertising? Reputation is generally an objective third-party assessment of the customer experience by credible influencers such as analysts and media observers. Credibility is similarly a transactional measurement of the company's trustworthiness and ability to generate consistent returns for its shareholders. The combined effects of reputation and credibility drive the ultimate "all in" valuation of the company.

14.7 Conclusion

In the case of innovation, while senior executives may very well be the official spokespeople for the organization, the various stakeholders within the ecosystem, most of the time, are in direct contact with middle management. So while middle management is making incompatible values live together within the organization's supervisory structure, middle managers are also a significant determinant of the authenticity of the company's public marketing statements. Can the company walk the talk? What are key stakeholders seeing?

Perhaps these are the middle managers whom should be groomed for greater roles to come; for these are the managers who drive significant value into their organizations by making innovation a reality—both for the company's transactional measurements and the intangibles that ultimately drive the stock price. They are doing the heavy lifting of transformation, yet are shielding day-to-day transactions from the strain. They are aligning systems within and without the organization to efficiently deliver what was not previously possible or available, and they are aligning the supporting culture(s) to make crossing the chasm of change a practical reality. Middle management stands where the rubber of transformation meets the road of transaction. It is a difficult job—one seemingly undervalued in American business.

Middle managers need to communicate their *indispensability* to senior executives (Zenger et al. 2011). They need to take the vision and goals of OI and cascade these ideas appropriately so that the front line teams become engaged and aligned with that vision. Davenport and Harding (2010) recommended that middle managers become *off-stage managers* who focus on managing the environment, not the people. The off-stage manager creates an enabling and empowering environment where front line teams and employees are empowered to identify issues and solutions and are provided with clear objectives and directions. Similarly, Kelly and Nadler (2007) suggest that middle managers should shift their focus to influencing processes and decisions, not controlling them. Sliding up to a governance role or down to an advocacy role, for example, could allow middle managers to use the variety of roles described in this chapter flexibly and systematically, especially during clutch time of innovation implementation (Hales 2006).

References

Arend, R.J.: Defending against rival innovation. Small Bus. Econ. **33**, 189–206 (2009)

Balogun, J., Johnson, G.: Organizational restructuring and middle manager sense making. Acad. Manage. J. **47**(4), 523–549 (2004)

Beck, T., Plowman, D.: The role of middle managers in animating and guiding organizational interpretation. Organ. Sci. **20**(5), 909–924 (2009)

Belasen, A., Eisenberg, B., Huppertz, J.: Mastering Leadership: A Vital Resource for Healthcare Organizations. Jones and Bartlett Learning, Boston, MA (2015)

Belasen, A., Frank, N.M.: The perceptions of human resource managers of the shifting importance of managerial roles in downsizing organizations. Int. J. Hum. Resour. Dev. Manage. **4**(2), 144–163 (2004)

Belasen, A.: The Theory and Practice of Corporate Communication: A Competing Values Perspective. Sage Publications, Thousand Oaks, CA (2008)

Belasen, A., Frank, N.M.: A peek through the lens of the competing values framework: what managers communicate and how. Atlantic J. Commun. **18**, 280–296 (2010)

Belasen, A., Rufer, R.: Innovation communication for effective inter-professional collaboration: a stakeholder perspective. In: Pfeffermann, N., Minshall, T., Mortara, L. (eds.) Strategy and Communication for Innovation, 2nd edn, pp. 227–240. Springer, Germany (2013)

Belasen, A., Lus, B.: Adaptive capacity and managing relationships in the supply chain. In: Paludi, M. (ed.) Implementing Best Practices in Human Resources, pp. 141–152. Praeger, Santa Barbara, CA (2013)

Brache, A.P.: Strategy and the middle manager, Training. Retrieve from: http://www.kepner-tregoe.com/pdfs/articles/StrategyMidMan.pdf (2004)

Chesbrough, H.: Open Innovation The New Imperative for Creating and Profiting from Technology. Harvard Business School Publishing Corporation, Boston (2003)

Chesbrough, H.W., Appleyard, M.M.: Open innovation and strategy. Calif. Manag. Rev. **50**(1), 57–76 (2007)

Christensen, C.: The Innovator's Dilemma: The Revolutionary Book That Will Change the Way You Do Business. HarperCollins Publishers, New York (2011)

Dager, J.: TRIZ—For People Who Struggle with Open Innovation. http://customerthink.com/triz-for-people-who-struggle-with-open-innovation/. Jan 2014

Davenport, T., Harding, S.: Manager Redefined: The Competitive Advantage in the Middle of Your Organization. Jossey-Bass, San Francisco (2010)

Day, J.: 3 Reasons to Go From Internal R&D to Open Innovation. http://www.business2community.com/business-innovation/3-reasons-go-internal-rd-open-innovation-01187775#M8EMsBCeLrEZ4xKS.97. 23 March 2015

De Wit, J., Dankbaar, B., Vissersn, G.: Open innovation: the new way of knowledge transfer? J. Bus. Chem. **4**(1), 11–19 (2007)

Dooley, L.: Managing within distributed innovation networks. Int. J. Innov. Manage. **11**(3), 397–416 (2007)

Floyd, S., Lane, P.J.: Strategizing throughout the organization: managing role conflict in strategic renewal. Acad. Manage. Rev. **25**(1), 154–177 (2000)

Giannopoulou, E., Ystrom, A., Ollila, S., Tobias, F., Elmquist : Implications of openness: a study into (all) the growing literature on open innovation. J. Technol. Manage. Innov. **5**(3), 162–180 (2010)

Hales, C.: Mowing down the line, the shifting boundary between middle and first-line management. J. Gen. Manage. **32**(2), 31–55 (2006)

Haneberg, L.: The High Impact Middle Manager: Powerful Strategies to Thrive in The Middle. ASTD Press, Alexandria, Virginia (2010)

Hemp, P.: Death by information overload, Harvard Bus. Rev. https://hbr.org/2009/09/death-by-information-overload. (2009)

Huang, T., Weng, W., Ken, Y.: Managing technology transfer in open innovation: the case study in Taiwan. Mod. Appl. Sci. **4**(10), 2–13 (2010)

Huy, Q.N.: In praise of middle managers. Harvard Bus. Rev. **79**(8), 72–9, 160 (2001)

Kelly, J., Nadler, S.: Leading from below: CEO's can't change companies on their own; the secret is to foster a leadership mentality throughout the ranks. Wall Street J, (3 March 2007)

Kuyvenhoven, R., Buss, C.: A normative view of the role of middle management in the implementation of strategic change. J. Manage. Mark. Res. **8**(1), 1–14 (2011)

Madlock, P.: The link between leadership style, communicator competence, and employee satisfaction. J. Bus. Commun. **45**(1), 61–78 (2008)

Mantere, S.: Role expectations and middle manager strategic agency. J. Manage. Stud. **45**(2), 294–316 (2008)

McCann, J., Seisky, J., Lee, J.: Building agility, resilience and performance in turbulent environments. People Strateg. **32**(3), 44–51 (2009)
Miles, R.E., Snow, C.C.: Organizational Strategy, Structure and Process. McGraw-Hill, New York (1978)
Mollick, E.R.: People and process, suits and innovators: the role of individuals in firm performance. SSRN: http://ssrn.com/abstract=1630546 or http://dx.doi.org/10.2139/ssrn.1630546 (2011)
Moore, G.A.: Escape Velocity: Free Your Company's Future from the Pull of the Past. HarperCollins Publishers, New York (2011)
Moore, G.A.: Strategy and your stronger hand. Harvard Bus. Rev. **83**(12), 62–75 (2005)
Morrison, P.D., Roberts, J.H., Von Hippel, E.: Determinants of user innovation and innovation sharing in a local market. Manage. Sci. **46**(12), 1513–1527 (2000)
Morgeson, F.P., DeRue, S., Karam, E.: Leadership in teams: a functional approach to understanding leadership structures and processes. J. Manage. **36**(1), 5–39 (2010)
Mumford, M.D., Scott, G.M., Gaddis, B., Strange, J.M.: Leading creative people: orchestrating expertise and relationships. The Leadersh. Quart. **13**(6), 705–750 (2002)
O'Reilly, C.A., Tushman, M.: The ambidextrous organization. Harvard Bus. Rev. **82**(4), 74–81 (2004)
Phillips, J.: Open innovation typology. Int. J. Innov. Sci. **2**(4), 175–183 (2010)
Quinn, R.E.: Beyond Rational Management. Jossey-Bass, San Francisco (1988)
Roland, D.: AstraZeneca Reinforces Pipeline with 'Open Innovation'. http://www.telegraph.co.uk/finance/newsbysector/pharmaceuticalsandchemicals/10775160/AstraZeneca-reinforces-pipeline-with-open-innovation.html. 28 April 2014
Rouleau, L., Balogun, J.: Middle managers, strategic sense making, and discursive competence. J. Manage. Stud. **48**(5), 953–983 (2011)
Shaw, K.: Getting leaders involved in communication strategy: breaking down the barriers to effective leadership communication. Strateg. Commun. Manage. **9**, 14–17 (2005)
Slowinski, G., Hummel, E., Gupta, A., Gilmont, E.: Effective practices for sourcing innovation. Res. Technol. Manage. **52**(1), 27–34 (2009)
Valentino, C.: The role of middle managers in the transmission and integration of organizational culture. J. Healthc. Manage. **49**(6), 393–404 (2004)
Vila, J., Canales, I.J.: Can strategic planning make strategy more relevant and build commitment over time? The case of RACC. Long Range Plan. **41**(3), 273–290 (2008)
Wallin, M., Krogh, G.: Focus on the integration of knowledge. Org. Dyn. **39**(2), 145–154 (2010)
Wei, Z., Khoury, A.E., Grobmeier, C.: How do leadership and context matter in R&D team innovation?—A multiple case study. Human Resour. Dev. Int. **13**(3), 265–283 (2010)
Wooldrige, B., Schmid, T., Floyd, S.: The middle management perspective on strategy process: Contributions, synthesis, and future research. J. Manage. **34**, 1190–1221 (2008)
Zenger, J., Folkman, J., Edinger, S.: Making yourself indispensable. https://hbr.org/2011/10/making-yourself-indispensable/ar/1. 10 Oct 2011
http://www.forbes.com/sites/brandindex/2014/10/11/crowdsourcing-campaign-appears-to-boost-brand-perception-for-lays/
http://www.innovationexcellence.com/blog/2011/11/18/legos-50-million-open-innovation-failure/
http://www.vtnews.vt.edu/articles/2015/04/042315-vbi-pg.html
http://www.wired.com/2014/04/how-ge-plans-to-act-like-a-startup-and-crowdsource-great-ideas/

Chapter 15
Innovation, Leadership, and Communication Intelligence

Ian C. Woodward and Samah Shaffakat

Abstract In this chapter, we consider the relationship between effective leadership, communication, innovation, and creativity within organizations and teams. In a dynamic business world where innovation is a critical driver for competiveness and growth, we argue that closing the gap between ineffective and effective leadership and communication approaches matters. To assist, we provide two interrelated "tools" that can improve effective leadership communication practices at every stage of the innovation cycle—from ideation through to implementation. These lead to clear, open, and compelling communication interactions that underpin innovation and engagement at inter and intra-organizational levels. Our focus is on increasing the chances of successful innovation outcomes by using effective leadership and communication approaches, combined with "communication intelligence" and "fair process" involvement.

15.1 Introduction

Why do highly innovative companies like Apple, Google, and Gore expend considerable effort and resources communicating the value and utility of their products and services? A simple answer—because by communicating effectively, they capture the minds and hearts of their customers. These firms emphasize effective communication that is clear, open, and compelling inside their businesses.

Effective communication means achieving the desired outcomes and objectives of communication exchanges, in a specific context or situation, which leads to shared understanding and satisfaction for the participants in those exchanges (Woodward et al. 2016). Effective communication underpins every phase of successful innovation: from tapping innovation resources and investment; through the ideation process; through implementing change; to marketing the innovative

I.C. Woodward (✉) · S. Shaffakat
INSEAD—The Business School for the World, 1 Ayer Rajah Avenue, Singapore 138676, Singapore
e-mail: ian.woodward@insead.edu

N. Pfeffermann and J. Gould (eds.), *Strategy and Communication for Innovation*,
DOI 10.1007/978-3-319-49542-2_15

products and services produced; and engaging in interactive customer feedback. On the other hand, ineffective communication or miscommunication cuts idea generation short, confuses creative exploration, wastes resources and effort, contributes to implementation disasters, and demotivates or disengages people. A key challenge for leading people toward innovation outcomes is how to close the gap between ineffective and effective communication interactions for all those involved.

Therefore, in this chapter, we discuss the relationship between innovation, creativity, leadership and effective communication. We examine the research on leadership and communication approaches as well as the behaviors and language that facilitate, engage and mobilize innovation, creativity, and collaboration within organizations and teams. We then present two interrelated "tools" that will substantially improve the potential for effective leadership communication in practice, in every phase of innovation. The fundamental "tool" is building "communication intelligence" among innovation stakeholders, to increase interaction that is clear, open and compelling. In addition, we recommend adopting the "*INVOLVE*"—"fair process" leadership communication practices across the various innovation phases to positively motivate and engage people.

15.2 Innovation, Creativity, and Work

Knowledge and information are two key constituents of dynamic innovation and change (Pfeffermann 2011). The focus on innovation and creativity is critical in a modern business world where organizations are under continuous pressure to perform and deal with the paradigm shift of knowledge work in a digital age (Dalkir 2013; Drucker 2009). To maintain competitive edge, organizations must meet this shift (Mayfield and Mayfield 2004), find efficient ways to promote innovation at different levels (Mayfield and Mayfield 2008), and understand the nature, opportunities and threats of disruption (D'Aveni 1999).

Moreover, to create and maintain continuous innovation flow, employees need to be motivated to innovate, and have the skills and capacity to do so (Mayfield and Mayfield 2008). Employee creativity lays the groundwork for organizational innovation (Oldham and Cummings 1996). Creative employees bring forth solutions to problems, defend their ideas and provide an action plan for how to put these ideas into practice (Gumusluoglu and Ilsev 2009a, b). Innovation is core to many established management approaches, such as total quality management (Osayawe Ehigie and Clement Akpan 2004), Kaizen (Imai 1986), and organizational learning (Senge 1990).

By its nature, innovation requires "out of the box thinking," doing new things or doing old things in new ways. This includes introducing novel and better ways of carrying out work tasks (West et al. 2003). Zaltman et al. (1973) see innovation as any "idea, practice, or material artifact" taken up by an individual, group, or organization in order to bring about change. From this perspective, types of innovation will differ depending on the *level* of focus (Amabile et al. 1996;

Mayfield and Mayfield 2008). At the organizational level, innovation concerns the domains of strategy, structure, organizational processes, and new market, product, or service selection. Innovation at the group level can include designing and creating new products, processes, and administrative routines. Individual level innovation includes idea generation, process effectiveness, and improving individual work (Stoker et al. 2001). At the leadership level, it can include creating, articulating, and persuading about innovative ideas or initiatives (Elkins and Keller 2003; Jung et al. 2003)

Creativity, the basis of innovation, is sometimes assumed to be limited to certain areas or professions, such as sciences or arts (Mumford et al. 1997). However, creativity is required in any jobs with tasks that pose complicated, unclear problems where effective performance is contingent on developing new and effective solutions (Ford 2000; Mumford and Gustafson 1988). This is the setting of modern business, where leaders and their teams need to deal with largely "adaptive" problems–problems that have no straightforward solution or quick fixes available. Solving adaptive problems requires innovation—and a transformation in beliefs, ideologies, values, and ways of working (Heifetz 1994).

Creative work, in general, comprises processes to both produce ideas, and to implement them (Vincent et al. 2002). The idea generation process (ideation) includes: defining a problem, collecting information, conceptual structure generation, and combing these concepts to form a new category (Mumford et al. 2003). For idea execution, the key processes include: idea evaluation, vetting and testing, and formulating and designing a plan (Lonergan et al. 2004). Creative work revolves around individuals who must actively look for, and manipulate knowledge and concepts (Byrne et al. 2009). This requires expertise, and years of experience (Qin and Simon 1990; Weisberg 1999). For a successful solution, problems will need expertise from several areas in various forms, which makes creative work collaborative (with communication exchanges), as well as individually focused (Cagliano et al. 2000).

Creative work constitutes several stages, consuming time, and energy. For example, successful idea execution requires continuous effort supported by a good amount of intrinsic motivation (Collins and Amabile 1999). Creative work demands organizational resources along with the time and commitment of several people and groups. As such, politics and persuasion are likely to come into play to secure resources for successful project completion (Dudek and Hall 1991). As innovation moves into implementation, leaders must also focus attention on "active monitoring and tailoring the plan" to cope with the inevitable challenges faced in the field (Byrne et al. 2009, p. 264).

Furthermore, creative work entails risk (Mumford et al. 2002), as the idea might not be generated at all or might not be sound enough, and as such the resulting product might not fit the market need (Cardinal and Hatfield 2000). Therefore, context plays a huge role and the leadership of creative efforts should consider not just the organizational strategy but sociotechnical aspects as well (Byrne et al. 2009). Cultivating an innovative environment fosters risk-taking and provides an opportunity to employ more creative techniques in the workplace (Gumusluoglu and Ilsev 2009a, b), which requires trust. Perceptions of trustworthy and engaging management are enhanced by leadership communication openness or transparency (e.g. Butler 1991; McCauley and Kuhnert 1992). To succeed, innovation, and creative work activities require effective leadership approaches and effective communication interactions at every point.

15.3 Innovation, Leadership Approaches, and Communication

Much research on leadership and innovation examines the results of leader behaviors on outcomes such as effectiveness or efficiency rather than the innovation outcomes (De Jong and Den Hartog 2007). A special issue of *The Leadership Quarterly* (2004, Vol. 15, No. 1)—"leading for innovation" examined the creative efforts of leaders. Mumford and Licuanan (2004) summed up the research by emphasizing the various communication roles played by leaders, such as facilitating problem definition, as well as encouraging open discussion on different concepts or ideas, that also allows followers to understand the source and meaning of these. They highlighted that many traditional leadership approaches might not fully fit the innovative leadership required into the future.

Leadership support and guidance are critical in facilitating innovation at the early stage, as these enable successful team processes (Tannenbaum et al. 1996). The skills, attitudes, and knowledge of a leader affect group climate and norms (Hackman 2002) and through monitoring, coaching and feedback, a leader creates a supportive environment, which helps the team to innovate (West et al. 2003) and perform successfully (McIntyre and Salas 1995).

Zaccaro et al. (2001) highlight a series of factors essential for team success, and they see leadership as the most critical. The degree to which the leader draws team objectives, and organizes and manages the team to make sure that these objectives are attained, adds significantly to team innovation (West et al. 2003). According to Yukl et al. (1990), leaders who clearly communicate instructions, such as deadlines, standards, and priorities, were more successful in leading innovative teams.

In examining the research on leaders of creative efforts, Byrne et al. (2009) noted that leaders are likely to structure the work environment "by creating groupings of technical expertise" (p. 259), and promoting effective communication between groups operating in a flat structure. They also, however, note the value of leadership coordination to assist the actors in the creative process. Earlier research noted the role of leaders structuring activities as well as fostering teams of diverse people who communicate effectively with one another (Mumford et al. 2007). The positive relation between innovation and effective, engaging leadership is confirmed by a number of studies in R&D settings (e.g., Keller 1992; Waldman and Atwater 1994).

Interestingly, Bel et al. (2015) examined the interplay between communication, leadership styles, and the probability of successful innovation, and found a positive link between innovation and firm size, regular communication, and result-oriented leadership. However, they also found that although organizations require "both strong leadership and sufficient communication to overcome inertia; frequent communication—particularly amongst strong managers and in larger firms—can cause leaders to pull the firm in different directions, resulting in disagreement and a failure to successfully innovate" (p. 1). This research suggests that as the organization size increases, it will be essential to achieve coordinated and collaborative communication with more emphasis on the effectiveness, rather than quantity, of communication.

Leaders also have a strong impact on employees' work behaviors, which also includes innovation behaviors (De Jong and Den Hartog 2007). The impact can be direct through identifying and addressing followers' intrinsic motivation essential to creativity (Tierney et al. 1999) or indirect by creating a safe climate for exploring different approaches (Amabile et al. 1996). Innovative behaviors in the workplace depend on interpersonal interaction (Anderson et al. 2004; Zhou and Shalley 2003). As Basadur (2004, p. 103) writes about the most effective leaders of tomorrow as the being the ones who "will help individuals (…) to coordinate and integrate their differing styles through a process of applied creativity that includes continuously discovering and defining new problems, solving those problems and implementing the new solutions." We argue that the leadership approaches deployed and underpinned by effective communication including appropriate language are crucial for innovation outcomes. So, what kind approaches are likely to contribute to this?

15.3.1 Organic, Transformational, and Charismatic Leadership

One contemporary leadership approach related to collective innovation is the so-called "organic" paradigm (Avery 2004). These are "leaderless" or "leaderful" organizations or teams, where leadership may not be "vested" in a single individual (p. 63); and leadership roles and tasks might shift among different people as teams self manage over time. It involves organizations where motivated people mutually work together and "sense-make" through collaborative communication. An analogy is a jazz quartet where rhythm, melody, and harmony flow dynamically through the improvisations of the different players to create a whole creative musical performance, and where the quartet members are shifting and signaling seamlessly between leading and supporting roles.

In organic work settings, effective communication exchange among organizational members is extensive to make sense of "rapidly changing circumstances," and to share vision, knowledge, and information (Avery 2004, pp. 63–64). Organic leadership is important, as this modern, ad hoc approach is process oriented, enabling people to quickly innovate, and adjust in a fast changing business environment while mutually solving adaptive problems. The growth of organic organizations (including matrix or latticed structures) is accelerating, particularly in the technology or entrepreneurial arenas (e.g., see Gore case study by Bell, in Avery 2004). Collaborative and transparent behaviors with active listening are essential for clear, open, and compelling communication in these organic situations.

By contrast, the two major leadership approaches that research over time demonstrates as particularly associated with innovation are: transformational and charismatic leadership. Both are built on the assumption of effective leadership communication and interaction.

By intellectually challenging followers, espousing innovation, and communicating a strong vision with a clear sense of emotional purpose, transformational leaders nurture a climate where employees are motivated to search for innovative approaches (Ling et al. 2008). Transformational leaders promote creative ideas, based on the "championing role" they adopt (Howell and Higgins 1990). In these settings employees can exceed performance expectations and be stimulated to take on innovative work techniques. Charismatic leadership with people engagement also influences the organizational climate (Koene et al. 2002). Charismatic leaders demonstrate innovative behaviors that deviate from the regular norms, and in doing so, they permit "out of the box" idea generation for those involved with technological innovations, such as R&D teams (Conger and Kanungo 1987).

There are many research examples of the impact of transformational leadership on creativity and innovation. For example, Sosik et al. (1998) in their research on 36 undergraduate students found that transformational leadership enhanced creativity in a "group decision support system" context. Research by Howell and Avolio (1992) on 78 managers in a Canadian financial institution found transformational leadership behaviors was positively related to the business-unit performance and this relationship also needed clear leadership support for innovation. Research by Gumusluoglu and Ilsev (2009a, b), shows a positive correlation between transformational leadership, followers' creativity, and organizational innovation also influenced by psychological empowerment, intrinsic motivation, and the perception of support for innovation. Likewise, Chen et al. (2014) in their study of 151 CEOs and matching senior management team members from Chinese Manufacturing firms, show transformational leadership to positively influence product innovation performance (conceptualized as the degree to which a new product or service achieves its market share, sales, investment return, and profit objectives).

Although transformational leaders at times can adopt a more directive communication style, they also actively seek followers' involvement by emphasizing the significance of collaboration in performing joint and collective tasks, offering a chance to share and learn, and delegating responsibility to their followers to perform any necessary work to ensure effective performance (Bass 1985). In doing so, they create an empowering environment, where followers pursue innovation in their work tasks. Amabile et al. (1996) show that autonomy enables employees to be more creative, as it enables them to believe that they have greater personal authority over how to go about accomplishing their tasks. Empowerment further caters to the intrinsic motivation of followers, which (as already discussed) contributes to innovative behaviors (Jung and Sosik 2002).

15.3.2 Leadership, Motivating Language, and Framing

Research studies show that employees' behaviors can be impacted by a leader's conscious use of speech (e.g. Mayfield et al. 1995). "Leader talk" enables leaders to seek, and gain, trust, and acknowledgement of subordinates (Reina and Reina 1999). Appropriate leader language skills in motivating and conveying vision are significant (Goleman et al. 2001). Transformational and charismatic leaders try to engage stakeholders around vision, ideas, innovation and change. Two effective communication attributes for these leadership approaches are the notions of motivating language and communication framing.

The motivating language framework involves classifying leadership communication language into three types: direction-giving language, empathetic language and meaning-making language. Direction-giving language occurs when leaders reduce uncertainty by elucidating roles, performance expectations, goals, and responsibilities. Empathetic language happens when leaders go beyond the mere economic exchange between them and their followers, to care for their peoples' emotional well-being. Meaning-making language takes place when leaders convey and express the organizational norms, culture, behaviors, and values that are unique and relevant for each organization building affiliation, supporting change management, and organizational socialization which Jablin (2001) refers to as "Entry and Assimilation."

Mayfield and Mayfield (2004) highlight motivating language as part of the innovation relationship between leader and followers, arguing that when leaders strategically communicate through motivating language, follower innovation increases. Direction-giving language, empathetic language, and meaning-making language encourage innovation through a combination of, for example: catering to followers' intrinsic motivation and understanding of what the task entails; and delineating reward policies and organizational goals, as well as risk-taking methods (Mayfield and Mayfield 2002). Additionally, this lays the groundwork for leadership training interventions to enhance workers' innovativeness (Zorn and Ruccio 1998) and provides ways to maximize employee outcomes such as satisfaction, performance, retention, and so on (Graen et al. 2004; Mayfield et al. 1995).

Fairhurst and Sarr (1996), who view leadership as a "language gamem," contend that framing is the most important skill in this game. "Just as an artist works from a palette of colors to paint a picture, the leader who manages meaning works from a vocabulary of words and symbols to help construct a frame in the mind of the listener" (Fairhurst and Sarr 1996, p. 100). They further explain how framing helps leaders to motivate actions and secure backing for their vision (Fairhurst and Sarr 1996; Fairhurst 2011). These visionary leaders frame the purpose in a way that is relevant and meaningful (Conger 1991). As Snow (2004) states, "Collective action frames, like picture frames, focus attention by punctuating or specifying what in our sensual field is relevant and what is irrelevant, what is "in frame" and what is 'out of frame,' in relation to the object of orientation" (p. 384). Particularly important framing methods for encouraging engagement with innovation are metaphors, stories, examples, and catchphrases.

Innovation literature also identifies the importance of framing. For example, Pfeffermann et al. (2008) note, "framing innovation/s for successful commercialization, innovation communication might be an important managerial function; understood as a firm's capital that tends to enhance competitive advantage" (p. 41). Such innovation communication (combining communication and innovation capital) would include framing innovations, "to facilitate the adoption process." (p. 41).

We argue that for innovation and creativity, an effective leader communicator will use a combination of language that motivates, and messages that are framed, to appeal to the innovation interests of participants. This supports effective communication that is clear, open, and compelling. A leader's language can give others a sense of direction and logic; harness emotive appeal; and place the rational and emotional basis in the context of why people should be doing or committing to something. Using words and phrases that resonate on these different levels in a relevant way should increase the motivation of stakeholders to engage in the innovation process and outcomes. Equally, getting innovation participants to describe objectives or ideas in their own language and choice of words should increase "ownership." It is not transmitting the message effectively (albeit this is crucial for a leader), but interacting, that builds commitment and understanding. Engaging in dialogue with open and constructive questioning, can clarify the vision, objectives and priorities. Besides, framing the messages using stories and examples should "bring to life" both the objectives of an innovation undertaking, and its potential impact. A resonating catchphrase or motivating language can encapsulate innovation intentions and culture. Just consider Apple's own "Think Different" campaign; the description of Toyota process innovation in "Lean Thinking"; or Elon Musk's quote, "Failure is an option here. If things are not failing, you are not innovating enough."

Our review highlights the need for leaders to encourage innovation and creativity, engage stakeholders in the process, engender commitment to change, and enable an environment of creative work and knowledge sharing. In all, effective communication is a key ingredient for the leadership approaches related to innovation outcomes. Some scholars view communication as an essence of leadership rather than a mere technique (Barge 1994; Macik-Frey 2007). Salacuse (2006, p. 23), maintains, "Indeed, leadership could not exist without communication." In a similar vein, Barge (1994, p. 21) comments "leadership is enacted through communication." Literature is replete with studies that highlight the significance of communication interactions in effective leadership (e.g., Den Hartog and Verburg 1997; Fairhurst 2011; Tourish and Jackson 2008). When leading for innovation, these communication interactions occur in different ways (e.g., speaking, listening, reading, writing, behaving, interpersonal relationships) and through different formats (e.g., face-to-face, or technology and media). We believe that the leader's role modeling of effective communication matters, and so does the environment for communication interaction and innovation work that they foster in their organization or team.

So, the question arises, how is this achieved and sustained? The answer lies in building on appropriate leadership approaches, behaviors, and motivating language or framing, by deploying effective leadership communication that is essentially clear, open, and compelling at its core. This will help to close the gap between ineffective and effective communication. To assist with increasing the capacity to do this, we present and recommend two interrelated "tools" that will substantially improve effective leadership communication at every phase of innovation. The fundamental "tool" is building "communication intelligence" among innovation stakeholders, especially when role modeled by leaders.

15.4 Innovation Leadership and Communication Intelligence

Effective leadership communication that reverberates with people is both relevant and comprehensible. The communication exchanges should resonate (such as with motivating language, vivid examples, or cooperative questioning). These exchanges are interactive, dynamic, and contextual. These build trust through the quality of the relationship experiences, and these relationships are stronger when communication contact is inclusive and accessible.

We contend that leaders should utilize their "communication intelligence" to engage stakeholders in innovation and creativity, by demonstrating effective leadership communication that resonates, clarifies, and connects. This will lead to communication that is clear (comprehensible and meaning based), open (inclusive and interactive) and compelling (motivating and relevant).

"Communication Intelligence" (CI) is a model that fully integrates eight elements to achieve effective leadership communication (Woodward 2015). CI combines four mindsets (the things people need to think about for effective communication); and four clusters of communication techniques and qualities (the ways people need to undertake communication activities to be effective). People with high levels of "communication intelligence" use all the mindsets and use techniques from across all four clusters, particularly those that are natural for them. Yet, they increase their communication effectiveness by learning, then using, techniques from other clusters that are less natural for them.

There are four CI Communication Mindsets (the "what," "why," "where," and "who"): Awareness, Message, Presence, Communication Formats; and Four CI Clusters of Communication Qualities and Techniques (the 'how'): rational, structural, expressive, and visual (see Fig. 15.1). For a leader's communication approach to be completely understood, these eight elements sit underneath a person's cultural

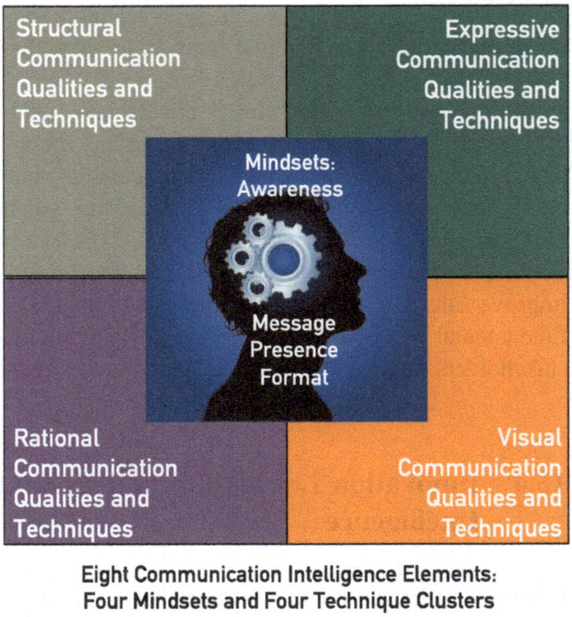

Fig. 15.1 Communication intelligence framework for leaders. *Source* Woodward (2015)

background (because different national cultures have unique communication characteristics), and their individual personality trait of extraversion/introversion—as all eight elements are present in all cultures (contextually adjusted); and are found in both extraverts and introverts (Woodward 2015) (see Fig. 15.1).

CI applies across all kinds of communication situations—public, group, interpersonal, and intrapersonal (with self), and are present across the various communication mediums used by leaders (from email to presentations; from personal conversations to team discussions; from blogs to video posts). Combining these CI elements produces effective interaction between leaders and people, both within and outside their organization or setting, and generates a platform for achieving relevant meaning, connection, and results. It also underscores the multiplicity and complexity of effective leadership communication interactions—especially for dealing with adaptive and creative processes.

The four CI "mindsets" for leaders, interrelated to innovation and creativity are:

- **Awareness**: of self, others, context and purpose. This refers to a person's ability to be deeply aware of the communication needs and preference styles of those involved in communication as well as oneself; and be aware of the situation, context, or purpose of the communication activity (e.g., informing, inquiring, influencing, persuading, entertaining, motivating, inspiring, brainstorming). For example, in innovation leadership communication, the "awareness mindset" is reflected in: a desire to involve others, and to take account of the diversity effects (e.g., culture, gender, generational cohort) that would influence the communication interactions (transmission and interpretation); understanding the likely motivators, language and themes that will be relevant to others; and considering cultural context or physical environment when planning activities, such as brainstorming or innovation evaluation discussions.

- **Message**: the core meaning and content of the communication, supported by structure and appropriate details, as well as message relevance and clarity. For example, in innovation leadership communication, the "message mindset" is reflected in: clear message framing and exchange among stakeholders, striving for comprehension, clarity, and relevance; opening up the free flow of ideas and analysis messaging including an appropriate balance of listening, inquiry, and advocacy (asking and telling) that leads to constructive dialogue and shared understanding; and innovation vision messages that resonate quickly.
- **Presence**: beyond the words—the nonverbal (e.g., body language and gestures), paraverbal (e.g., sound/tone of voice) and visual, symbolic or expressive features. For innovation leadership communication, the visible presence of leaders and followers during their interactions will influence the engagement and environment for creative and open thinking. This is reflected in, for example, open body postures; respectful and measured, yet expressive voice tone; and visual communication devices (charts and technology collaboration tools) that are stylistically owned by the participants in the innovation process. Moreover, some form of visualization and "personal energy" is an essential ingredient in creative brainstorming and ideation activity.
- **Format**: the choice and use of communication formats, media and repertoire that are "fit for purpose and situation" (e.g., behaviors, spoken, written, listening, thinking/reflection, novels, text, email, instant messaging, video, and the like). For example, in innovation leadership communication, the "format mindset" is reflected in continuously adopting or adjusting communication activities, media, and technologies that are available, accessible, appropriate, and useful for each stage of the innovation process and assist "ease of collaboration."

The four CI clusters of communication techniques and qualities for leaders interrelated to innovation and creativity are:

- **Rational**: techniques and qualities that affect the logic, factuality, knowledge level, intellectual substance, idea clarity, and simplicity of language for comprehension in communication. For innovation, "rational" qualities would include: being objective; using verifiable evidence and key facts; suppressing and recognizing bias in thoughts and words (especially when separating idea generation from analysis in the brainstorming and decision stages); clarifying complex ideas and concepts into simple words for understanding; and providing precise summaries of action items and priorities for innovation implementation.
- **Structural**: techniques and qualities that affect the language or sound clarity, consistency, order/flow, construction, thoroughness, levels of detail and accuracy in communication. For innovation, "structural" qualities would include: methods for agenda setting, organizing, disseminating, and exchanging

information; discussion preparation; sequencing participative debate; utilizing deliberate "unstructured" times for communication exchange to allow free-low dialogue and openness without power control; using rhetorical tools, such as "catchphrase," repetition, and triads (lists in three for summation); and ensuring innovation implementation plans are appropriately and accurately documented.
- **Expressive**: qualities that affect the expression, emotion, interactivity, personalization, and authenticity of communication. In innovation leadership communication, "expressive" qualities would include: storytelling; using inspiring, and motivating language; displaying appropriate expressive nonverbal and paraverbal communication (such as body gestures and voice tone) in support of ideas; active listening (where mind, verbal and nonverbal communication are focused); demonstrating personal commitment and enthusiasm; and exhibiting behaviors engendering a sense of trustworthiness, risk-taking, openness, and collaboration.
- **Visual**: qualities that affect the appearance, visuality, conceptuality, creativity, and symbolism of communication. In innovation leadership communication, "visual" qualities would include: active idea generation; producing graphical, design, or visual representations of ideas and messages; demonstrating future facing messaging to allow people to imagine success after problem solving and ideation; and articulating the "big picture." [Adapted from Woodward (2015) and Woodward et al. (2016).]

Combining the CI elements creates clear, open, and compelling communication. This is valuable for all leadership approaches to engage people with innovation and to encourage everyone to have "communication intelligent" interactions. One additional "tool" will help to increase involvement levels in innovation processes, as well as build innovation capability and confidence over time. This is "*INVOLVE*"—the "fair process" leadership communication practices, which are usable across different phases of innovation engagement.

15.5 INVOLVE—Fair Process Leadership and Communication

Our earlier discussion on leadership approaches raised the fundamental issue of successfully managing processes for innovation and creativity. "Fair process leadership" (FPL) is one framework that does this successfully at individual, team, and organizational levels. FPL is an integrative framework that supports effective leadership, particularly in situations emphasizing process engagement and transparency with stakeholders, as well as objective evaluation (Van Der Heyden 2013). These characteristics are important for innovation and creativity cultures, as "fair process" promises a high level of commitment and trust, which are necessary ingredients for high performance (Kim and Mauborgne 1997; Van Der Heyden and Limberg 2007).

In simple terms, "fair process" exist where the participants in any decision-making process understand the process that will be followed, as well as the associated rules and modes of engagement and communication, and perceives these to be fair with respect to all participants. Fair process principles provide a "means," rather than an "end," toward more engaged decision-making and improved implementation (Van Der Heyden and Limberg 2007), which we contend are essential elements for successful innovation.

Van Der Heyden et al. (2005) developed a "*Fair Process Leadership*" model with *a process description*, consisting of five steps (the *5 "E"s*) and a description of *fair play behaviors* that leaders need to demonstrate throughout these steps (the *5 "C"s*). This model as further espoused by Van Der Heyden (2013) represents an interlinked cycle for decision-making, implementation and continuous review (see circular model within Fig. 15.2). It can be directly applicable for teams and organizations seeking innovation built on deep engagement and commitment including organic teams. In additional work, Woodward et al. (2016) posited three communication practices to enhance "fair process" as an actionable concept, these are the '*INVOLVE*' practices (also see Fig. 15.2). How might fair process principles translate into leading, working and communicating within innovation processes? We adapt and cite from the relevant research below.

Adapted from: Van Der Heyden (2013) and Woodward et al. (2016)

Leaders and teams can adopt the "*INVOLVE*" communication practices to enact "fair process" to facilitate innovation, creativity, and engagement. The core

Fig. 15.2 INVOLVE—fair process effective leadership communication

principle is a simple and compelling message: "*INVOLVE*." This is the deep belief and conviction to encourage effective communication that demonstrates fair process for all those involved. Three specific communication practices put this principle into action at each stage of the innovation cycle:

- **Active Listening and Constructive Dialogue** [productive behaviors for interaction, deliberation, option generation, analysis, decision-making, explanation, and evaluation; as well as an appropriate balance of inquiry (asking) and advocacy (telling) with demonstrable active listening];
- **Agreed Communication Protocols** [mutually developed and transparent communication rules with commitment and follow-through that are culturally appropriate—these should not be bureaucratic, but guidelines to coordinate the development and exchange of knowledge as the creative ideas emerge and are taken forward—e.g., the expected rules or ways of behaving for doing unstructured or structured activities, or the expected norms for using technology collaboration systems]; and
- **Accessible Communication Connection** [useful, convenient, and readily available communication activities, formats, and media to facilitate participation and engagement with internal and external stakeholders—these are the communication format choices that make sense at any point in time, and are subject to change during the innovation cycle to foster collaboration]. (Adapted from Woodward et al. 2016.)

The five complementary and mutually reinforcing behavioral characteristics of FPL identified by Van Der Heyden (2013), and to which the "*INVOLVE*" communication practices relate, are:

- **Communication**: the ability to give all actors a "voice" without fear or pressure of retaliation once that "voice" is exercised;
- **Clarity**: the transparency of behaviors, interactions, and exchanges by the actors of the process;
- **Consistency**: the uniformity in the treatment of actors, issues, and steps, including over time;
- **Changeability**: the possibility of "correction," changing actors' beliefs, and possibly changing the chosen course, as a function of new evidence; and
- **Culture**: the commitment to 'do the fair thing' not just superficially, but deeply and authentically. [Adapted from Van Der Heyden (2013).]

With these behaviors in place, all participants can dynamically enact the five stages of the FPL model (Van Der Heyden 2013) for innovation and creativity—deploying the "*INVOLVE*" practices along the way:

- *Engage*: Establish an innovation process to involve relevant people; seek inputs to framing issues and generating ideas; seek constructive challenge to views; make contributions to the process design and priorities before the decisions are actually made, when influencing this is still a possibility.

- *Explore*: Generate and explore all options and their potential outcomes thoroughly and comprehensively. Allow an open and dynamic ideation process, by not closing options early and keeping idea generation separated from analysis. Then through constructive debate and analysis eliminate those options that are neither promising nor capable of successful implementation, and take forward the most prospective options.
- *Explain*: Make a clear innovation decision, where the leadership (or group in an organic setting) explains its rationale. Effective communication will take sufficient time and energy to develop understanding, especially for those impacted outside the decision-making group. All the innovation participants should be thoroughly briefed, fully committed, and hold clear and compelling messages for stakeholders outside the decision group. Roles, responsibilities and priorities for successful implementation and execution are articulated clearly; and the challenges, expected benefits, rewards, or appropriate sanctions for poor execution are enunciated.
- *Execute*: Ensure all relevant individuals implicated by the innovation decision are clear on what they are supposed to do and their focus for implementation. Adjust and adapt if outcomes are not according to plan, while informing and involving others to sustain coordination in execution; and maintain rewards (or sanctions) in line with expectations formed and announced previously.
- *Evaluate*: Seek critical feedback from relevant stakeholders on the decision, the plan, and the process followed to get there; share lessons learnt based on the evidence; utilize this knowledge for future innovation process work. [Adapted from Van Der Heyden et al. (2005), Van Der Heyden (2013) and Woodward et al. (2016).]

We argue that when individuals, teams, and organizations demonstrate "fair process" and the "*INVOLVE*" practices for innovation processes there will be transparent, respectful, constructive, and objective communication behaviors. This is characterized by communication described as: "open and authentic rather than hidden or opaque; inclusive rather than dictating; and clear rather than confused" (Woodward et al. 2016). Such communication encourages creative thought with open minds and comprehensible articulation, that is, "communication intelligent," and clear, open, and compelling.

15.6 Conclusion

Today's world is volatile, uncertain, complex, ambiguous, and diverse (VUCAD). It is intensely competitive, with change as the "increasing constant." Innovation (from process engineering and new products, to technology creation, and new business models) is an imperative for contemporary business. In global organizations, there are initiatives to increase learning, share knowledge, and to develop new capabilities for leaders to engage people and ideas. These are intended to deliver positive and dynamic business outcomes (Beechler and Woodward 2009). This is especially so in entrepreneurial and emerging organizations which are seeing rapid growth in the twenty-first century (Koryak et al. 2015); as well as global business opportunities for technology innovation, transfer, and investment (Audretsch et al. 2014). Effective communication interactions are essential to empower these innovation initiatives and exchanges. This applies equally to positional leaders and their followers, and where leadership is dispersed and organic. Leaders are required to champion innovation by: planning, implementing, and assessing innovation, shaping and managing various capabilities; and assembling resources at inter and intra-organizational levels (Zerfass et al. 2004). Communicating by inviting and responding to creative ideas is the first step for engaging employees and defining innovation objectives.

As such, we believe that effective leadership communication is evolving in a VUCAD world as an essential ingredient for successful innovation—whether in tapping innovation resources and investment, contributing to ideation, implementing innovation decisions, or interacting with customers. By communicating effectively leaders can increase their ability to nurture innovation, and translate complex innovations in a way that others comprehend, accept, and then embrace. In the innovation workspace, leadership communication is embedded in concrete actions, language, processes and interpersonal relations, and in the depth and breadth of idea dialogue amongst "aware" and motivated participants.

We contend that leadership communication built on the two "tools" ("communication intelligence" and the "*INVOLVE*" communication practices of "fair process"); combined with appropriate innovation based leadership approaches, collaborative behaviors, motivating language, and framing; can positively inspire and engage people toward innovation outcomes and support innovation cultures. We believe these should make a substantial contribution to closing the gap between ineffective and effective leadership communication for innovation. These are part of the solution to: ensuring ideation is not cut short; increasing the commitment levels to innovation decisions; clarifying the focus and understanding of innovation implementation issues and priorities; improving articulation of the benefits of the innovation for internal and external stakeholders; and learning from evaluation and knowledge exchange that is objective and constructive.

Furthermore, "communication intelligence" with involvement-based "fair process" will encourage trust, risk-taking, creativity, and collaboration. This supports an environment where people are more likely to contribute and commit to the changes, new directions or initiatives; and see these to fruition. This builds capacity and confidence for innovative and creative work into the future. Effective leadership communication for innovation will be framed to appeal to emotion and rationality. It will be replete with relevant messages, visuality, expressive examples, interactive engagement, listening, motivating language, engagement processes, and "communication intelligence." Such innovation communication will be clear, open, and compelling.

Bibliography

Amabile, T.M., Conti, R., Coon, H., Lazenby, J., Herron, M.: Assessing the work environment for creativity. Acad. Manag. J. **39**(5), 1154–1184 (1996)

Anderson, N., De Dreu, C.K., Nijstad, B.A.: The routinization of innovation research: a constructively critical review of the state-of-the-science. J. Organ. Behav. **25**(2), 147–173 (2004)

Audretsch, D., Lehmann, E., Wright, M.: Technology transfer in a global economy. J. Technol. Transf. **39**, 301–312 (2014)

Avery, G.C.: Understanding Leadership: Paradigms and Cases. Sage Publications, London (2004)

Barge, J.: Leadership: Communication Skills for Organizations and Groups. St. Martin's Press, New York (1994)

Basadur, M.: Leading others to think innovatively together: creative leadership. Leadersh. Quart. **15**(1), 103–121 (2004)

Bass, B.M.: Leadership and Performance Beyond Expectations. Free Press, New York, NY (1985)

Beechler, S., Woodward, I.C.: The global "war for talent". J. Int. Manage. **15**(3), 273–285 (2009). doi:10.1016/j.intman.2009.01.002

Bel, R., Smirnov, V., Wait, A.: Team composition, worker effort and welfare. Int. J. Ind. Organ. **41**, 1–8 (2015)

Butler, J.: Toward understanding and measuring conditions of trust: evolution of a conditions of trust inventory. J. Manage. **17**, 643–663 (1991)

Byrne, C.L., Mumford, M.D., Barrett, J.D., Vessey, W.B.: Examining the leaders of creative efforts: what do they do, and what do they think about? Creativity Innov. Manage. **18**(4), 256–268 (2009)

Cagliano, R., Chiesa, V., Manzini, R.: Differences and similarities in managing technological collaborations in research, development, and manufacturing: a case study. J. Eng. Technol. Manage. **17**(2), 193–224 (2000)

Cardinal, L.B., Hatfield, D.E.: Internal knowledge generation: the research laboratory and innovative productivity in the pharmaceutical industry. J. Eng. Tech. Manage. **17**(3), 247–271 (2000)

Chen, Y., Tang, G., Jin, J., Xie, Q., Li, J.: CEOs' transformational leadership and product innovation performance: the roles of corporate entrepreneurship and technology orientation. J. Prod. Innov. Manage. **31**(S1), 2–17 (2014)

Collins, M.A., Amabile, T.M.: Motivation and creativity. In: Sternberg, R.J. (ed.) Handbook of Creativity, pp. 297–312. Cambridge University Press, Cambridge, UK (1999)

Conger, J.A.: Inspiring others: the language of leadership. Executive **5**(1), 31–45 (1991)

Conger, J.A., Kanungo, R.N.: Toward a behavioral theory of charismatic leadership in organizational settings. Acad. Manag. Rev. **12**(4), 637–647 (1987)

D'Aveni, R.A.: Strategic supremacy through disruption and dominance. Sloan Manag. Rev. **40**(3), 127–135 (1999)

Dalkir, K.: Knowledge Management in Theory and Practice. Routledge, London, UK (2013)

De Jong, J.P.J., Den Hartog, D.N.: How leaders influence employees' innovative behavior. Eur. J. Innovative Manage. **10**(1), 41–64 (2007)

Den Hartog, D.N., Verburg, R.M.: Charisma and rhetoric: communicative techniques of international business leaders. Leadersh. Quart. **8**(4), 355–391 (1997)

Drucker, P.F.: Managing in a Time of Great Change. Harvard Business Press, Boston, MA (2009)

Dudek, S.Z., Hall, W.B.: Personality consistency: eminent architects 25 years later. Creativity Res. J. **4**(3), 213–231 (1991)

Elkins, T., Keller, R.T.: Leadership in research and development organizations: a literature review and conceptual framework. Leadersh. Quart. **14**, 587–606 (2003)

Fairhurst, G.T.: The Power of Framing: Creating the Language of Leadership. Jossey Bass, San Francisco, CA (2011)

Fairhurst, G.T., Sarr, R.A.: The Art of Framing: Managing the Language of Leadership. Jossey-Bass, San Francisco, CA (1996)
Ford, C.M.: Creative developments in creativity theory. Acad. Manag. Rev. **25**(2), 284–289 (2000)
Goleman, D., Boyatzis, R., McKee, A.: Primal leadership: the hidden driver of great performance. Harvard Bus. Rev. **79**(11), 42–53 (2001)
Graen, G.B., Hui, C., Gu, Q.: A new approach to intercultural cooperation. In: Graen, G.B. (ed.) New Frontiers of Leadership, pp. 225–246. Information Age Publishing, Greenwich, CT (2004)
Gumusluoglu, L., Ilsev, A.: Transformational leadership and organizational innovation: the roles of internal and external support for innovation. J. Prod. Innov. Manage. **26**(3), 264–277 (2009a)
Gumusluoglu, L., Ilsev, A.: Transformational leadership, creativity, and organizational innovation. J. Bus. Res. **62**(4), 461–473 (2009b)
Hackman, J.R.: Leading Teams: Setting the Stage for Great Performances. Harvard Business School Press, Boston, MA (2002)
Heifetz, R.A.: Leadership Without Easy Answers. Belknap/Harvard University Press, Cambridge, MA (1994)
Howell, J.M., Avolio, B.J.: The ethics of charismatic leadership: submission or liberation? Executive **6**(2), 43–52 (1992)
Howell, J.M., Higgins, C.A.: Champions of technological innovation. Adm. Sci. Q. **35**(2), 317–341 (1990)
Imai, M.: Kaizen. Random House Business Division, New York (1986)
Jablin, F.M.: Organizational entry, assimilation, and disengagement/exit. In: Jablin, F.M., Putnam, L.L. (eds.) The New Handbook of Organizational Communication: Advances in Theory, Research, and Methods, pp. 732–818. Sage Publications, Thousand Oaks, CA (2001)
Jung, D.I., Sosik, J.J.: Transformational leadership in work groups the role of empowerment, cohesiveness, and collective-efficacy on perceived group performance. Small Group Res. **33**(3), 313–336 (2002)
Jung, D.I., Chow, C., Wu, A.: The role of transformational leadership in enhancing organizational innovation: hypotheses and some preliminary findings. Leadersh. Quart. **14**(4), 525–544 (2003)
Keller, R.T.: Transformational leadership and the performance of research and development project groups. J. Manag. **18**(3), 489–501 (1992)
Kim, W.C., Mauborgne, R.: Fair process: managing in the knowledge economy. Harvard Bus. Rev. **75**(4), 65–75 (1997)
Koene, B.A., Vogelaar, A.L., Soeters, J.L.: Leadership effects on organizational climate and financial performance: local leadership effect in chain organizations. Leadersh. Quart. **13**(3), 193–215 (2002)
Koryak, O., Mole, K., Lockett, A.: Entrepreneurial leadership, capabilities and firm growth. Int. Small Bus. J. **33**, 89–105 (2015)
Ling, Y.A.N., Simsek, Z., Lubatkin, M.H., Veiga, J.F.: Transformational leadership's role in promoting corporate entrepreneurship: examining the CEO–TMT interface. Acad. Manag. J. **51**(3), 557–576 (2008)
Lonergan, D.C., Scott, G.M., Mumford, M.D.: Evaluative aspects of creative thought: effects of idea appraisal and revision standards. Creativity Res. J. **16**, 231–246 (2004)
Macik-Frey, M.: A Communication-Centered Approach to Leadership: The Relationship of Interpersonal Communication Competence to Transformational Leadership and Emotional Intelligence. University of Texas, Arlington (2007)
Mayfield, J., Mayfield, M.: Leader communication strategies: critical paths to improving employee commitment. Am. Bus. Rev. **20**(2), 89–94 (2002)
Mayfield, M., Mayfield, J.: The effects of leader communication on worker innovation. Am. Bus. Rev. **22**(2), 46–51 (2004)
Mayfield, M., Mayfield, J.: Leadership techniques for nurturing worker garden variety creativity. J. Manage. Dev. **27**(9), 976–986 (2008)

Mayfield, J., Mayfield, M., Kopf, J.: Motivating language: exploring theory with scale development. J. Bus. Commun. **32**(4), 329–344 (1995)
McCauley, D., Kuhnert, K.: A theoretical review and empirical investigation of employee trust in management. Publ. Adm. Quart., 265–284 (1992)
McIntyre, R.M., Salas, E.: Measuring and managing for team performance: emerging principles from complex environments. In: Guzzo, R., Salas, E. (eds.) Team Effectiveness and Decision Making in Organizations, pp. 9–45. Jossey-Bass, San Francisco, CA (1995)
Mumford, M.D., Gustafson, S.B.: Creativity syndrome: integration, application, and innovation. Psychol. Bull. **103**(1), 27–43 (1988)
Mumford, M.D., Licuanan, B.: Leading for innovation: conclusions, issues and directions. Leadersh. Quart. **15**(1), 163–171 (2004)
Mumford, M.D., Whetzel, D.C., Reiter-Palmon, R.: Thinking creatively at work: organizational influences on creative problem-solving. J. Creative Behav. **31**(1), 7–17 (1997)
Mumford, M.D., Scott, G.M., Gaddis, B., Strange, J.M.: Leading creative people: orchestrating expertise and relationships. Leadersh. Quart. **13**(6), 705–750 (2002)
Mumford, M.D., Baughman, W.A., Sager, C.E.: Picking the right material: cognition processing skills and their role in creative thought. In: Runco, M.A. (ed.) Critical and Creative Thinking, pp. 19–68. Hampton Press, Cresskill, NJ (2003)
Mumford, M.D., Eubanks, D.L., Murphy, S.T.: Creating the conditions for success: best practices in leading for innovation. In: Conger, J.A., Riggio, R.E. (eds.) The Practice of Leadership: Developing the Next Generation of Leaders, pp. 129–149. Jossey-Bass, San Francisco, CA (2007)
Oldham, G.R., Cummings, A.: Employee creativity: personal and contextual factors at work. Acad. Manag. J. **39**(3), 607–634 (1996)
Osayawe Ehigie, B., Clement Akpan, R.: Roles of perceived leadership styles and rewards in the practice of total quality management. Leadersh. Organ. Dev. J. **25**(1), 24–40 (2004)
Pfeffermann, N.: The scent of innovation: towards an integrated management concept for visual and scent communication of innovation. In: Hulsmann, M., Pfeffermann, N. (eds.) Strategies and Communications for Innovations, pp. 163–181. Springer, Berlin, Germany (2011)
Pfeffermann, N., Hüulsmann, M., Scholz-Reiter, B.: Framing Innovations to Grasp Stakeholders' Attention: A Dynamic Capability-Based Conception of Innovation Communication (2008)
Qin, Y., Simon, H.A.: Laboratory replication of the scientific process. Cogn. Sci. **14**(2), 181–312 (1990)
Reina, D.S., Reina, M.L.: Trust and Betrayal in the Workplace. Berrett-Koehler, San Francisco, CA (1999)
Salacuse, J.: Leading Leaders: How to Manage Smart, Talented, Rich, and Powerful People. American Management Association, New York, NY (2006)
Senge, P.: The Fifth Discipline: The Art and Practice of the Learning Organization. Doubleday, New York (1990)
Snow, D.A.: Framing processes, ideology and discursive fields. In: Snow, D.A., Soule, S.A., Kriesi, H. (eds.) The Blackwell Companion to Social Movements, pp. 380–412. Blackwell Publishing, Oxford, UK (2004)
Sosik, J.J., Kahai, S.S., Avolio, B.J.: Transformational leadership and dimensions of creativity: motivating idea generation in computer-mediated groups. Creativity Res. J. **11**(2), 111–121 (1998)
Stoker, J.I., Looise, J.C., Fisscher, O.A.M., de Jong, R.D.: Leadership and innovation: relations between leadership, individual characteristics and the functioning of R&D teams. Int. J. Hum. Resour. Manage. **12**(7), 1141–1151 (2001)
Tannenbaum, S.I., Salas, E., Cannon-Bowers, J.A.: Promoting team effectiveness. In: West, M.A. (ed.) Handbook of Work Group Psychology, pp. 503–529. Wiley, Sussex, UK (1996)
Tierney, P., Farmer, S.M., Graen, G.B.: An examination of leadership and employee creativity: the relevance of traits and relationships. Pers. Psychol. **52**(3), 591–620 (1999)
Tourish, D., Jackson, B.: Guest editorial: communication and leadership: an open invitation to engage. Leadership **4**(3), 219–225 (2008)

Van Der Heyden, L.: Setting a tone of fairness at the top. J. Bus. Compliance, 19–29 (2013). Retrieved from https://scholar.google.com.sg/scholar?hl=en&q=setting+a+tone+for+fairness+at+the+top&btnG=&as_sdt=1%2C5&as_sdtp=#0

Van Der Heyden, L., Limberg, T.: Why fairness matters. Int. Commer. Rev. **7**(2), 94–102 (2007)

Van Der Heyden, L., Blondel, C., Carlock, R.S.: Fair process: striving for justice in family business. Fam. Bus. Rev. **XVIII**, 1–21 (2005)

Vincent, A.S., Decker, B.P., Mumford, M.D.: Divergent thinking, intelligence, and expertise: a test of alternative models. Creativity Res. J. **14**(2), 163–178 (2002)

Waldman, D.A., Atwater, L.E.: The nature of effective leadership and championing processes at different levels in an R&D hierarchy. J. High Technol. Manage. Res. **5**(2), 233–245 (1994)

Weisberg, R.W.: Creativity and knowledge: a challenge to theories. In: Sternberg, R.J. (ed.) Handbook of Creativity, pp. 226–251. Cambridge University Press, Cambridge, UK (1999)

West, M.A., Borrill, C.S., Dawson, J.F., Brodbeck, F.C., Shapiro, D.A., Haward, R.: Leadership clarity and team innovation in health care. Leadersh. Quart. **14**(4–5), 393–410 (2003)

Woodward, I.C., More, E.A., Van der Heyden, L.: "Involve": The Foundation for Fair Process Leadership Communication (INSEAD Working Paper Series No. 2016/17/OBH/TOM/EFE). Retrieved from http://ssrn.com/abstract=2747990 (2016)

Woodward, I.C.: Understanding Communication Preference Styles Guidebook, 2nd edn. INSEAD (2015)

Yukl, G., Wall, S., Lepsinger, R.: Preliminary report on validation of the managerial practices survey. In: Glark, K.E., Glark, M.B. (eds.) Measures of Leadership, pp. 223–238. Leadership Library of America, West Orange, NJ (1990)

Zaccaro, S.J., Rittman, A.L., Marks, M.A.: Team leadership. Leadersh. Quart. **12**(4), 451–483 (2001)

Zaltman, G., Duncan, R., Holbeck, J.: Innovations and Organizations. Wiley, New York, NY (1973)

Zerfass, A., Sandhu, S., Huck, S.: Innovationskommunikation-Strategisches Handlungsfeldfür Corporate Communications. In: Bentele, G., Piwinger, M., Schönborn, G. (eds.) Kommunika-tionsmanagement, pp. 1–30. Luchterhand, Neuwied, Germany (2004)

Zhou, J., Shalley, C.E.: Research on employee creativity: a critical review and directions for future research. In: Martocchio, J. (ed.) Research in Personnel and Human Resources Management, pp. 165–217. Elsevier, Oxford, UK (2003)

Zorn Jr., T.E., Ruccio, S.E.: The use of communication to motivate college sales teams. J. Bus. Commun. **35**(4), 468–499 (1998)

Chapter 16
Redefining Collaborative Innovation in the Digital Economy

Eric Viardot

Abstract The continuous growth of the digital economy is redefining the way innovative companies are collaborating with external partners. We analyze the case of the technology driven companies which are at the forefront of this evolution that impact both the inbound phase of the collaborative innovation, to get new ideas, knowledge as well as technology, and the outbound phase of commercialization and adoption of innovative products and services.

16.1 Introduction

With more than 3.2 billion people connected to the Internet, via computer or mobile device, in November 2015, the digital economy is now a fact of life and it keeps growing fast: the number of Internet users reached one billion in 2005, two billion in 2010, and three billion in 2014 (InternetLive stats 2015). Meanwhile an increasing amount of information, sometimes referred as "big data" is now available on the Internet. At the end of 2015, the total traffic over the Internet was close to one zettabyte, i.e. one billion gigabytes (Cisco 2016), while about 90% of the world's data has been generated in the past two years alone (Dragland 2013).

The digital economy is characterized by an increase in digitization of businesses and the intensification of e-commerce, as the new generation of Internet users is getting even more mobile, interaction and transaction. Another consequence of the digital economy has been the fast emergence of new players which have disrupted traditional business while reshaping markets, such as Amazon, Apple, Google and Facebook. In addition, many other lesser known digital companies, including private companies valued at $1 billion or more like Xiaomi, Palantir or Flipkart—commonly called "unicorns" (Fortune 2016) as well as a score of other well-funded start-ups have also disrupted traditional business. Bradley et al. (2015) have investigated the impact of digital disruption for 12 industries and 13 countries and

E. Viardot (✉)
EADA, c.Aragó 204, 08011 Barcelona, Spain
e-mail: eviardot@eada.edu

N. Pfeffermann and J. Gould (eds.), *Strategy and Communication for Innovation*,
DOI 10.1007/978-3-319-49542-2_16

have concluded that digital disruption will displace approximately 40% of incumbent companies within five years.

In such a fast moving digital environment, innovation has become a key priority for companies (Wagner et al. 2014). However, innovating is not an easy path to market success. One reason is that it is difficult to anticipate the market potential of an innovation as many inventions proceed to solve a specific problem but often turn out to have unexpected uses in unexpected conditions (Klein and Tornatzky 1982). Another cause is that very often, the impact of an innovation relies on complementary inventions, which contribute to a full system solution that will add to its performance and, consequently, its demand (Chesbrough and Teece 1996). An extra source of uncertainty is that development time for these complementary innovations can fluctuate very significantly (Viardot 2011).

But the failure to innovate is much riskier than the alternative of doing nothing while a successful innovation can be the source of a unique and sustainable differentiation which provides a competitive edge and generates significant profitability.

This is especially notable in technology driven industries such as telecommunications, electronics, or information technologies and services, where the pace of technology innovation is quite substantial. The Technology Products and Services sector is actually the first industry prone to be impacted by digital disruption but it is also the emblematic sector where innovation is the best way to attain success in the business.

In the latest survey done by the Boston Consulting Group (2015), four technology companies are cited as the top five most innovative companies by a poll of more than 1500 executives. Since the first edition of this survey in 2004, technologies constantly dominate the top-ten list. Apple has been number one every year since 2005 and Google has been number two every year since 2006. Microsoft has been in the top-ten every year since 2005 and has been joined by Samsung since 2013. IBM and Sony have also been in the top-ten nearly every year since 2005. Interestingly, in 2014 Facebook was listed for the first time as one of the most innovative companies which illustrate the accrued importance of the social technologies in the digital economy. Innovation has permitted to those companies and others to achieve a "winner-takes-all" position (Frick and Torres 2002) for a given innovative product category with a very dominant market share as illustrated in Table 16.1.

16.2 The Rise of Collaborative Innovation

An analysis of the various case studies shows that all those winning companies rely less on technology than on their marketing capacities to transform a successful idea into a product or a service which is valuable to customers (Viardot 2004). Among the marketing skills which those firms have developed, communication is extremely

Table 16.1 Dominant technology companies in selected innovative product categories[a] Year 2015

Product/service category	Market share(%)	Names of the dominant players
Operating systems (PC)	93.5	Microsoft
DRAM chips	93	Samsung, Hynix, Micron
Mainframe	90	IBM
Browser	90	Google, Microsoft, Mozilla
Social networks	88	Facebook, Google
PC microprocessors (notebooks)	82	Intel
Operating systems (Smart phones)	78.5	Android
PC microprocessors (desktops)	75	Intel
Relational database software	71	Oracle, IBM
Desktop search engine	69	Google
Chat application	69	WhatsApp, FbMessenger, QQMobile, Wechat
Wireless local area network (WLAN)	68	Cisco, Aruba
Servers	55	HP, Dell, IBM
Custom semiconductors	53.7	TSMC
Personal computer	51	Lenovo, HP, Dell
Tablets	42.5	Apple, Samsung

Source Companies annual reports, press releases, Bloomberg, IDC, Gartner group
[a]Companies are associated with its major successful innovation, even though the company may be diversified in other businesses. For example, Google is still closely related to web search engine, as Microsoft is with PC software, or Apple and Samsung with smartphones

important in order to create awareness for the new product in the market and to convince "early adopters" to buy the innovation (Frattini et al. 2013).

This ability to communicate externally has become even more fundamental as in the recent years, there has been a major paradigm shift in the innovation process with the rise of collaborative innovation (Baldwin and von Hippel 2009). In the past few years the proportion of large innovative firms that rely heavily on external support for innovation has increased dramatically because top executives believe that their organizations will no longer succeed alone when faced with the complexity of the world and they have to engage and collaborate with the external world system of customers, partners, governments and institutions (IBM 2011). A study about collaborative innovation in "Digital Europe" found that 62% of respondents are already making more than a quarter of their revenues through collaborative innovation (World Economic Forum 2015).

The logic of collaborating with customers and other partners to innovate is not particularly new, but the trend towards open innovation has been dramatically accelerated with the development of the digital economy technology, where real-time communication fosters external and internal learning networks by establishing and enhancing the quantity and quality of communications (Inauen and Schenker-Wicki 2012).

Various studies have shown the value generating effects of integrating a broad range of external parties which are bringing a large range of resources, skills, as well as technical and commercial competences in the innovation process (Love and Roper 1999; Tether and Tajar 2008). Other works have underlined that companies relying on external parties have better innovation performance than endocentric companies (Miotti and Sachwald 2003; Nieto and Santamaría 2007). A recent research has also shown that companies that emphasize innovation are more likely to create radical innovations while firms pursuing closed innovation are more likely to exhibit a higher incremental product innovation performance (Bigliardi et al. 2012). When engaging in collaborative innovation companies look forward also to accelerating time to complete their research projects while mitigating the risks and reducing the costs per project.

In this chapter, we will detail how the collaborative innovation process is changing with the development of the digital economy. We will use the typology of Gassmann and Enkel (2004) which is built along the flows between a company and the many external innovation stakeholders. There is the "inbound" process where ideas, knowledge, and technology are obtained from the outside and brought into the company at the level of research and development of new solutions. There is the "outbound" process which refers to the external exploitation of the internal knowledge of a firm in order to develop or commercialize the innovation. Thus in the first part of this chapter we will analyze the recent evolution of the collaboration in the inbound innovation process of successful technology firms. Then we will examine the changes in the practice of collaboration for the outbound phase of innovation management.

16.3 The Impact of the Digital Economy on Collaboration in the Inbound Innovation Phase

The digital economy is drastically changing the way companies are collaborating with their environment for developing innovative solutions. Firms are now trying to pull all potential contributors to their innovation process in order to get new ideas, feedback or technologies. There is a wide variety of potential external partners available for companies which are looking to initiate collaborative innovation. They are the customers, the suppliers, the competitors, the universities the private research institutes, the government research organizations, the "complementors" that provide the product and services around the technology, the consultants, acting as carriers of the innovation or facilitators to the markets.

As the world requires greater levels of digital collaboration and cooperation among participants, the various partners have been increasingly organized into wide and dynamic ecosystems around large technology companies. There is also an increasing interest to collaborate with young, dynamic, small start-up firms. Finally, the expansion of the collaboration goes beyond businesses with crowdsourcing which aims to enrol everyone in creating innovation.

16.3.1 The Ecosystem as the New Paradigm for Collaborative Innovation in the Digital Economy

Some companies are pushing aggressively the forming of an external innovation network with various partners in order not only to get new ideas but also to develop the products or services. Ecosystem innovation is made possible through digitally enabled platforms, which enable companies of all sizes to create, craft, develop and market new products and solutions. Constant feedbacks for improvement are constantly searched from the network participants and they are nurtured by a digital communication flow which is forcefully managed by the conducting firm. Moreover, the addition of more participants to a group creates an incentive for others to join in. Such a snowball effect may provide the necessary momentum to make an innovation successful enough to become a de facto standard and eliminate other competitive solutions.

Apple offers a good example of ecosystem innovation with its iTunes service or its iPhones. ITunes was launched in 1998 as a simple music player to support the iPod, an MP3 player, but over time it has developed into a sophisticated multimedia content manager available for all the Apple devices including the iPhone and the iPad with the collaboration of music artists, publishers, movies and TV shows producers, etc. Similarly, a large part of the success of the iPhone is based on the open innovation of millions of software applications developers. In July 2015, there were more than 1.5 million application software available for iPhone users (Statista 2015). In this innovating ecosystem, the collaborative innovation is symbiotic (Thomas and Wind 2013) as the partners, the application developers on one side and Apple on the other side, need each other to be successful.

Apple is not the only case in the technology industry. For instance, Microsoft, SAP, or IBM have made and forged entire ecosystems around their solutions, namely Windows, Hana, and Notes, with application developers, system integrators, trainers, and hardware companies working together to provide solutions to end users.

- The Microsoft Partner Ecosystem included about 640,000 partners including software developers, all types of distributors, telecommunication companies, and Internet hosting services. Microsoft is spending around more than USD 5 billion a year to manage such an ecosystem, including channel incentives, partner marketing business and investment fund. One of the first decision of the new CEO of Microsoft in 2015 has been to reemphasize the importance of a collaborative ecosystem strategy for the company while playing down the intent made by his predecessor to grow a standalone phone business (Rubin 2015).

- SAP, the leader in ERP software for business-to business applications, had more than 13,300 partners all over the world in 2015—which the company describes as the SAP Ecosystem—working with and around its software solutions to offer an extensive range of industry specific solutions for its professional customers. In addition to its 14 Development centres known as SAP Labs and its 21 research locations worldwide, SAP has set up 13 Co-Innovation and Living Labs worldwide and it has invested in more than 130 Information Technology startups through its venture capital structure named Sapphire Ventures (SAP 2015)
- In the same line, IBM is now offering to its customers, business partners, regional government and academia to connect directly with its 13 worldwide IBM Collaborative Innovation Centers. Their role is to develop in-demand skills, to accelerate research innovations into markets and to drive economic development at regional level (IBM 2015).

Another innovation driven company, Samsung has recently decided to put the principles of Open Innovation into operation in addition to its existing overseas research centres. In July 2013, Samsung has opened its Open Innovation Centre (OIC) with 4 main activities: commercial partnerships with third party, investments in start-ups, acquisitions of small tech firms and innovation incubation with 2 accelerators in Palo Alto and New York City. The explicit objective of the OIC is to develop innovative software and services that touch the core software products that are on all Samsung flagship devices: phones, TVs, tablets, cameras (Penenberg 2015).

Another recent example of technologies companies which are actively nurturing an innovation ecosystem is Huawei. In May 2015 Huawei launched LiteOS, an operating system designed for running connected appliances and machines which is said to be 20% faster than other systems and which is open-source so that developers can modify the code and use it in a wide range of devices. Huawei is now actively working with domestic and overseas partners to make an ecosystem centred on LiteOS.

The development of the digital economy encouraged "open collaboration" which originated at the end of the 80s with the development of new "open source" application software such as Linux, Apache or Mozilla for instance. They were developed in a collaborative manner with a free access to an end product's design and implementation details as well as a free redistribution. The success of "open-source" software was achieved by making the software architecture widely available for free, so that it could benefit from the value co-creation by the complementors, the customers and any other third party.

However, the "open-source" approach has also some limitations, especially regarding the full compatibility of new software versions over time. From a single software project emerged different versions of the original, because of the split between various development teams with no coordination or control. The most famous example is the multiple versions of the original UNIX computer operating system which was developed in the 70s by AT&T's Bell Labs but is now sold in many different and often incompatible versions, including HP/UX. AIX (IBM), Berkeley BSD, SINIX (Siemens), Solaris (Sun), Inx (Silicon Graphics), etc. Consequently, an application developed originally for the UNIX market could run only on one of the versions and required a substantial adaptation to run on another version.

One key lesson from the "open source" software is that opening the innovation process does not guarantee the full compatibility of an innovation over time. It requires an aggressive stand from a company to make sure that this will happen and will last in order to make the innovation widely available to external parties who will adopt and fine-tune it.

16.3.2 Opening to Smaller Companies Young Dynamic Innovative Firms

The growing digital economy is also giving path to an accrued interest for large firms to collaborate with young companies in order to innovate more aggressively. Especially, young firms are frequently closer to new categories of users or potential users of novel products and services in growth-oriented markets. They are usually organized around the development of really innovative and potentially disruptive solutions. They can also experiment more rapidly than large firms allowing them to adapt more quickly to any change in customers' needs or desires. Conversely, recently established firms are habitually lacking the financial, marketing, and organizational resources required to scale their business and increase growth. So there is a sound rationale of collaborative innovation between large and small firms to make the most of their complementary proficiencies.

Traditionally there has been 3 ways for collaboration: Corporate venture, incubators (also known as accelerators) and joint innovation.

- **Corporate venture** is when a large firm—or its investment entity—invest equity into a privately held startup. This kind of collaboration is especially interesting for large firms which are willing to cut on the financial risk of betting only on internal research and/or which are looking for the next "cutting edge" technology or product innovation. Nearly half of the top 100 companies in the Fortune 500 ranking have a corporate venturing unit with Google Venture, Intel Capital, Qualcomm Ventures Salesforce venture and Comcast Ventures being the most active in 2015 in the US with investments in Internet, mobile and cybersecurity companies. About one-third of the total of corporate venture funds in the US are invested in digital companies (Global Corporate Venturing 2015).
- **Incubators** are fixed-term programs physically locating a start-up firm in one work space with many other start-up companies. Incubators are supporting start-up firms with consulting, mentoring, prototype creation and other services.

They are run independently by large companies or jointly with other actors such as universities or governments for example. Quite often, the start-ups in these incubators can be also venture funded by large companies at the same time. For example, the Microsoft Innovation Centers, present in over 100 locations worldwide, and embodying partnership with local government, universities and industry partners, are incorporated within Microsoft Ventures. They offer three years of free software, developer tools and free Cloud Services.

- Recently incubators have been completed with startup "**accelerators**" which are quite similar but with some distinct differences as the goal is to help to jump start a business as fast as possible (Deering et al. 2014) while there is no time limit for a company to stay in an incubator. The time in an incubator space is typically limited to 3–4 months and the cash investment from the accelerator itself is very minimal compared to an incubator. The main benefit of an accelerator is to improve the chances of raising venture capital from a third-party entity after graduating from the program, as the best accelerators significant educational and mentorship programs that culminate in a demo day (Solomon 2015).
- **Joint innovation** or co-creation is defined as an agreement between a large firm and one or many partners to work together for the development and/or the marketing of a common and new product, service or technology. It may take the form of development funding, development of new Intellectual Property (IP), purchasing agreement. Joint innovation usually covers a specific part of the innovation lifecycle (ideation, prototyping, development, production and commercialization) and usually requires allocation of resources limited to a project scope.

But the digital economy is reshaping the way large and small firms are collaborating for innovation as illustrated by a recent survey from Accenture (2015) about the status of the previously mentioned three types of relationships in 2015 and in the next 3 years both for large companies and young start-ups.

Corporate venturing is losing its top spot for large firms as the number of companies relying on it will be cut by half between 2015 and 2018, from 36% of users to 19%. While corporate venturing comes only second for start-ups in 2015, it will also lose its appeal in the same proportion with only 13% of entrepreneurs considering it as important in 2018 against 23% in 2015. The relative status of incubators will stay almost the same for large firms as 34% consider them as important in 2015 and 36% believe they will still have an essential role in 2017 for collaborative innovation. In parallel, a growing number of small firms expect that incubators will play an important role from 19% in 2015 to 25% in 2018, indicating that startups are still looking for all the supportive services that provide effective incubators and accelerators as a means of progress towards joint innovation.

But the most significant shift is in the evaluation of the importance of Joint Innovation as a way to collaborate with external partners for getting new technology, skills, talent and resources. The use of Joint Innovation will grow by 50% between 2015 and 2018 regarding large companies as the share of firms actively involved will reach 38% in 2018 from 26% in 2015. In parallel, start-ups are also considering increasing their involvement in Joint Innovation from 26% in 2015 to 36% in 2018, representing a rise of 38% (Fig. 16.1).

While such a development of collaborative innovation is emblematic of the need from more co-operation for innovation between large and small firms, it does not come that easy because of different organization, practice and culture. While small startups are more prone to disruptive innovations with a fast prototyping and a quick market probing, big companies tend to prefer incremental innovations with a longer sequential "gate stage" prototyping and selection process, and a structured and slower route to market implementation along their distribution channels. In terms of organizations, startups are usually flat, informal, decentralized and risk oriented while established firms have a tendency to be hierarchical, more formal in terms of communication and decision-making, centralized and risk averse. When considering the investment in any innovation project, young startups are obviously considering first the short-term impact on the free cash flow while the large companies are sensitive to the Net Present Value (NPV) over the long term.

Added to the challenges coming from those structural imbalances is the difference in perceived commitments from each of the 2 parties in Joint Innovation. According to the same survey by Accenture, 29% of small startups were considering that their bigger partner was not committed to the collaboration while only 7% of the largest firms were thinking the same from their smaller partner. Conversely, only 24% of the small firms were confident that their larger partner was very

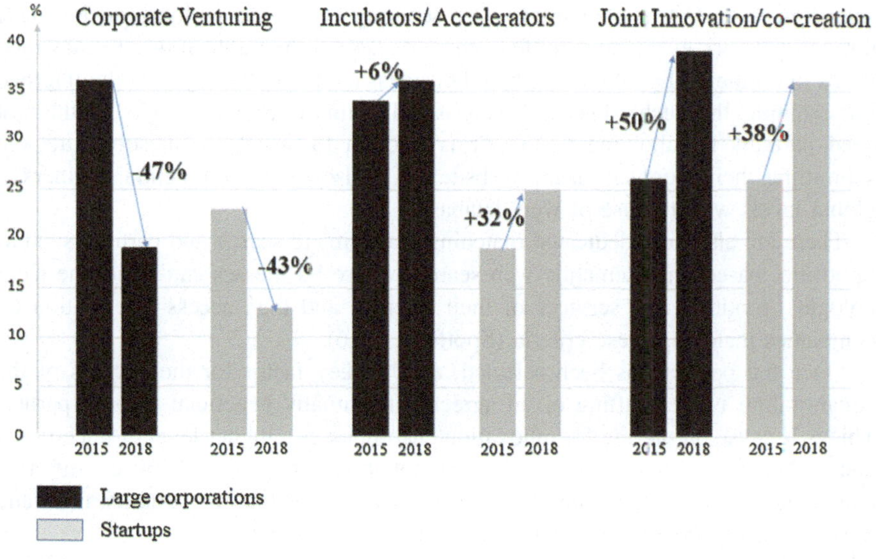

Fig. 16.1 Evolution of the percentage of companies considering the importance of the three major tools for collaborative innovation between 2015 and 2018. *Source* Accenture (2015)

committed while 41% of large firms were thinking the same about their smaller associate.

In order to overstep those potential roadblocks, there are 3 important steps to follow in order to set up a successful collaborative innovation between large and small firms particularly in the digital economy: The first one is to make a thorough preparation, both internally and in the selection of the potential partner. The second phase is to carefully craft the partnership, and the last stage is to ensure the sustainability of the collaboration for innovation.

The **preparation** phase is often critical for the future success of a C collaborative innovation I initiative but it is often overlooked by companies, large or small. It starts with a clear delineation of the objectives of the future collaborative innovation project answering questions such as: which area of innovation do we want to reach? Which technology? How the collaboration will contribute to our corporate strategy? Is collaboration better than a mere acquisition?

Another important element is preparing the organization to support in terms of structure, culture and incentives. A key element of collaborative innovation is the rapid and continuous flow of information from the company to the partner. This cannot happen in a structure with a defensive structure prone to defend only internal ideas; similarly a complex decision making process—either because it is too hierarchical or too much consensus-based can slow up the communication and frustrate a more agile partner. Also, it is very common for incentives to be modified in order to reward collaborative mindset, risk taking mentality, entrepreneurial attitude, while cross-organizational and cross-functional teams have to be set up.

The ultimate element of the preparation phase to collaborative innovation is the selection of the adequate partner. The scouting for future partner can be made through network events and conferences but also via the publication of innovation needs in offline or digital industrial and business media or directly over the Internet. For example, the Technology to Business (TTB) unit of Siemens looks for the best sited partners by allowing young firms through to answer a questionnaire and submitting their proposal on the website. TTB also monitors potential partners at global levels with the use of web databases.

There are also now dedicated matching platforms to search and partner, such as Spotfolio, based on a semantic web search, where businesses can select the technologies, products and services of their interest, and then access information on companies matching these criteria (Spotfolio 2016).

Once the partner has been selected, another key factor for the success of the collaboration is the crafting of an agreement mutually beneficial to both parties. This is because partnering for innovation encompasses a higher level of uncertainty and a longer timeline than the typical collaboration project between different companies. Here the firms involved must adequately define the benefits, risks and governance aspects of the collaborative innovation project.

The most important element in a collaborative innovation project in the digital economy is probably the issue of Intellectual Property (IP) rights where there is very often a huge gap between start-ups and large firms in terms of know-how and resources. IP rights are at the heart of collaborative innovation in technology industries as very often companies bring existing IP to the collaboration or will develop new IP through the partnership. Therefore, it is fundamental to have agreement in advance about the exact use and protection of IP to build a relationship based on trust in order to extract the full value of collaboration. More specifically large firms have to be careful not to dodge too much the IP of innovative start-ups because they may refrain to team up and thus jeopardize the future of the innovation partnership. They have to be prepared to accept not to have the full-ownership and control of IP.

Large firms also have to pay attention to be flexible enough in term of control, legal compliances or due diligence in order to keep younger and faster startups motivated. Conversely small firms have to be ready to pay for legal and expert advice at the same level as a large firm and to finance significant upfront IP protection costs. They must also be able to enlarge their timeframe when negotiating with a bigger partner.

The availability of a common platform for sharing and diffusing innovation can also be important to facilitate the access to new and smaller companies to join the innovation ecosystem of a larger partner, as seen previously. But in some cases the collaboration is more than digital and may entice a physical infrastructure like AT&T Foundry or the Verizon Innovation Center to allow start-ups to test their proof of concepts in reality.

The final step to ensure a fruitful CI is to sustain the partnership overtime to achieve a continuous benefit of mutual parties as the context evolves. This is not a small feat to achieve as we have underlined the physical and cultural differences between small and large firms regarding the management of innovation, in the unstable environment of the digital economy.

In order for collaborative innovation to succeed in the long term, experience shows that firms, large and small, have to:

- share knowledge and integrate the results systematically across their product lines,
- develop ongoing mutual benefits and safeguarding incremental IP in a transparent manner,
- incentivizing team and employee support for collaboration with external partners
- make sure that lines of communication are always functioning between partners
- honest assessing of the challenges and risks of failure for the future (World Economic Forum 2015).

In some cases, it may imply to manage the partnership outside the sales organization so that collaborative innovation outputs may blossom without being subject to short-term business demands.

The building of an ecosystem of external innovators is based on the setting of strong ties which involve a strong degree of trust and are characterized by frequent contacts over a longer period. Those interpersonal ties that are built through frequent communication can lead to more effective interactions (Uzzi 1997) but conversely they may provide redundant information, especially as they occur among a small group of people in which almost everyone knows what the others know. That is the reason why some firms are also trying to enlarge their reach of innovation contributors far outside their regular business environment in getting ideas from everyone with crowdsourcing.

16.3.3 The Expansion of the Ecosystem to Everyone: Crowdsourcing

Crowdsourcing can be defined as outsourcing a task in the problem solving process to an informal group of people, and not a designated agent in the form of an open call to contribution (Afuah and Tucci 2012).

Based on crowdsourcing principles and using the reward structures of tournaments (Morgan and Wang 2010), a challenge is posted to a large public and the contributions are evaluated by a jury to select and reward the winning ideas. For instance, Cisco has launched an external innovation competition called the I-Prize to help the company identify promising business platforms for future growth and with a prize of $250,000 prize for the winner (Jouret 2009). Dell is using a website called IdeaStorm where customers can submit ideas to improve the company's products or services and vote on the ideas of others. Since its launching in 2007 this platform has dealt with more than 24,000 customer ideas and had implemented 549 ideas by February 2016 (Dell 2016).

Crowdsourcing has been used since a long time with the Longitude Prize offered by the British government in 1714 to anyone able to design a method to measure a ship's longitude (Dawson and Bynghall 2012). But the recent rise of crowdsourcing as a tool for innovative collaboration is directly related to the development of the digital economy as crowdsourcing as well as the rise of the Internet and the social applications is tremendously easing the way for companies to engage and collaborate with mainstream users or contributors by posting their call for ideas on line (Bilgram 2013).

The key benefit of crowdsourcing is to reach out to an undefined mass and a wider variety of user types than the traditional external partners; it provides a more heterogeneous background favourable to highly creative and "out of the book" suggestions, with a more heterogeneous background favourable to highly creative and "out of the book" suggestions. Furthermore, crowdsourcing is based on mostly weak ties with the participants of the network which are built on loose emotional tendencies and are maintained via infrequent communication. This kind of relationship is considered to increase the probability of stimulating creativity because they bridge otherwise disconnected groups and individuals (Tsai 2001), they are providing access to original information (Granovetter 1973), and they encourage autonomous thinking (Perry-Smith 2006).

The effective way of using crowdsourcing for collaborative innovation requires first to be able to design a contribution structure that allows for large participatory efforts while respecting the diversity of opinions (Seltzer and Mahmoudi 2013). Second, it necessitates the capacity to capture the valuable contribution from heterogeneous individuals or group of participants (Howes 2008). As a consequence, the digital platform which supports crowdsourcing is very important with its look and feel in order to make it easy to use but also with the rules for participation and interaction among the participants. Zhao and Zhu (2014) have shown that the platforms are enablers and drivers for extracting the maximum value of the open calls. With the constant evolution of digital technology, those platforms are getting increasingly more efficient in mediating the relationship between a firm issuing a call and the multitude of contestants.

The use of crowdsourcing is not yet widely used by companies compared to other sources of open innovation but it is slowly making its way (Chesbrough and Brunswick 2013). This is notably due to the limitation in time and scope of the contribution from crowdsourcing. But a recent study by Xu et al. (2015) showed that Chinese firms in the telecommunication industry which use outsourcing technologies to capture the knowledge of the customers and transform it in innovation competences, can obtain a significant advantage in business performance compared to their competitors.

In recent years, private companies are also using specialized platform to externalize crowdsourcing such as Ideaken, which is specialized in open innovation, or Zoopa where companies can launch advertising contests.

16.4 The Impact of the Digital Economy on Collaboration in the Outbound Innovation Phase

It is not enough to have good new ideas: first and foremost, an innovation must be adopted by the market. Without market success, it is just a useless invention whose failure will dent the profitability of the vendor or may even lead it to bankruptcy. Still a large percentage of innovations are still failing to be successful in the market and to be profitable (Castellion and Markham 2013). The outbound activities of the innovation process have become significant priorities for companies especially in the technology sector.

The adoption of innovation follows the S-curve defined by Bass et al. (2004) when considering the number of users over time. The shape of the curve is determined by the limited number of *early adopters* of an innovation which reaches a tipping point where there is an acceleration of the adoption of the innovation by an *early majority* of *mainstream adopters* followed by a *late majority* up to a tripping point where the innovation is now massively adopted and starts plateauing. Early adopters and the majority of adopters have different expectations when it comes to adopting an innovation.

Early adopters are focusing on performance, the technical credibility of a provider and the quality of the innovation. Also known as "lead users", they are often ready to contribute to work with the provider to test the innovation and to improve it. Early adopters have also been shown to not be very sensitive to the price of an innovation. They are much more sensitive to delayed market introduction and may be less tolerant if the providers fail to deliver on time an innovation that has been announced.

Mainstream innovation adopters are the ones who can break or make an innovation because they represent the biggest number of potential customers. The significant failure rate of market acceptance of innovations reflects the difficulty that some companies have to convince the mainstream adopters to adopt an innovation. They are very sensitive to the credibility of the innovation provider and the quality of the solution because they are looking more for safety than performance. They also prefer easy-of-use innovation and consider the organizational and technical infrastructure that exists to support the use of the innovation. They also follow the social influence of peers and "opinion leaders" who recommend a particular innovation or technology. They also pay a lot of attention to the reliability of the vendor as they do not want to spend money or time with the repair of a potentially poor solution. Consequently, they prefer to go with innovation which has already been recognized as a kind of standard reference. Also, they go with vendors which already have an established brand and a good reputation in order to minimize the risk of malfunction.

In order to facilitate the adoption of innovation, successful companies have figured out that it was important to work well with the early adopters in order to achieve a critical mass of users able to convince a larger majority of the value of the innovation. Those firms had also understood the importance of setting up strong technology standards and developing a solid brand image to attract and convince the majority of users to adopt an innovation.

But the digital economy has also redefined and changed how innovative companies are collaborating in the commercialization phase of the innovation process. More specifically, it concerns first the new interaction with early adopters through crowdsourcing, second the accrued role of a new category of "enabling" standards, and finally the renewed importance of branding.

16.5 A More Intense Collaboration with Early Adopters: Crowdfunding

Among the early adopters, the "lead users" have always played an essential role on accelerating the diffusion of an innovation (Salah et al. 2010). Lead users are dissatisfied users ahead of the market trends who are willing to develop their own solutions or to collaborate with the provider (Franke and Piller 2003) because they enjoy the problem-solving techniques (Bilgram et al. 2008). Some innovative companies have managed to develop a "toolkit approach" (von Hippel and Katz 2002) that transfers most of the product and service development tasks from the research and development department to pre-qualified lead users (Piller and Walcher 2006) which are participating directly in most of the stages of the product development process (Prűg and Schreier 2006). Such a method facilitates the development of new products that are accepted by the market (Henkel and von Hippel 2005).

Actually, lead user contributes to an accelerated rate of diffusion in comparison with the traditional internal innovation method (van Oast et al. 2009) with acceptance performance superior to any other external agents including the external product development partners (Al-Zu'bi and Tsinopoulos 2012). One of the reasons is that lead users are also powerful evangelizers and opinion leaders in their industry. Being among the first adopters of new products, they represent a reliable source of information and their word-of-mouth power is strong enough to influence the behaviour of other people in terms of search, purchasing and usage of new products (Goldsmith et al. 2003).

A recent and even more intimate way to associate lead users to innovation comes with the development of crowdfunding, which allows the general public to invest investment in businesses through a dedicated website. In 2015, crowdfunding had surpassed Venture Capital as a way to finance innovation with a total of $34.4 billion invested against $30 billion with the US leading the way before Asia and Europe: while crowdfunding can be applied to any cause, business and entrepreneurship are the most popular categories ahead of social causes. The most famous crowdfunding sites include Kickstarter and Indiegogo for instance but in 2016 there were more than 1250 active platforms in the world (Massolution 2015) with technology dedicated platform such as Appbackr, a crowdfunding site for application developers. Crowdfunding is now strongly established as a new way for collaborative innovation as Mollick and Kuppuswamy (2014) have found that 90% of the projects funded on Kickstater remained active companies 4 years after their (reward-based) campaign while 32% of them had yearly revenues of over $100,000.

They are two main forms of crowdsourcing: the "All or Nothing", where no money is collected if the pre-announced amount of money pledged is not achieved, and the "Keep it All", where the money is handed over anyway for the project. The most popular model is the reward-based crowdfunding where providers received rewards such as the products created, personalized incentives, copies of a creative work, or promotional products, etc. Other models are relying on donations, lending (debt crowdfunding), or equity in the company (equity crowdfunding). Finally, a donation can be made once or continuously if the content is delivered on a regular basis such as Bountysource, another crowdfunding site for open-source developers, which is accepting one-time or monthly donations (Bountysource 2016).

Crowdfunding is naturally a type of crowdsourcing (see above) as it allows the company to get additional ideas from its fund providers for developing the product or service as well as the opportunity to test prototypes directly with crowdfunders. But crowdfunding is also a very effective way to enhance the commercial application of collaborative innovation because the crowdfunders are usually the first to buy the new product they have supported either because they believe in the idea or the product or because they receive it as part of the reward. It allows the subscribing companies to use their fund providers as early customers to who they can presell their products or services, which are often non-existing yet. Beyond the assistance in the (pre)-commercialization phase, the crowdfunders are also usually contributing actively to the diffusion of the new offer through an enthusiastic and positive word-of-mouth campaign among their network, and especially their digital network communities. With such a backing some projects have received millions of dollars in crowdfunding, the most famous being "Star Citizen" a space combat video game which is the first crowdfunded project to have received more than $100 million and which should be launched in 2016 after 4 years of development (Cieslak 2016).

16.6 The Accrued Role of a New Category of "Enabling" Standards

Standards have always played an important role in facilitating the acceptance and the diffusion of an innovation, especially among the late majority of adopters as well as the late adopters because they represent a guarantee of quality. They make life easier to the customers who know that they are buying a solution which has been acknowledged by the majority of the vendors within an official standard committee, often with the backing of national and/or international governments and administrations. In some cases, the "dominant design" which is recognized as a standard is not coming from a committee but from a powerful company which has managed to turn its innovation into a de facto market standard, such as Microsoft with Windows, Google with its eponym web search engine, or Sony with its Blu-Ray technology. But for the users, the results are the same: they are sure that they are buying a solid and stable solution with proven facilitating conditions of use and support. Standards perform the fundamental functions of quality assurance, information and measurement, and they contribute to the prevention of the occurrence of a risk and/or to the reduction of a potential loss should a risk materialize. In telecommunication for instance, standards reduce the managerial and financial risks in joint venture and mergers or in acquisitions of networks (Sherif 2006).

In the technology sector, the need for standards has always been reinforced by the need for compatibility between different categories of equipment and service suppliers. For instance, in the computer industry, compatibility is required to ensure that computers, software, modems, printers and other peripherals interface easily. Similarly, in the mobile telecommunications market, compatibility demands a common set of technological standards for the design of cellular base stations, digital switches and handsets to ensure maximum geographical coverage for users.

Traditionally, the discussions about compatibility have taken place in the various standardization committees like the International Telecommunication Union (ITU), the Institute of Electrical and Electronics Engineers (IEEE), or the European Telecommunications Standards Institute (ETSI). This compatibility approach was very effective when the market was mostly dominated by the large suppliers. For instance in the 90s the European mobile telecom vendors and operators companies managed to agree on one compatible technology, the Global System for Mobile Communications (GSM) developed by the ETSI while there were four different and non-compatible technologies in the US. The value for the cellular phone users clearly was much bigger in Europe than in the US and the cellular phone caught on more quickly in Europe than in the US. The GSM can be described as "anticipatory" standards which is looking forward to solve interoperability and compatibility issues. It is usually opposed to the "responsive" standard which comes at the end of the technology development and which officializes a dominant design of a product or service.

Between those two categories, the pressing need for more collaborative innovation in the digital economy is pushing for the development of "enabling" standards (Egyedy and Sherif 2010) to facilitate the surge of acceptance of an innovation. Actually, once an innovation has passed the initial acceptance of early users and is looking for the early majority of adopters, there is a risk of fragmentation of the offer: This is because new competitors, enticed by market growth and improvement in the innovation, may enter the business and offer alternative solutions based on another technology. Such a fragmentation is denting the credibility of the innovation vendors because different competing technologies or solutions are confusing the prospective early majority of adopters. Accordingly, they prefer to wait for a dominant model to emerge before making a decision. Additionally, market fragmentation prevents economy of scale and the associated decrease of production costs which cannot be translated in the price. As the late majority of adopters are sensitive to price this is another limitation to the adoption of an innovation. Finally enabling standards signal to the customer that competition is shifting to areas that are not covered by the standard (cost, quality of implementation, service support, etc.).

Thus in the context of open, collaborative innovation, large companies are pushing for "enabling standards" which support the offer around a common set of characteristics in order to contribute to the emergence of a prevailing model of product or service and to prevent the fragmentation of the market. But those enabling standards are dynamics because they integrate the feedback from the market as they are influenced by competitive forces and the need to reduce production costs. An interesting illustration is the HyperText Markup Language a.k.a. HTML, which is the standard language used to create web pages. It was originally created in 1990 by T. Berners-Lee which moved it to the World Wide Web Consortium (W3C), the main international standards organization for the World Wide Web. Since that time HTML has been continually evolving with technology and the market and it is currently in its fifth version while being the most adopted solution for designing web page application for computers and mobiles.

Enabling standards are by definition flexible standards as opposed to the rigidity and uniformity which is associated with responsive standards. Enabling standards demonstrate that consensus-based standards do not necessarily mean rigidity because they encourage collaboration as well (Choi et al. 2011), and that standardization is also a powerful way to enforce the acceptance of an innovation (Viardot et al. 2016). Actually, open collaboration with other partners has been positively correlated with the proclivity to join standardization activities (Blind et al. 2012).

16.7 The Renewed Importance of the Brand Image

The running of collaborative innovation in the digital economy has also reinforced the importance for a company to have a strong brand image and to communicate it forcefully and effectively outside in order to rally the maximum of external parties around an innovation (Corkindale and Belder 2009).

Actually, a major issue for the adoption of an innovation is the anxiety of the majority of customers, developers or external parties in front of the uncertainties of any novelty (Boyd and Mason 1999). Some are intimidated by the task of learning how to use the innovation, some are risk averse to any novelty, and others are afraid that the innovation will become obsolete quickly; all are always postponing their decision to take it on. What is true for consumers is also true for organizations. Many managers fret about innovations and try to assess the balance on the risk/return relationship of such investment more than considering the sheer novelty of an innovation.

Consequently, a well-known and familiar brand helps to reassure individuals or industrial buyers when they consider the purchase of an innovative solution which represents always a leap into the unknown. In that case, one of the main criteria that determines a customer's choice is confidence in a company and its products (Temporal and Lee 2000).

A brand is a name, a set of words, a sign, a symbol, a design, or a combination that identifies a seller's goods or services (Keller 1993). A strong brand facilitates the identification of the innovation while attaching a quality image and a personality that establish a bond with the customers and facilitate their loyalty (Urde 1999). For instance, Google is perceived as a clean, friendly but credible path to accessing the tremendous wealth of the Internet. Cisco's image is associated with being a visionary and an expert in Internet telecommunication as well as a partner with its clients. Huawei, its Chinese competitor has managed to carve out an image of being extremely innovative at the global level. And the Apple brand personality is about lifestyle, imagination, innovation, passion, and aspirations. It suggests also power-to-the-people through innovation thanks to simplicity and the removal of complexity from people's lives (Marketingminds 2013).

Successful technology companies which are relying on collaboration for innovation have also invested in the building of a strong brand, which has put them on the top of the rankings of the most valuable brand in the world in 2015, ahead of more traditional sectors. As illustrated in Table 16.2, 15 technology corporate

Table 16.2 Different rankings of the top most valuable marketing brands in 2015

Company	Value ($M)	Industry	Millward Brown[a]	Forbes[b]	Interbrand[c]
Apple	247.0	Technology	1	1	1
Google	173.7	Technology	2	3	2
Microsoft	115.5	Technology	3	2	4
IBM	94.0	Technology	4	5	5
Visa	*92.0*	*Payments*	*5*		
AT&T	89.5	Telecom provider	6	12	
Verizon	86.0	Telecom provider	7	21	
Coca Cola	*83.8*	*Soft drinks*	*8*	*4*	*3*
McDonalds	*81.2*	*Fast food*	*9*		*9*
Malboro	*80.4*	*Tobacco*	*10*		
Tencent	76.6	Technology	11		
Facebook	71.1	Technology	12	10	23
AlibabaGroup	66.4	Retail online	13		
Amazon	62.3	Retail online	14	13	10
China Mobile	59.9	Telecom provider	15		
Wells Fargo	*59.3*	*Regional banks*	*16*		
GE	59.3	Conglomerate	17	9	8
UPS	*34.1*	*Logistics*	*18*		
Disney	*43.0*	*Entertainment*	*19*	*11*	*13*
MasterCard	*40.2*	*Payments*	*20*		
Baidu	40.0	Technology	21		
ICBC	*38.8*	*Regional banks*	*22*		
Vodafone	38.5	Telecom provider	23		
SAP	38.2	Technology	24		
American Express	*38.1*	*Payments*	*25*	22	25
Samsung	37.9	Technology		7	7
Toyota	*37.8*	*Automotive*		*8*	*6*
Louis Vuitton	*28.1*	*Luxury*		*14*	*20*
Cisco	27.6	Technology		15	15
BMW	*27.5*	*Automotive*		*16*	*11*
Oracle	26.8	Technology		17	16
NIKE	*26.3*	*Apparel*		*18*	*17*
Intel	25.8	Technology		19	14

(continued)

Table 16.2 (continued)

Company	Value ($M)	Industry	Millward Brown[a]	Forbes[b]	Interbrand[c]
Walmart	*24.7*	*Retail*		*20*	
Honda	*22.6*	*Automotive*		*23*	*19*
Mercedes-Benz	*22.5*	*Automotive*		*24*	*12*
Budweiser	*22.3*	*Beverages*		*25*	
HP	23	Technology			18

Brand value represents the fractions of intangible corporate earnings of a company which is attributable to the brand multiplied by an earnings multiple, depending on the brand market valuation and the brand growth

Source [a]http://www.millwardbrown.com/BrandZ/2015/Global/2015_BrandZ_Top100_Chart.pdf

[b]www.forbes.com/powerful-brands/list

[c]http://interbrand.com/best-brands/best-global-brands/2015/

brands dominates the top of the ranking by MillwardBrown with a brand value in a range of $37.8M to $247M. Two other rankings by Forbes and Interbrand come with a slightly different order but the overall result is that technology brand dominates the branding panorama.

This preeminent position of innovation driven brands does not come by accident. One may argue that their value reflects their market success. Actually part of their hit performance has been achieved through a strong forceful branding strategy which they have started very early in their corporate life.

Some of the companies listed in the ranking above certainly are now spending huge amounts of money to promote their brand. For example, in 2014 Samsung spent a total of $14 billion on marketing, the biggest marketing budget ever, bigger than Iceland GNP (Miyoung 2013) but scaled it down the following year. In 2015, Microsoft spent almost US$2 billion, while Apple devoted US$1.6 billion, in advertising.

But the building of a strong brand image for an innovation does not always require big amounts of money. Some highly successful innovative companies have managed to achieve recognition essentially through creativity, quality and word-of-mouth. In the 80s many successful innovators such as Intel, Microsoft, Compaq, Cisco, and others were first mentioned in the pages of the Wall Street Journal, the Financial Times, Business Week, Forbes and Fortune magazines. Only once their brand image was made, then they spent money in advertising to maintain their image and notoriety. More recently in the 2000s, a new generation of web-based firms such as Google, EBay, Amazon, Apple or Facebook also achieved top of mind recognition on a low advertising budget and by generating "buzz" over the Internet. Now than they have matured, they are also relying on larger and more expensive traditional marketing campaigns. It is now to a new generation of "companies—such as Snapchat, Okta or MongoDB—to generate excitement and passion over social networks and the web in hope that this will transform into sales afterwards.

Collaborative innovation provides additional ways to enhance brand awareness and brand image, such as developing and tightening deep relationships when hundreds of consumers spend significant amounts of time interacting with companies and their brands on idea contest platforms (Nambisan and Nambisan 2008). Furthermore, consumers share their experiences with a brand over the social network and spread positive word-of-mouth in their networks extending the reach of simple open collaboration platforms or initiatives (Füller et al. 2010). For instance, Microsoft has created the **Imagine** Cup to give the opportunity for students to turn their ideas into reality, as well as to solve challenges and problems provided to them. On average, more than 165,000 students are participating every year.

When promoting an innovation, the use of branding is not exclusive to private companies. It has been used very effectively by some alliances to promote an innovation in order to make it a standard. Take for an example HDMI, a compact audio/video interface which was initiated in 2002 by a handful of companies, and has now over than 1700 adopters (HDMI 2013). Another successful example is Bluetooth, a short-range networking protocol for connecting different types of digital devices (mobile phone, computer, GPS, etc.) or accessing the Internet by wireless signals within a 35-ft or 10-m range. In 1998, five companies founded the Bluetooth Special Interest Group (SIG), Ericsson, IBM Corporation, Intel Corporation, Nokia and Toshiba Corporation. Its goal was to promote the development of the new protocol as the standard solution for wireless connections. In the early stages the decision was made to develop a strong brand so as to communicate with the end consumers in order to accelerate its recognition and to step up its adoption by other industrial companies. Today, the Bluetooth SIG has more than 25,000 member companies in 2014 (Bluetooth.org 2014).

16.8 Conclusion

In this chapter, we have analyzed how the digital economy is redefining the strategy of open collaboration of innovative companies both in the inbound and outbound part of the innovation process. Though we have used different examples, the reality is that successful innovative companies in the technology sector are coupling the two processes—inbound and outbound—when working with external partners. An interesting illustration is the case of SAP which is also relying on crowdsourcing with programs such as the SAP HANA Idea Incubator and the SAP Idea Place to facilitate collaborative inbound innovation, in addition to its SAP Labs, its Co-Innovation Labs and its venture capital units that we have presented in 1.1. While on the outbound side of innovation, SAP is not directly involved in crowdfunding projects because of its size and the nature of its business (including its industrial customers), but it has developed its own set of sixteen flexible standards for key operations processes to guarantee a better use of its solution operations. Finally, in 2015, SAP has made the major decision to rebrand all the companies acquired under the unified SAP Brand which represent the underlying innovations and integration that SAP offers to its customers.

Collaborative innovation does not come automatically and requires an effective management because collaborative innovation is redesigning completely the flows of information—and power—inside the company. As illustrated with the case of innovative companies from the technology sector, collaborative communication leads to the emergence of a new category of managers, called the "network orchestrators" (Fung et al. 2007) who are able to deal with a large diversity of contributors. They have to develop a specific set of management skill, because their role is not the same as managing internal collaboration. It requires a more fluid approach with a network-centric perspective and not only a firm or a market centric viewpoint. Network orchestrators must also have considerable communication skills and they must be able to communicate clearly, simply, effectively and consistently with all innovation partners in order to identify, and then keep them motivated and engaged (Thomas and Wind 2013).

Finally in the globalized word of technology, the digitalization of the economy is also turning internationalization as a new key success factor with collaborative innovation. **Innovative technology companies are increasingly going abroad to interact with their most demanding customers and locating the most competent or lowest priced suppliers. They are also seeking ideas or knowledge with leading research environments which are getting more geographically dispersed and searching for new markets for their technologies** while promoting their brand at a global level. **Consequently, the proportion of corporate R&D centres performed outside domestic countries is increasing rapidly** (Herstad et al. 2008) *while some companies are even relocating their headquarters in order to be more collaborative, such as* SAP which has moved the R&D head office to the Silicon Valley in the US. Of particular interest is the growing presence of Chinese and Indian innovative technology companies—including Alibaba, Huawei, TCS or Infosys among many others—which have managed to increase significantly their business outside of their native countries in recent years. There is no doubt that globalization is clearly the current challenge for companies which want to develop collaborative innovation in the digital economy.

Bibliography

Accenture: Harnessing the power of entrepreneurs to open innovation. Retrieved at https://www.accenture.com/us-en/insight-b20-digital-collaboration (2015)

Afuah, A., Tucci, C.L.: Crowdsourcing as a solution to distant search. Acad. Manag. Rev. **37**(3), 355–375 (2012)

Al-Zu'bi, Z.M.F., Tsinopoulos, C.: Suppliers versus lead users: examining their relative impact on product variety. J. Prod. Innov. Manag. **29**(4), 667–680 (2012)

Baldwin, C.Y., von Hippel, E.: Modeling a paradigm shift: from producer innovation to user and open collaborative innovation. MIT Sloan Research Paper, No. 4764-09, pp. 1–34 (2009)

Bass F.: Comments on "a new product growth for model consumers durables": the bass model. Manag. Sci. **50**(12 Supplement), 1833–1840 (2004)

Bigliardi, B., Dormio, A.I., Galati, F.: The adoption of open innovation within the telecommunication industry. Eur. J. Innov. Manag. **15**(1), 27–54 (2012)

Bilgram, V.: Performance assessment of co-creation initiatives: a conceptual framework for measuring the value of idea contest. In: Brem, A., Viardot, E. (eds.) Evolution of Innovation Management: Trends in an International Context, pp. 32–51. Palgrave Macmillan, UK (2013)

Bilgram, V., Brem, A., Voigt, K.-I.: User-Centric innovations in new product development-systematic identification of lead user harnessing interactive and collaborative online tools. Int. J. Innov. Manag. **12**(3), 419–458 (2008)

Blind, K., de Vries, H.J. Mangelsdorf, A.: External knowledge sourcing and involvement in standardization—evidence from the community innovation survey. In: 2012 IEEE International Technology Management Conference (ITMC), pp. 1–9. IEEE, New York (2012)

Bluetooth.org: Retrieved at www.bluetooth.org/ (2014)

Boston Consulting Group: The most innovative companies in the world retrieved from https://www.bcgperspectives.com/most-innovative-companies-2015/ (2015)

Bountysource: Support for open-source software. Retrieved at http://www.bountysource.com/ (2016)

Boyd, T.C., Mason, C.H.: The link between attractiveness of "Extrabrand" attributes and the adoption of innovations. J. Acad. Mark. Sci. **27**(3), 306–319 (1999)

Bradley J., Loucks J., Macaulay J., Noronha A., Wade M.: Digital vortex. How digital disruption is redefining industries. http://global-center-digital-business-transformation.imd.org/vortex/ (2015)

Castellion, G., Markham, S.: New product failure rates: influence of *Argumentum ad Populum* and self-interest. J. Prod. Innov. Manag. **30**(5), 976–979 (2013)

Chesbrough, W., Teece, D.J.: When is virtual virtuous. Harvard Bus. Rev. **74**(1), 65–73 (1996)

Chesbrough, W., Brunswick S.: Managing open innovation in large firms. Retrieved at http://openinnovation.gv.at/wp-content/uploads/2015/08/Fraunhofer-2013-studie_managing_open-innovation.pdf (2013)

Choi, D.G., Lee, H., Sung, T.: Research profiling for "standardization and innovation." Scientometrics **88**(1), 259–278 (2011)

Cieslak, M.: Star citizen: will big budget space game satisfy its backers? BBC. Retrieved at http://www.bbc.com/news/technology-35555086 (2016)

Cisco: The zettabyte era—trends and analysis. Retrieved at http://www.cisco.com/c/en/us/solutions/collateral/service-provider/visual-networking-index-vni/VNI_Hyperconnectivity_WP.html (2016)

InternetLive Stats: Internet Users http://www.internetlivestats.com/internet-users/ (2015)

Corkindale, D., Belder, M.: Corporate brand reputation and the adoption of innovations. J. Prod. Brand Manag. **18**(4), 242–250 (2009)

Dawson, R., Bynghall, S.: Getting Results from Crowds. Advanced Human Technologies, Incorporated (2012)

Deering, L., Cartagena, M., Dowdeswell, C.: Accelerate: Founder Insights Into Accelerator *Programs*. FG Press, Boulder (2014)

Dell: IdeaStorm can help take your idea and turn it into reality. Retrieved at http://www.ideastorm.com/ (2016)

Dragland, A.: Big data—for better or worse. http://www.sintef.no/en/corporate-news/big-data–for-better-or-worse/ (2013)

Egyedy, T.M., Sherif, M.H.: Standards dynamics through an innovation lens: next-generation Ethernet networks. In: Proceedings of the first ITU-T Kaleidoscope Academic Conference, 'Innovations in NGN', Geneva, 12–13 May 2008, pp. 127–134. ITU, Geneva (2010)

Facebook: Microsoft imagine cup. Retrieved at www.facebook.com/microsoftimaginecup (2013)

Fortune: The unicorns list. Retrieved at http://fortune.com/unicorns/ (2016)

Franke, N., Piller, F.T.: Key research issues in user interaction with configuration toolkits in a mass customization system. Int. J. Technol. Manag. **26**(5/6), 578–599 (2003)

Frattini, F., Colombo, G., Dell'Era, C.: Exploring the role of early adopters in the commercialization of innovation. In: Brem, A., Viardot, E. (ed) Evolution of Innovation Management: Trends in an International Context, pp. 151–182. Palgrave Macmillan (2013)

Frick, K.A., Torres, A.: Learning from high-tech deals. McKinsey Quart. **1**, 112–124 (2002)

Füller, J., Mühlbacher, H., Matzler, K., Jawecki, G.: Consumer empowerment through internet-based co-creation. J. Manag. Inf. Syst. **26**(3), 71–102 (2010)

Fung, V.K., Fung, W.K., Wind, Y.: Competing in a Flat World: Building Enterprises for a Borderless World. Wharton School Publishing, Upper Saddle River, NJ (2007)

Gassmann, O., Enkel, E.: Towards a theory of open innovation: three core process archetypes. In: Proceedings of the R&D Management Conference, Lisbon, 6–9 July (2004)

Global Corporate Venturing: The 117 most active corporate VC firms of the last year. Retrieved at https://www.cbinsights.com/blog/active-corporate-vc-firms-h1-2015/ (2015)

Goldsmith, R.E., Flynn, L.R., Goldsmith, E.: Innovative consumers and market mavens. J. Mark. Theory Pract. (Fall), 54–64 (2003)

Granovetter, M.: The strength of weak ties. Am. J. Sociol. **78**(6), 1360–1380 (1973)

HDMI: Retrieved at http://www.hdmi.org/learningcenter/adopters_founders.aspx (2013)

Henkel, J., von Hippel, E.: Welfare implications of user innovation. J. Technol. Transfer **30**(1/2), 73–87 (2005)

Herstad, S., Bloch, C., Ebersberger, B., de Veld, E.V.: Open innovation and globalisation: theory, evidence and implications, Visionera.net. Retrieved at https://www.tem.fi/files/25709/Open_innovation_and_globalisation.pdf (2008)

Howe, J.: Crowdsourcing: Why the Power of the Crowd is Driving the Future of Business. Crown Publishing Group, New York, NY (2008)

IBM: From Stretched to Strengthened. IBM Global Business Services, Somers, NY. Retrieved at http://www-935.ibm.com/services/us/cmo/cmostudy2011/cmo-registration.html (2011)

IBM: Collaborative Innovation Centers (CIC). Retrieved at http://www.research.ibm.com/university/cic/index.shtml1ad7f6442045/entry/maximize_your_relationship_with_ibm32?lang=en (2015)

Inauen, M., Schenker-Wicki, A.: Fostering radical innovations with open innovation. Eur. J. Innov. Manag. **15**(2), 212–231 (2012)

Jouret, G.: Inside Cisco's search for the next big idea. Harvard Bus. Rev. **87**(9), 43–45 (2009)

Keller, L.K.: Conceptualizing, measuring and managing customer-based brand equity. J. Mark. **57**(1), 1–22 (1993)

Klein, K.J., Tornatzky, L.G.: Innovation characteristics and innovation adoption implementation: a meta-analysis of findings. IEEE Transm. Eng. Manag. **29**(11), 28–45 (1982)

Love, J.H., Roper, S.: The determinants of innovation: R&D, technology transfer and networking effects. Rev. Ind. Organ. **15**(1), 43–64 (1999)

Marketingminds: Apple's branding strategy. Retrieved at http://www.marketingminds.com.au/branding/apple_branding_strategy.html (2013)

Massolution: 2015CF—Crowdfunding Industry Report. Retrieved at http://www.crowdsourcing.org/editorial/global-crowdfunding-market-to-reach-344b-in-2015-predicts-massolutions-2015cf-industry-report/45376 (2015)

Miotti, L., Sachwald, F.: Co-operative R&D: why and with whom? An integrated framework of analysis. Res. Policy **32**(8), 1481–1499 (2003)

Miyoung, K.: Samsung's marketing splurge doesn't always bring bang-for-buck. Reuters. Retrieved at http://in.reuters.com/article/samsung-marketing-idINDEE9AQ0EC20131128 (2013)

Mollick, E.R., Kuppuswamy, V.: After the campaign: outcomes of crowdfunding, 9 Jan 2014. UNC Kenan-Flagler Research Paper No. 2376997. Retrieved at http://dx.doi.org/10.2139/ssrn.2376997 (2014)

Morgan, J., Wang, R.: Tournament for ideas. Calif. Manag. Rev. **52**(2), 77–97 (2010)

Nambisan, S., Nambisan, P.: How to profit from a better "Virtual Customer Environment". MIT Sloan Manag. Rev. **49**(3), 53–61 (2008)

Nieto, M.J., Santamaría, L.: The importance of diverse collaborative networks for the novelty of product innovation. Technovation **27**(6–7), 367–377 (2007)

Penenberg, A.: Dancing giants: Samsung's Open Innovation Center. Retrieved from https://pando.com/2014/04/22/dancing-giants-11-samsungs-1b-open-innovation-center/ (2015)

Perry-Smith, J.E.: Social yet creative: the role of social relationships in facilitating individual creativity. Acad. Manag. J. **49**(1), 85–101 (2006)

Piller, F.T., Walcher, D.: Toolkits for idea competitions: a novel method to integrate users in new product development. R&D Manag. **36**(3), 307–318 (2006)

Prűg, R., Schreier, M.: Learning from leading-edge customers at the Sims: opening up the innovation process using toolkits. R&D Manag. **36**(3), 237–250 (2006)

Rubin, R.: Microsoft has hit the ecosystem wall. Fast Company. Retrieved from http://www.fastcompany.com/3048603/microsoft-has-hit-the-ecosystem-wall (2015)

Salah, S.H., Maha, M., Tolba, A.H.: Conceptualising the influence of lead users and opinion leaders on accelerating the rate of innovation diffusion. Int. J. Technol. Mark. **5**(3), 203–218 (2010)

Samsung: Open innovation is a Samsung initiative to identify and grow the technologies and infrastructure of the future. Retrieved at http://www.samsung.com/global/business/semiconductor/aboutus/business/open-innovation/overview (2013)

SAP: SAP: run simple—the world's largest provider of enterprise application software. Retrieved from http://www.sap.com/bin/sapcom/en_us/downloadasset.2015-10-oct-20-01.SAP-Corporate-Fact-Sheet-en-2015-10-20-pdf.html (2015)

Seltzer, E., Mahmoudi, D.: Citizen participation, open innovation, and crowdsourcing: challenges and opportunities for planning. J. Plann. Lit. **28**(1), 3–18 (2013). Retrieved at http://jpl.sagepub.com/cgi/doi/10.1177/0885412212469112

Sherif, M.H.: Managing Projects in Telecommunication Services, Chapter 10. Wiley-IEEE Press (2006)

Solomon, B.: The best startup accelerators of 2015. Forbes. Retrieved at http://www.forbes.com/sites/briansolomon/2015/03/17/the-best-startup-accelerators-of-2015-powering-a-tech-boom/#3e0fb85934e4 (2015)

Spotfolio: Innovation meets industry. Retrieved from http://spotfolio.com/en/ (2016)

Statista: Number of apps available in leading app stores as of July 2015. Retrieved from http://www.statista.com/statistics/276623/number-of-apps-available-in-leading-app-stores/ (2015)

Temporal, P., Lee, K.C.: Hi-Tech Hi-Touch branding: creating brand power in the age of technology. Wiley, New York (2000)

Tether, B., Tajar, A.: Beyond industry—university links: sourcing knowledge for innovation from consultants, private research organisations and the public science-base. Res. Policy **37**(6–7), 1079–1095 (2008)

Thomas, R.J., Wind, Y.: Symbiotic innovation: getting the most out of collaboration. In: Brem, A., Viardot, E. (ed.) Evolution of Innovation Management: Trends in an International Context, pp. 1–31. Palgrave Macmillan (2013)

Tsai, W.: Knowledge transfer in intra-organizational networks: effects of network position and absorptive capacity on business unit innovation and performance. Acad. Manag. J. **44**(5), 996–1004 (2001)

Urde, M.: Brand orientation: a mindset for building brands into strategic resources. J. Mark. Manag. **15**(1–3), 117–133 (1999)

Uzzi, B.: Social structure and competition in interfirm networks: the paradox of embeddedness. Adm. Sci. Q. **42**, 35–67 (1997)

Van Oast, E., Erhaegh, S., Oudshoorn, N.: From innovation community to community innovation: user-initiated innovation in wireless Leiden. Sci. Technol. Hum. Values **34**(2), 182–205 (2009)

Viardot, E.: Successful Marketing Strategy for High-Tech Firms. Artech House, 3rd edn (2004)

Viardot, E.: Achieving market leadership for innovation through communication. In: Hülsmann, M., Pfeffermann, N. (eds.) Strategies and Communications for Innovations: an Integrative Management View for Companies and Networks, pp. 243–256. Springer, Berlin (2011)

Viardot, E., Sherif, M.H., Chen, J.: Innovation with standardisation. An introduction to recent trends and new challenges. Technovation, **48–49** (2016)

Von Hippel, E., Katz, R.: Shifting innovation to users via toolkits. Manag. Sci. **48**(7), 821–833 (2002)

Wagner, K., Taylor, A., Zablit, H., Foo, E.: The Most Innovative Companies 2014: Breaking Through is Hard to Do. BCG Perspectives. Retrieved from https://www.bcgperspectives.com/Images/Most_Innovative_Companies_2014_Oct_2014_tcm80-174313.pdf (2014)

World Economic Forum: Collaborative Innovation Transforming Business, Driving Growth. Retrieved on 11 Nov 2015 from http://www3.weforum.org/docs/WEF_Collaborative_Innovation_report_2015.pdf (2015)

Xu, Y., Ribeiro-Soriano, D.E., Gonzalez-Garcia, J.: Crowdsourcing, innovation and firm performance. Manag. Decis. **53**(6), 1158–1169 (2015)

Zhao, Y., Zhu, Q.: Evaluation on crowdsourcing research: current status and future direction. Inf. Syst. Front. **16**(3), 417–434 (2014)

Chapter 17
Customer-Centricity in the Executive Suite: A Taxonomy of Top-Management–Customer Interaction Roles

Noel Capon and Christoph Senn

Abstract The quest for customer-centricity drives top-management relationships with customers in business-to-business (B2B) markets. But the impact of executive engagement varies greatly across supplier–customer relationships. Based on exploratory field research, this paper develops a taxonomy of top-management–customer interaction roles. We also provide suggestions for leveraging senior executives for both supplier and customer benefit.

17.1 Introduction

The fundamental core of this paper emphasizes the importance of the marketing and sales functions to firm success. Whereas many firm functions provide significant value, marketing and sales differ qualitatively from most others. Specifically, the purpose of marketing and sales is to attract, retain, and grow customers. If these functions fail, the firm does not make profits, the organization does not survive, and shareholder value atrophies. And from the perspective of employees, no one gets a paycheck!

These assertions are not particularly revolutionary. After all, a half century ago, Peter Drucker stated, "If we want to know what a business is, we have to start with its purpose. There is only one valid definition of business purpose—to create a customer." As just noted, that is the job of the marketing and sales functions. Indeed, Drucker continued, "Because it is [the purpose of a business] to create a customer, [the] business enterprise has two—and only these two—basic functions: marketing and innovation." (Drucker 1954).

N. Capon (✉)
Columbia Business School, 3022 Broadway, New York, NY 10027, USA
e-mail: nc7@gsb.columbia.edu

C. Senn
Universität St.Gallen (HSG), Dufourstrasse 50, CH-9000 St.Gallen, Switzerland
e-mail: Christoph.Senn@unisg.ch

N. Pfeffermann and J. Gould (eds.), *Strategy and Communication for Innovation*,
DOI 10.1007/978-3-319-49542-2_17

In recent years, two factors have highlighted the prescient nature of Drucker's observations. First is the evolution from considering customers as important firm assets in general, to placing a specific monetary value on individual customers. Indeed, during the past decade, both academics and practicing managers alike have become enamored of the customer lifetime value (CLV) concept. CLV can be complicated mathematically, but essentially equates to the anticipated stream of future profits a customer delivers to the firm, factored by the probability of retaining that customer year-to-year, and the firm's discount rate (Gupta and Lehmann 2004). Hence, management can calculate the value (individually) of current and potential customers, and make more informed resource allocation decisions about customer retention and acquisition.

The second factor is application of Pareto's law to the firm's customer base. Many firms have realized that a disproportionately small number of customers is responsible for a large portion of the firm's sales revenues. Indeed, the 80:20 rule states that 80% of firm revenues derive from 20% of firm customers. Of course, this ratio is not exactly 80:20; it may be 70:30, 90:10, or even more skewed. The basic point is that all customers are not equal; rather, some customers are more equal than others. Hence, a relatively small number of customers is disproportionately important to firm health.

As a direct result of this realization, during the past quarter century, many firms have developed strategic/key account programs. In these programs, the firm pays special attention to a relatively small number of its most important customers. These programs come in several shapes and sizes, but are typically staffed by specially trained strategic/key account managers (SAMs/KAMs), frequently backed up by a program office, and employing formal systems and processes like annual planning and periodic supplier/customer reviews. More recently, globalization has led firms to initiate global account programs staffed by even more highly trained global account managers (GAMs), to address customers with operations in many countries/continents around the world.

The foregoing places the following question squarely on the table. Given the critical importance for the firm of attracting, retaining, and growing customers; given the firm's new ability to calculate the economic value of individual customers; and given the large investment many firms have made in strategic/key/global account management programs to optimize relationships with their most important customers; what is top-management's responsibility to become involved in this endeavor?

In most corporations, the CEO and executive team are busy people. Executive team members direct and manage areas of functional expertise like accounting, finance, human resources, operations, marketing, R&D, sales. The CEO leads and directs the executive team and guides the firm to achieve its objectives in the face of increasingly difficult environmental challenges, and demanding stakeholders. And from time-to-time, top-management intervenes at the frontline in attempting to secure major deals. Consider for example the case of Accenture, IBM, and Merck: A few years ago, along with many other large corporations, pharmaceutical giant Merck decided to outsource its data-processing system. Middle managers developed an RFP and evaluated submissions from several potential suppliers. The consensus decision: Accenture had earned the contract. Then IBM CEO Sam Palmisano came calling on Merck's CEO: IBM won the contract!

A major reason for this success was IBM's strongly developed focus on *Integrated Accounts*. High-caliber strategic account managers (Account Managing Directors) are responsible for securing revenue—and—profit growth from IBM's most important customers. Supported by global team resources and a systematic executive sponsorship program, IBM implemented an *Embedded*-type of customer management (Capon and Senn 2010). This program allowed IBM to take action at a critical time when strategic customer Merck was about to award its contract to IBM's competitor.

Perceptive senior executives understand that customers are crucial to organizational survival and growth, and that success with *strategic* customers can have a significant impact on revenues, profits, and shareholder value. Many firms have initiated programs similar to IBM to supplement ongoing sales efforts. But while some research and received industry wisdom suggests that greater top-management involvement tends to be positively related to revenue generation and profitability (Homburg et al. 2002; Workman et al. 2003), we suggest three important caveats:

- Using the CEO as a persuader of last resort is not sustainable. Even *rock-star* CEOs will not be able to turn around customers' closed (and almost closed) deals with competitors on a regular basis.
- CEOs and executive team members face ongoing time constraints. They must carefully balance time they allocate to their day-by-day responsibilities with time they direct to addressing customer issues (Lafley 2009). Spending large amounts of time with customers on a regular basis is not sustainable (Quelch 2008).
- Many executive–customer interactions focus overly on securing immediate sales revenues, rather than discuss how the firm and its customer may shape the future together (Senn 2006).

17.2 About the Research

A half century of research on management time has identified a broad range of general insights and specific recommendations/guidelines on how best to allocate this scarce resource (Drucker 1966; Mintzberg 1973; Kotter 1999). But findings on top-management relationships with customers remain mostly anecdotal (Guesalaga and Johnston 2010; Shi et al. 2005; Bossidy and Charan 2002). Nor do these studies offer practical advice for senior executives who care seriously about customer–supplier relationships (Kohli and Jaworski 1990; Jaworski and Kohli 1993; Narver and Slater 1995; Homburg et al. 2000; Day 2006). Regardless, recent research confirms increasing supplier appetites for top-level customer relationships, leading to deeper strategic dialogues and greater influence on customer decisions (IBM Corporation 2013; Senn et al. 2013).

In our research, we set out to answer three related questions:

1. How, if at all, do senior executives relate to customers?
2. What are the impacts of various relationship types?
3. How may firms leverage these relationships in the best way?

The taxonomy we develop results from a multi-year study embracing 30 personal interviews with top managers, and 15 workshops with more than 300 participants. The main preliminary findings are:

Interviews: CEOs and top managers with sales backgrounds interface with customers more broadly, deeply, and frequently than their peers. They also view spending time with customers as a strategic necessity, rather than as an agenda-filler.

Workshops: Four direct and one indirect role capture the ways in which senior executives interface with customers; each role offers rewards and risks. For strategically important customers, the *Growth Champion* role provides the most benefits.

Loose cannon	Senior executives make visits to customers without decent prior and/or post meeting interaction with account managers. Often leads to bad impression and/or unrealistic promises
Social visitor	Senior executive meet with customers, but interactions focus on relationship building. Firm–customer interactions do not involve substantive business issues, but create a certain level of trust
Deal maker	Senior executives (including the CEO) get involved in significant revenue opportunities when the customer is about to make (or has just taken) the supplier decision. Strong signal of commitment
Growth champion	The firm builds senior executive involvement into relationships with strategic customers via regular meetings and/or an executive sponsor program

17.3 Top-Management Customer-Focused Roles

Our research identified five relationship roles that CEOs and top managers actually play in addressing customers. To be more precise—four customer-focused roles plus an absence of customer interaction. We start with absence and label this role—***not my problem***.

17.3.1 Not My Problem

The CEO of a large manufacturing firm addressed one of the authors: "I typically do not see customers. That's why we have a sales force. Our products and solutions are world-class and we have one of the strongest R&D groups in the industry. If only our account managers would do a better job in selling our value to customers!" A competitor has since acquired this firm and the former CEO is no longer with the organization.

It is not uncommon for senior managers to adopt this perspective. Indeed, the motto, *Let the sales force do its job*, sounds eminently sensible. After all, as noted above, each top manager has his/her own set of responsibilities; competitive pressures continue to increase and most executives have to do more with less. The perspective underlying this role is quite straightforward: The human resource function should do its job—hire the best talent to direct and manage the selling effort. In turn, senior sales executives should use the tools at their disposal to effectively populate the sales force and ensure salespeople reach their goals. If sales force results do not meet firm objectives, then required actions are quite straightforward: Make the appropriate personnel changes. The common mindset is simply stated: *We don't expect top marketing and operations managers to get involved in finance; why should any functional leader, or the CEO, get involved with sales?*

The underlying assumption of ***not my problem*** is that, at a high level, all functions are equally important. Of course, most functional areas are critical to firm success, but as our observations of companies around the world show, and noted above, sales (and marketing) is perhaps a little more equal than others. More than any other function, sales is a boundary role—the critical interface between the firm and its customers. If sales underperform, revenues do not reach expectations, and everyone's budget suffers. If top managers can assist sales in doing its job, then all functions benefit.

17.3.2 Four Direct Customer-Focused Roles

Not my problem assumes that senior managers cannot add value to the firm's selling effort. This judgment is typically erroneous. Indeed, each executive team member has significant ability to improve firm–customer relationships, but there are pitfalls. Some approaches can be positively damaging to customer relationships. In our interviews and workshops, we found evidence for four separate roles. We organize these roles along two dimensions—*revenue seeking* and *relationship building*. Each dimension has two levels—*low* and *high* (Exhibit 17.1).

Loose Cannon

The ***loose-cannon*** role can be positively damaging for the firm. Top managers must clearly understand the potential negative consequences from ***loose-cannon*** behavior, and implement eradication processes.

Relationship Building	High	**Social Visitor**	**Growth Champion**
	Low	**Loose Cannon**	**Deal-Maker**
		Low	High
		Revenue Seeking	

Exhibit 17.1 Top-management–customer relationship roles

A U.S.-based account manager for an Indian technology outsourcing firm complained to one of the authors during an executive education program: "I had been working my account for two years, gaining trust and making steady progress, and then it all fell apart. A senior Hyderabad-based executive made an appointment with my customer's top-management without letting me know, and without any briefing or debriefing. He had no idea what was going on with the customer, and we just could not honor he promises he made. The meeting was a disaster and set us back at least one year." It was later revealed that the Indian executive's daughter was at college in the U.S., and the customer visits were an excuse to charge the trip to his company expense account.A top-level executive at a European chemical firm met with a major customer. The executive was unfamiliar with both customer challenges and his firm's initiatives—he left a negative impression. This executive also made unrealistic promises the firm could not fulfill; he did not tell the account manager about the meeting nor about the promises. The result: A badly damaged relationship that remains unrepaired despite heroic efforts by the account manager and her team.

The critical question is straightforward: Are the CEO and/or senior executives acting as ***loose cannons***? ***Loose-cannon*** behavior typically occurs when a senior executive meets with a major customer but has no (or minimal) prior interaction with the account manager. Sometimes the account manager only finds out long afterwards—often from customer executives! Not only was the senior executive not (or improperly) briefed, the executive commits the firm to actions without full awareness of intricacies in the firm–customer relationship. And the senior executive may fail to advise the account manager of these agreements.

Loose-cannon interactions with customers are typically short-lived and not particularly focused on revenue generation, at least not in the short term. But they seem to be ubiquitous. We have met few account managers who could not identify ***loose-cannon*** behavior and its negative consequences from one or more of their senior executives.

Account managers have another name for ***loose cannons—seagulls***. The ***seagull*** flies in, leaves a deposit, and flies off, rarely returning to the same spot! Regrettably, many readers probably have *seagulls* in their firms. Consider what the senior sales executive at a major Canadian financial institution told us when we discussed our evolving framework with him: "We don't have isolated senior executives acting as ***seagulls***; we have a whole flock of them!"

If such interactions are so damaging, why do they occur? Our research revealed that, in many cases, senior executives thought they were being helpful. They saw their organizational positions as door openers to customer executive suites. Typically, they were not mistaken. But lack of preparation and failure to develop the appropriate working relationships with relevant account managers led to disaster.

Fortunately, most sample firms had sufficiently strong processes to avoid ***loose-cannon*** behavior most of the time. They employed some form of strategic account management and executive sponsor program to interface with major customers. These firms clearly defined roles and responsibilities as a central program element.

Social Visitor

Social-visitors also have a low revenue focus, but a long-term relationship focus. The role's purpose is to demonstrate supplier commitment, and to create trust through long-term personal relationships. Though less negative than ***loose cannons***, the ***social visitor*** has its own set of risks and rewards.

> The CEO of a European-based engineering firm frequently met with customer CEOs at social events and trade shows. Attempting to take these conversations to a more business-oriented level, the CEO accepted an invitation to visit a major customer's U.S.-headquarters. The CEO advised the account manager of the visit but decided to go alone. The customer CEO, used to high-level visits from high-performing suppliers, awaited the supplier CEO with a senior executive entourage. After a moment's surprise at seeing the supplier CEO alone, the customer CEO asked: "Good to see you, but where on earth is your account manager and his team?" The supplier CEO responded: "Well, I thought we were going to discuss confidential issues." "Sure," was the response, "but who's going to take the notes?"

Whereas ***loose cannons*** are positively destructive for the firm–customer relationship, the ***social visitor's*** impact ranges from mildly positive to mildly negative. The ***social visitor*** specializes in relationship building via meeting and greeting. Typically, the firm arranges meetings for senior customer executives—educational events on company premises, cocktail parties at trade shows, and/or hosting trips to sporting events. The ***social visitor*** works the crowd, engages in conversation about industry issues, and builds personal relationships with individual customer executives.

These interactions rarely involve deep discussion of business issues; indeed, spouses may attend these events. Relationships are often long-term in nature and occur at regular intervals—the annual duck hunt, U.S. Open Tennis Championships, U.S. Golf Masters, or Formula 1 Grand Prix races in Europe or Asia. Customer executives often look forward to these events.

There is nothing particularly undesirable with social interactions between senior firm and customer executives—everyone has a good time, and leaves with pleasant feelings; these meetings enhance personal relationships. Rarely do such social interactions cause positive harm. But customer executives may wonder why these interactions are all they see of senior supplier executives. Indeed, they may feel mildly negative about lack of relationship depth, and be frustrated about missed opportunities for meaningful discussion. Negative feelings maybe especially strong if other suppliers make concerted efforts to engage in deeper business relationships.

Customer expectations can be easily misconstrued. The *lonely* CEO and his firm described in the opening vignette learned a big lesson, and learned it quickly. Today, the firm prohibits executives from visiting customers alone, except under special circumstances, and only with proper briefing and debriefing. A centrally managed tracking system ensures no top-management visit occurs without evaluating potential alternatives. Results from recent meetings are built into briefing packages for upcoming customer discussions.

Deal Maker

Deal Makers have high revenue focus but low relationship focus. The purpose of this role is to stabilize shaky relationships or secure deals. Sam Palmisano's actions as described earlier exemplify a senior executive fulfilling the ***deal-maker*** role.

> A mid-sized coating systems manufacturer was trying to supply eco-friendly, cost-saving production technology to a leading German automotive firm. Contrary to the experiences during lab- and field-tests, a stable operational solution at the customer's main production line could not be reached within the contractually agreed timeframe. The supplier CEO personally intervened. The purchase-decision delay he secured enabled the firm to demonstrate technical ability to solve customer problems in a full-scale production environment.

Senior executives can have a major positive impact as ***deal makers***. Sophisticated suppliers with good sales pipeline systems know what deals customers are deciding on a week-by-week basis. For many opportunities, the critical deciding issue is not so much the value propositions that supplier account managers offer, but customer beliefs that the supplier will/or will not live up to its commitments. Account manager promises go so far, but only a senior executive, often the CEO, can fully commit firm resources. The ***deal maker*** is very revenue focused.

In many customer organizations, middle managers negotiate deals and make purchase recommendations, but senior managers have the final word. As our opening vignette showed, Merck's CEO overruled internal advice to outsource its data-processing system to Accenture when IBM's CEO came calling. Indeed, we have observed several cases where top-management intervened in the buying process reversed previously made purchase decisions. On some occasions, customers even assumed penalties to void signed contracts.

Senior executives, especially CEOs, are uniquely placed to fulfill the ***deal-maker*** role. Only senior executives can truly commit the firm to allocate necessary resources to serve a customer. But there are dangers: Too many deals escalated to the executive suite create a culture of upward delegation. The senior executive team can certainly help the sales force close deals, but too frequent rush meetings on customer schedules may be unsustainable. And although customers may be happy to see senior supplier executives engaged in real business issues (versus meet-and-greet ***social visitors***), they may also wonder where senior executives are the rest of the time. Could they not have been involved earlier in the process?

Notwithstanding potential success, ***deal makers*** are no panacea. Consider what one account manager told us: "The division president decided to accompany me to my customer to close an important deal. Shortly after the meeting started, the customer pushed for significant additional price concessions. The division president was so focused on winning the deal, he was about to agree on a massive discount that would have cost us more than $2 million. I didn't know how to stop him, so I faked a heart attack. The meeting ended and the negotiations were put on hold. A few weeks later I negotiated the deal at a much better price!"

Growth Champion

Growth champions have a strong revenue focus and a long-term relationship focus. Unlocking new growth opportunities and serving as a role model to inspire others are the main purposes of this role. Thus, growth champions represent the most positive customer-facing behavior for senior executives.

> Cisco's former long-time CEO John Chambers fit the profile of a truly customer-centric top manager. Chambers accompanied strategic account managers on business trips and demanded feedback on how to improve his participation. And Chambers regularly interacted with customer CEOs by leveraging Cisco's tele-presence technology. eBay's CEO John Donahoe also plays a major role in developing and sustaining retailer partnerships that drive substantial revenue growth for eBay.

Senior executives acting as ***growth champions*** play a critical role in driving revenue and profit growth. Our research suggests that well-planned, well-organized, and well-executed executive involvement at customers can pay huge dividends. Findings from our preliminary research indicates that executives who combine revenue seeking and relationship building create significantly more favorable conditions for customer-centricity than other approaches.

Two items merit further consideration: First, some of the most successful executives we identified commenced their customer-facing behavior as ***social visitors*** or ***deal makers***; later, they evolved into ***growth champions***. Second, senior executives should take customer behavior into account when planning customer relationship efforts. When customers behave in a transactional manner, ***growth-champion*** behavior is often a waste of resources. Conversely, if customers desire long-term collaborative relationships, behaving as a ***deal maker*** or ***social visitor*** delivers sub-optimal results.

Perhaps the most significant corporate initiative that firms take to implement ***growth-champion*** behavior is installing an executive sponsor program.

Executive Sponsor Programs. Over and above periodic and episodic interactions between top supplier and customer executives, many suppliers are appointing executive sponsors for important customers (Capon and Wallner 2012). Executive sponsors are senior managers who work with individual strategic account managers at specific customers. The executive sponsor is a *permanent* senior executive who interacts with senior customer executives. Executive sponsors can add significant value to the supplier–customer relationship.

Suppliers with these programs typically draw executive sponsors from the C-suite. These firms work through a serious matching process to ensure positive chemistry between and among the strategic account manager, senior customer executives, and the executive sponsor. Strategic account managers know that some senior firm executives make good executive sponsors—others do not; sometimes divorce is necessary. DHL gives strategic account managers the right to suggest changing executive sponsors without penalty, provided they back up proposals with solid evidence. The strategic account manager–executive sponsor relationship is particularly tricky because, at some level, the executive sponsor *works for* the account manager. Frequently, the CEO is executive sponsor for the firm's most important customers: J.W. (Bill) Marriott Jr. was executive sponsor for major Marriott customer Accenture; Oracle's executive sponsor at GE is Larry Ellison. At Siemens, the entire Executive Board participates in the executive sponsor program—members, including the CEO, average seven customer sponsorships.

Most firms introduce executive sponsor programs for customer-facing reasons. But executive sponsors may offer significant benefits within the firm. It is one thing for the account manager to discover customer needs and propose value propositions; it is quite another for the firm to actually deliver promised value. Powerful executive sponsors can cut through supplier bureaucracies when account managers face internal roadblocks. Some executive sponsors offer direct support to account managers:

In a significant downsizing, IBM released an engineer with considerable expertise in transaction processing. IBM's strategic account manager at Merrill Lynch realized IBM's position would be severely weakened by this loss; he contacted his executive sponsor. The sponsor said he could not reverse the layoff, but authorized hiring the engineer on a consulting contract and agreeing to pay on his budget. The engineer remained as a consultant for the next dozen years, helping IBM to secure business with a strategic customer during economically difficult times.

Senior executives acting as ***growth champions*** play a critical role in driving revenue and profit growth. Our research suggests that well-planned, well-organized, and well-executed executive involvement at customers can pay huge dividends. Findings from our preliminary research indicate that executives who combine revenue seeking and relationship building create significantly more favorable conditions for customer-centricity than other approaches.

17.4 Conclusion

Leveraging top-management–customer interactions requires a clear view of the firm's strategic direction. Greater customer focus by top-management and appropriate executive relationships with customers are key differentiators for many firms. Senior executives have critical functional roles, but they may also play an important additional role in helping the firm reach sales and profit objectives. Forming meaningful relationships and becoming more engaged in customer affairs allows senior executives to provide higher degrees of customer-centric behavior, and hence greater value for both the supplier and its customers. Peter Drucker was very clear about the purpose of a business. Active and well-managed participation in fulfilling that purpose should become an increasingly important role for senior managers, led by the CEO.

Bibliography

Bossidy, L., Charan, R.: Execution: The Discipline of Getting Things Done. Crown Business, New York (2002)

Capon, N., Senn, C.H.: Global customer programs: how to make them really work. Calif. Manage. Rev. **52**(2), 32–55 (2010)

Capon, N., Wallner, C.: Key account management at siemens: the executive relationship program. In: Capon, N., Senn, C. (eds.) Case Studies in Managing Key, Strategic, and Global Customers. Bronxville, NY, Wessex (2012)

Day, G.S.: Aligning the organization with the market. Sloan Manage. Rev. **48**(1), 41–49 (2006)

Drucker, P.F.: The Practice of Management, pp. 37–38. Harper and Row, New York (1954)

Drucker, P.F.: The Effective Executive. Harper & Row, New York (1966)

Guesalaga, R., Johnston, W.: What's next in key account management research? building the bridge between the academic literature and the practitioners' priorities. Ind. Mark. Manage. **39** (7), 1063–1068 (2010)

Gupta, S., Lehmann, D.R.: Managing Customers as Investments. Wharton, Philadelphia, PA (2004)

Homburg, Ch., Workman, J., Jensen, O.: fundamental changes in marketing organization: the movement toward a customer-focused organizational structure. J. Acad. Mark. Sci. **28**(4), 459–478 (2000)

Homburg, C., Workman, J.P., Jensen, O.: A configurational perspective on key account management. J. Mark. **66**, 38–60 (2002)

IBM Corporation: The customer-activated Enterprise—insights from the global c-suite study. Somers, NY (2013)

Jaworski, B., Kohli, A.K.: Market orientation: antecedents and consequences. J. Mark. **57**(3), 53–70 (1993)

Kohli, A.K., Jaworski, B.J.: Market orientation: the construct, research propositions and managerial implications. J. Mark. **54**(2), 1–18 (1990)

Kotter, J.P.: What executive general managers really do. Harvard Bus. Rev. **77**, 145–159 (1999)

Lafley, A.G.: What only the CEO can do. Harvard Bus. Rev. **87**, 54–62 (2009)

Mintzberg, H.: The Nature of Managerial Work. Harper & Row, New York (1973)

Narver, J.C., Slater, S.F.: Market orientation and the learning organization. J. Mark. **59**(3), 63–74 (1995)

Quelch, J.A.: How CEO's should work with customers. Harvard Business Review Blog Network. Retrieved at http://blogs.hbr.org/2008/09/how-ceos-should-work-with-cust/ (2008). 22 Sept 2008

Senn, Ch.: The executive growth factor—how siemens invigorated its customer relationships. J. Bus. Strat. **27**(1), 27–34 (2006)

Senn, Ch., Thoma, A., Yip, G.S.: Customer-centric leadership: how to manage strategic customers as assets in B2B markets. Calif. Manage. Rev. **55**(3), 27–59 (2013)

Shi, L.H., Zou, S., White, J.C., McNally, R.C., Cavusgil, S.T.: Executive insights: global account management capability: insights from leading suppliers. J. Int. Mark. **13**(2), 93–113 (2005)

Workman, J.P., Homburg, C., Jensen, O.: Intra-organizational determinants of key account management effectiveness. J. Acad. Mark. Sci. **31**(1), 3–21 (2003)

Chapter 18
What Is Innovation Communication? A Dynamic Capability View

Nicole Pfeffermann

Abstract What is the inherent source of enterprise-generated future cash flows? While no firm wins forever in a market, dynamic capabilities allow an entrepreneurially managed firm to renew and leverage its difficult-to-imitate resources so as to gain competitive advantage in high-velocity environments. This chapter aims at presenting innovation communication as a dynamic capability from a strategic management perspective. After a brief literature on innovation communication, the dynamic capability of innovation communication is defined and its construct dimensions explained in detail. Second, the direct and indirect effects of innovation communication capability on company value are illustrated and propositions are set up to further develop theory. Finally, innovation communication is described as one of four communication fields in information–innovation management practice.

18.1 Introduction

> Apple has the ability to innovate … and create magic. This isn't something you can just write a check for.
>
> Tim Cook, Apple CEO 2013

The basic idea of transformation goes far beyond the 'strategic fit' approach that has been applied in the organizational-change view (Teece 2014). In business practice, change managers often tend to emphasize the importance of adapting to external factors, such as new technologies or processes, but do not focus on the building of new capabilities for creating durable competitive advantage. For example, a firm decides to invest in new software programs or a co-creation platform and is not necessarily capable of transforming this investment into a critical

---

N. Pfeffermann (✉)
Pfeffermann Strategy and Research, Im Medienhafen,
D-40221 Düsseldorf, Germany
e-mail: np@npfeffermann.de

N. Pfeffermann and J. Gould (eds.), *Strategy and Communication for Innovation*,
DOI 10.1007/978-3-319-49542-2_18

driver of business growth and value-driven performance. Or a startup business with a great new business model is not capable of building up a competitive resource set for the implementation and execution of this business model and for transforming this business model into a sustainable market success.

Enhancing corporate innovation and successfully leveraging open innovation initiatives require corporations to build up and expand capabilities and resources, such as (adapted from Weiblen and Chesbrough 2015)

- To screen, identify, work with, and monitor larger numbers of startups and talents on a more dispersed global level
- To manage different types of relationships at the same time on a company-wide, global level
- To understand and evaluate business model and strategy including business model innovation
- To dive into the entrepreneurial world to understand the ecosystem and speak the same language
- To invest in new tools and platforms (apps) to connect, co-create, and manage engagement activities
- To be more flexible/agile in structure, culture, and processes for re-integration of startup solutions, higher outcomes of commercialization activities, talent recruiting and competitive resource building.

What are capabilities and specifically dynamic capabilities? According to Teece (2014), a capability, ordinary or dynamic, is a firm's ability to accomplish its goals, which is attained in part through learning and innovation, organizational resources, and its organizational heritage over time. Whereas ordinary capabilities focus on technical efficiency in business functions, the purpose of dynamic capabilities is to achieve congruence with changing customer needs and with technological and business opportunities for doing the right things. Dynamic capabilities are important at the top management level and culture level; for instance, the collective ability to quickly implement a new business model. Dynamic capabilities govern other organizational activities understood as entrepreneurial 'asset orchestration.' From a managerial view, this 'asset orchestration' function encompasses (1) coordination/integration (= combining various resources in an entrepreneurial fashion); (2) reconfiguration (recombining existing resources); and (3) learning (= outcome of practice and experimentation). While no firm wins forever in a market, dynamic capabilities allow them to renew and leverage their difficult-to-imitate resources so as to retain the 'pivot' potential of a 'lean startup,' i.e., dynamic capabilities are an essential driver of sustainable competitive advantage due to their combination of entrepreneurially managed actions and organizational routines, especially in high-velocity environments.

The social web and new digital technologies have changed information exchange worldwide. For example, crowdsourcing platforms have become a new corporate innovation tool that engages stakeholders in knowledge acquisition and transfer processes (Boudreau and Lakhani 2013). In this context, *innovation is understood as a progress at a corporate level* to survive in dynamic changing environments and achieve competitive advantage. The use of various social media platforms and crowd-powered innovation tools in corporate innovation, however, leads to new challenges in communication management. Communication management is no longer just a task of marketing and corporate communication in the traditional logic, but rather a central task in marketing, corporate communication, idea/innovation management, and corporate strategy and entrepreneurship. The fragmentation in marketing and corporate communication, involving "silos" mainly related to different traditional communication fields (employee communication, finance communication, product communication, etc.), can no longer exist. Instead, management practice shows that cross-functional activities are key to succeed in open corporate innovation (Keyword: *Open Communication View*). From a management practice view, there is a pressing need for new communication management approaches related to open corporate innovation.

This chapter aims at presenting a new understanding of innovation communication as a dynamic managerial capability of a company on the corporate level to govern ordinary capabilities and orchestrate assets in entrepreneurial fashion for achieving congruence with changing customer needs and with technological and business opportunities. This includes the direct and indirect of innovation communication on company value and the linkage to other communication fields/abilities in a corporate innovation process.

18.2 Innovation Communication: A Review

Three main streams of research can be identified in the field *communication of innovations*: (1) Marketing of innovations/innovation marketing in marketing research; (2) Communication in marketing diffusion research; and (3) Innovation communication linked to innovation management research.

First, research in marketing including consumer behavior and psychology encompasses scientific investigations regarding the antecedents and consequences in marketing of innovations. Marketing is an essential part in the innovation process (Crosby and Johnson 2006). Communication can inform consumers about the advantages and characteristics of an innovation using mass media and individual communication throughout the adoption process (Hofbauer et al. 2009). Theoretical findings and managerial implications provide essential information concerning strategies and mechanisms to introduce innovations successfully. Marketing of innovation includes both the commercialization of radical innovations, technologies and services (e.g., Mohr et al. 2009; Sandberg 2008; Sowter 2000) and strategic innovation marketing (e.g., Talke 2005; Trommsdorff and Steinhoff 2007). Various useful definitions are provided in the literature, for instance, "innovation marketing encompasses all market-oriented activities of innovation management—that is, all strategic and operative decisions for marketing new products" (Steinhoff and Trommsdorff 2011).

Second, "diffusion research seeks to understand the spread of innovations by modeling their entire life cycle from the perspective of communications and consumer interactions" (Peres et al. 2010: 91). Several innovation diffusion models have been introduced mainly in the marketing diffusion literature (e.g., Mahajan et al. 2000; Peres et al. 2010) related to specific industries, adopter groups, or steps in the adoption process (e.g., Arndt 1967; Hesse 1987; Mahajan et al. 1990, 1995; Pae and Lehmann 2003; Rohlfs 2001). Research interest has shifted in its focus from the forecasting focus to the managerial diagnostic focus in order to provide answers in marketing management (Peres et al. 2010). Three social influence factors are mentioned as drivers of communication in innovation diffusion: (1) word-of-mouth communication (e.g., Martilla 1971; Mazzarol 2011), (2) network externalities (e.g., Rohlfs 2001; Tomochi et al. 2005), and (3) social signals (e.g., van den Bulte and Stremersch 2004; Berger and Heath 2008). These social influence factors, referred to as *interdependencies among consumers*, "affect various market players with or without their explicit knowledge" (Peres et al. 2010: 91) and, thus, have to be considered in marketing of innovations. Further research in this field requires the consideration of online communities, web services, and complex types of product–services categories in innovation diffusion (Peres et al. 2010, see also Rodriguey-Montemayor 2014).

Third, researchers have focused on innovation communication and its impact on the innovation process from idea to launch as a part of corporate communication and constitutive element in innovation management (e.g., Huck 2006; Mast and Zerfaß 2005; Mast et al. 2005, 2006; Zerfaß 2009; Zerfaß and Ernst 2008; Zerfaß and Möslein 2009; Zerfaß et al. 2004). Three communication fields are as follows: (1) internal communication; (2) external communication; and (3) *public relations* (innovation journalism: Nordfors 2009, Nordfors and Ventresca 2006). *Innovation communication* is defined as "symbolic interaction between organizations and their stakeholders about innovative products, services, technologies, and ideas" (Huck 2006: 3). Zerfaß (2009) defines innovation communication as a systematic initiation of communication processes with internal and external stakeholders to support technical, economic, and social novelties through (a) the interest-led construction, revision, and destruction of socially dependent conceptual patterns and communication resources, including professional promotion, and (b) by stimulating content-related catalysts for the development of novelties. The object of communication is primarily the innovation itself, but in many cases it is also the organization behind the innovation (Zerfaß 2009: 42 translated into English).

"**Integrated innovation communication** is a process that aims to identify internal and external contact points in the development and implementation process of an innovation, as well as to develop and implement communicative measures that guarantee the alignment of these interaction points in order to achieve a maximum level of development efficiency and effectiveness internally as well as optimal market saturation externally" (Bruhn and Ahlers 2014, p. 144).

18.3 Innovation Communication Capability

The dynamic capability of innovation communication is an individual and organizational ability of managing transactional procedures of transmitting information related to (1) ideas, business models, concepts, prototypes, practices, etc., or a combination of them, referred to as a cluster; perceived as new; (2) context-issue(s) for (1); (3) the innovation capability; considering openness, interrelation, and time/timing used to create value through activating and modifying knowledge schemata, improving management of strategic assets, and intensifying corporate reputation (adapted from Pfeffermann 2011, 2014; see Pfeffermann 2016/17 for a comparison of dynamic capabilities and innovation communication capability).

The eight dimensions of the innovation communication capability, understood as key constructs to develop and test theory and, thus, measure the multi-dimensional construct "innovation communication capability," are described as follows (Fig. 18.1):

Information transmission is defined as an organizational ability to transfer information between organizations and *stakeholders* in terms of introducing ideas, business models, concepts, prototypes, practices, etc., or a combination of them, referred to as a cluster, which is perceived as new by recipient; generating and highlighting *context-issues* related to perceived novelties; and presenting *innovation capability* considering openness, interrelation, and time/timing.

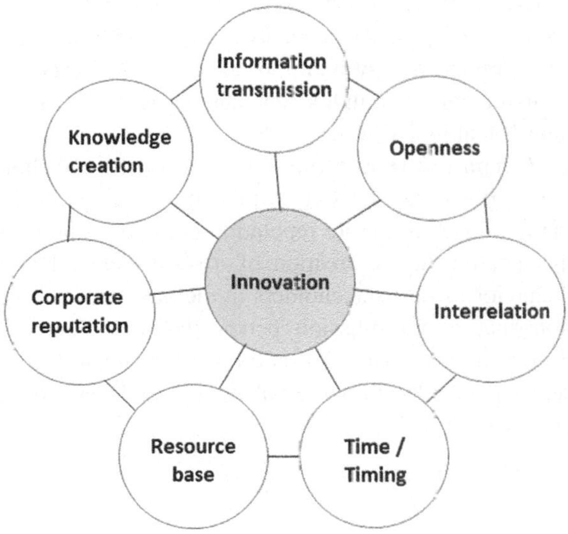

Fig. 18.1 Eight dimensions of the innovation communication capability. *Source* Adapted from Pfeffermann (2011, 2014)

Stakeholders are groups related to a company, for instance, financial analysts, customers, suppliers, media, or investors (Freeman and McVea 2005; Freeman 1999). First developed in the mid-1980s by Freeman (1984), the stakeholder approach is linked to strategy for strategy formulation and implementation (Freeman and McVea 2005).

A context-issue is defined as a "frame of reference" of an innovation or innovation cluster, which integrates the innovation or cluster into a topic of concern to grasp stakeholders' attention and to lead to a better understanding of an innovation (Huck 2006).

Innovation capability reflects human capital, social capital, and the cognition of managers involved in the creation, use and integration of market knowledge and innovation resources in order to match and create market and technological change (adapted from Bruni and Verona 2009).

Openness is defined as the ability to plan, coordinate, execute, monitor, and evaluate the use of purposefully information transmission across boundaries (e.g., cross-functional, cross-cultural and cross-disciplinary, mindsets, and comfort zones) to connect with stakeholders through information transmission (open dialog).

Interrelation is defined as the ability to plan, coordinate, execute, monitor, and evaluate connected information transmission (instruments and activities) addressing three markets: Resource market, sales market, and communication market.

Time/Timing encompasses the ability to plan, coordinate, execute, monitor, and evaluate time-related information transmission considering the linkage among past-related, present-related, and future-related information transmission.

Knowledge creation is defined as the ability to plan, coordinate, execute, monitor, and evaluate learning processes of organizations and individuals. *Learning* means revising existing knowledge and activating or modifying new knowledge schemata (Miller 2005). A schema, interrelated with other schemata, will be activated and developed if, for instance, stakeholders have made their first experiences with a new product or received information about it that they perceive as a new situation. New information or experiences change and develop existing knowledge domains into a complex schemata system (e.g., Bruhn 2009; Miller 2005; Brewer and Nakamura 1984).

Corporate reputation is defined as the stakeholders' collective positive judgments of a company's innovativeness over time (based on Barnett et al. 2006). The definition of corporate reputation consists of both (1) the stakeholder relationship perspective in the creation of trustful stakeholder relationships (the enterprise's behavior toward stakeholders in the past, present, and expected future) and (2) the information transmission perspective (the degree of informative transparency). Information transmission is crucial for enhancing trust/credibility and stakeholder satisfaction (de la Fuente Sabate and de Quevedo Puente 2003).

Management of strategic assets is defined as the planning, coordination, execution, monitoring, and evaluation of operationally managing specific resources and capabilities aimed at creating, extending and/or modifying a **unique resource base of an organization** (VRIN resources) for achieving competitive advantage. A company coordinates and implements its strategic assets in concert with other specific resources and capabilities, which leads to the inherent value of strategic assets (McGee et al. 2005).

Innovation: Ideas, business models, concepts, prototypes, practices, etc., or a combination of them, referred to as a cluster, which is perceived as new by recipient (adapted from Rogers 2003).

18.4 Innovation Communication Effects

Figure 18.2 illustrates the direct and indirect effects of innovation communication capability (ICOMM) on company value [measured by Economic Value Added (EVA)]. Beginning from the left-hand side, ICOMM has four main effects on value creation (see conceptual definition), which are divided into two direct effects and one indirect effect: the *building and modifying* function (the *knowledge creation* direct effect—LEARN), the *improving* function (the *management of strategic assets* direct effect—INNOVATE), and the *intensifying* function (the *corporate reputation* indirect effect and dependent mediator variable—CONNECT). In order to explain the positive effects of the three main functions of ICOMM on company value CV/EVA arrows are drawn from the left-hand side to the right-hand side.

The following propositions can be made based on the conceptual definition and a literature review, as detailed in Fig. 18.2 from left to right:

- P1 (general proposition): ICOMM has a positive impact on company value through knowledge creation, management of strategic assets (resource base), and corporate reputation.
- P2: ICOMM positively influences knowledge creation by building up new knowledge schemata or modifying existing ones which leads to knowledge adoption and innovation adoption that directly affects company value [measured by Economic Value Added].
- P3: ICOMM positively influences the management of strategic assets by improving management of resources and capabilities (VRIN resources/competitive resource base) which leads to advancement of operational management that directly affects company value [measured by Economic Value Added].
- P4: ICOMM positively influences intensifying corporate innovation reputation (IREP), leading to corporate reputation, which indirectly affects company value [measured by Economic Value Added].

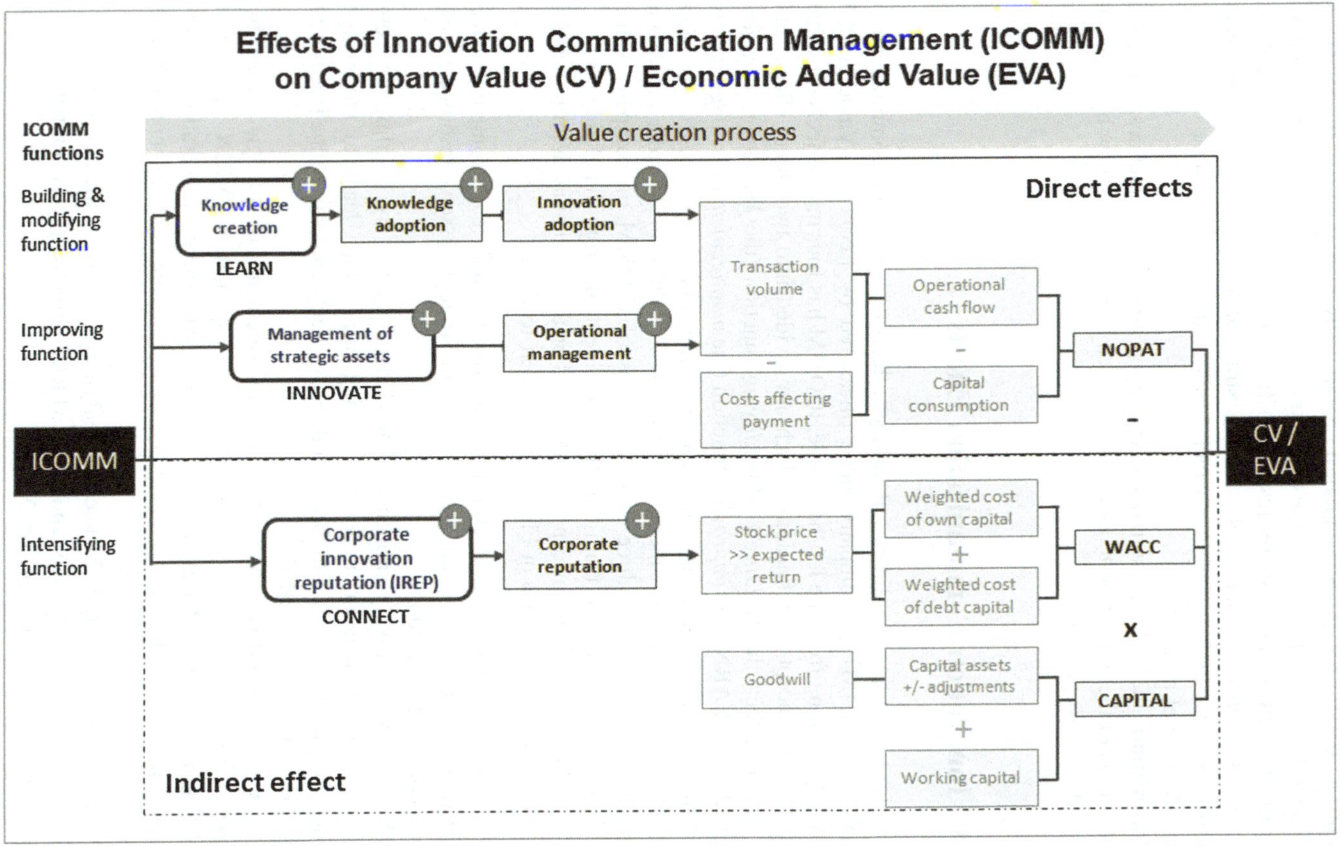

Fig. 18.2 Effects of innovation communication on company value. *Source* By the author

18.5 Innovation Communication in Management Practice

From the perspective of an innovation process, innovation communication is one of four communication fields in management practice. In fact, each innovation process can be divided into three generic phases: (1) Idea generation and selection; (2) idea visualization; and (3) idea realization and market launch, as illustrated in Fig. 18.3 (right-hand side). From the information view (data: input–processing–output), each innovation process has three phases: (1) Gathering and identification; (2) strategizing/decision-making; and (3) application (Fig. 18.3, left-hand side). The information–innovation view is linked to communication which represents the dialog level of any innovation process (see Daschkovska et al. 2010): Connecting, productive thinking, and re-connecting (Fig. 18.3, center).

Lean communication—Connection: Focus on reaching out to people and identifying the right data for a better understanding of customer needs and market specifications (lean business and communication process design).

Mental model communication—Productive thinking: Focus on connecting data (new information) with mental model(s) of reality for effective strategizing practice and sound decision-making.

Strategy communication—Re-connection: Focus on translating strategy into execution and re-connecting to customers and markets to successfully apply (big) data and turn idea into market success.

The three steps of an information–innovation process are undergrid by an "invisible" layer, which focuses on both dynamic strategizing (data view) and renewal and creating the *New* (innovation view): **Innovation Communication**. As a dynamic capability, innovation communication is a *secret* layer of a firm that, linked to good strategy, can govern ordinary capabilities and renew and shape

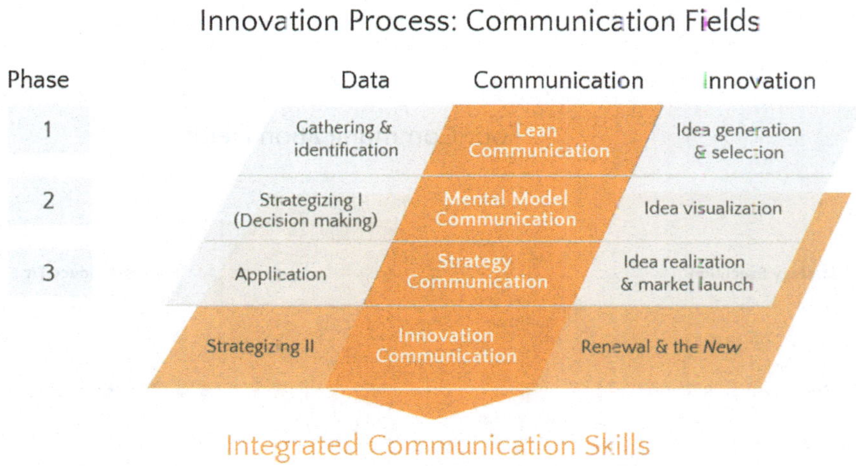

Fig. 18.3 Innovation process: communication fields. *Source* By the author

difficult-to-imitate resources (VRIN resources), for instance, data application, strategizing, and idea visualization, resulting in *doing the right things* to enhance corporate innovation and achieve competitive advantage.

Innovation communication—Growth zone: Focus on value creation and enlarging a company's growth zone, i.e., the creation driver for sustainable innovation.

Dynamic capabilities reside, in part, with individual managers and the top management team (Teece 2014). From this perspective, the four integrated communication fields can be divided into *top management skills* and *entrepreneurial thinking skills*, as illustrated in Fig. 18.4:

Strategy Communication: Top management ability to effectively translate strategy into execution and accomplish goals in specific environments.

Innovation Communication (dynamic capability): Top management ability to shape innovation culture and build up a firm's resource base for accelerating growth and creation.

Lean Communication: Entrepreneurial thinking ability to disseminate ideas and achieve traction for faster desired outcomes and outputs using lean communication process design.

Mental Model Communication: Entrepreneurial thinking skill to connect data (inner dialog) and manage information–interaction designs for facilitating innovation adoption and sound decision-making.

Innovation communication and mental model communication are defined as interrelated strong dynamic capabilities in the digitalized information age. Linked to good strategy, they tend to accelerate any information–innovation process for creating the new and achieving long-term market success.

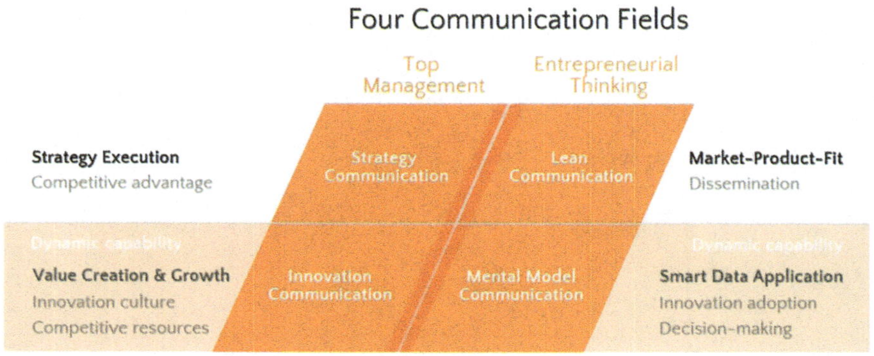

Fig. 18.4 Four communication fields. *Source* By the author

18.6 Conclusions

From an open corporate innovation view, this chapter presented a new understanding of innovation communication as a dynamic managerial capability of a company. The following eight dimensions of the innovation communication capability were defined: Information transmission, openness, interrelation, time/timing, corporate reputation, knowledge creation, resource base (= management of strategic assets), and innovation.

As one of four communication fields in an information–innovation process, innovation communication capability tends to collect information, sense opportunities in a strategizing process and innovate and transform. Innovation communication is a dynamic capability both at the individual and corporate levels. In management practice, innovation communication is a top management ability to shape innovation culture and build up a firm's resource base for accelerating growth and creation. In fact, managers, entrepreneurs, and innovators can become the instruments that help achieve the shrewd allocation of company resources to drive business growth and innovation in the long-run.

Bibliography

Arndt, J.: Role of product-related conversations in the diffusion of a new product. J. Mark. Res. **4**, 291–294 (1967)

Barnett, M., Jermier, J., Lafferty, B.: Corporate reputation: the definitional landscape. Corp. Reputation Rev. **9**(1), 26–38 (2006)

Berger, J., Heath, C.: Who drives divergence? identity signaling, outgroup dissimilarity, and the abandonment of cultural tastes. J. Pers. Soc. Psychol. **95**(3), 593–607 (2008)

Boudreau, K.J., Lakhani, K.R.: Using the crowd as an innovation partner. Harvard Bus. Rev. 60–69 (2013)

Brewer, W.F., Nakamura, G.V.: The nature and functions of schemas. In: Wyer, R.S. Jr., Srull, T.K. (eds.) Handbook of Social Cognition, pp. 119–160. Lawrence Erlbaum Associates, Hillsdale, NJ (1984)

Bruhn, M.: Integrierte Unternehmens- und Markenkommunikation: Strategische Planung und operative Umsetzung, 5th edn. Schäffer-Poeschel, Stuttgart, Germany (2009)

Bruhn, M., Ahlers, G.M.: Integrated communication in the innovation process: an approach to integrated innovation communication. In: Pfeffermann, N., Minshall, T., Mortara, L. (eds.) Strategy and Communication for Innovation, pp. 139–160. Springer, Heidelberg (2014)

Bruni, D.S., Verona, G.: Dynamic marketing capabilities in science-based firms: an exploratory investigation of the pharmaceutical industry. Br. J. Manage. **20**, 101–117 (2009)

Crosby, L.A., Johnson, S.L.: Customer-centric innovation. Mark. Manage. **15**(2), 12–13 (2006)

Daschkovska, K., Pfeffermann, N., Meyer-Larsen, N., Hülsmann, M.: Adoption framework for innovations in logistic transport chains. In Kersten, W., Blecker, T., Lüthje, C. (eds.) Pioneering Solutions in Supply Chain Management. A Comprehensive Insight Into Current Management Approaches, pp. 347–364. Erich Schmidt Verlag, Berlin (2010)

de la Fuente Sabate, J.M., de Quevedo Puente, E.: The concept and measurement of corporate reputation: an application to Spanish financial intermediaries. Corp. Reputation Rev. **5**(4), 280–301 (2003)

Freeman, R.E.: Strategic Management: A Stakeholder Approach. Pitman, Boston (1984)
Freeman, R.E.: Divergent stakeholder theory. Acad. Manage. Rev. **24**, 123–237 (1999)
Freeman, R.E., McVea, J.: A stakeholder approach to strategic management. In: Hitt, M.A., Freeman, R.E., Harrison, J.S. (eds.) Handbook of Strategic Management, pp. 189–207. Blackwell, Malden (2005)
Hesse, H.-W.: Kommunikation und Diffusion von Produktinnovationen im Konsumgüterbereich : Prognose d. Erstkäufe bei kauffördernder und kaufhemmender Kommunikation mit Diffusionsmodellen. Duncker & Humblot, Berlin, Germany (1987)
Hofbauer, G., Körner, R., Nikolaus, U., Poost, A.: Marketing von Innovationen. Strategien und Mechanismen zur Durchsetzung von Innovationen. Kohlhammer, Stuttgart, Germany (2009)
Huck, S.: New perspectives on innovation communication: findings from Germany's survey INNOVATE 2006. Innovation Journalism **3**, 3–20 (2006)
Mahajan, V., Muller, E., Bass, F.M.: New product diffusion models in marketing: a review and directions for research. J. Mark. **54**, 1–26 (1990)
Mahajan, V., Muller, E., Bass, F.M.: Diffusion of new products: empirical generalizations and managerial uses. Mark. Sci. **14**(3), 79–88 (1995)
Mahajan, V., Muller, E., Wind, Y.: New-product diffusion models: from theory to practice. In: Mahajan, V., Muller, E., Wind, Y. (eds.) New-product Diffusion Models, pp. 3–27. Springer, Düsseldorf (2000)
Martilla, J.A.: Word-of-mouth communication in the industrial adoption process. J. Mark. Res. **8** (2), 173–178 (1971)
Mast, C., Zerfaß, A. (2005). Neue Ideen erfolgreich durchsetzen. Das Handbuch der Innovationskommunikation. F.A.Z.-Institut für Management-, Markt- und Medieninformation, Frankfurt am Main, Germany
Mast, C., Huck, S., Zerfass, A.: Innovation communication. outline of the concept and empirical findings from Germany. Innovation Journalism **2**(7), 1–14 (2005)
Mast, C., Huck, S., Zerfaß, A.: Innovationskommunikation in dynamischen Märkten. Empirische Ergebnisse und Fallstudien. Lit Verlag, Berlin, Germany (2006)
Mazzarol, T.: The role of word of mouth in the diffusion of innovation. In: Hülsmann, M., Pfeffermann, N. (eds.) Strategies and Communications for Innovations, pp. 117–132. Springer, Berlin, Germany (2011)
McGee, J., Thomas, H., Wilson, D.: Strategy: Analysis and Practice. McGraw Hill, New York (2005)
Miller, K.: Communication Theories: Perspectives, Processes, and Contexts, 2nd edn. McGraw-Hill, New York (2005)
Mohr, J., Sengupta, S., Slater, S.: Marketing of High-Technology Products and Innovations, 3rd edn. Pearson, New Jersey (2009)
Nordfors, D.: Innovation journalism, attention work and the innovation economy. A review of the innovation journalism initiative 2003–2009. Innovation Journalism **6**, 1–46 (2009)
Nordfors, D., Ventresca, M.: Innovation journalism: towards research on the interplay of journalism in innovation ecosystems. Innovation Journalism **3**(2), 1–18 (2006)
Pae, J.H., Lehmann, D.R.: Multi-generation innovation diffusion: the impact of intergeneration time. J. Acad. Mark. Sci. **31**(1), 36–45 (2003)
Peres, R., Muller, E., Mahajan, V.: Innovation diffusion and new product growth models: a critical review and research directions. Int. J. Res. Mark. **63**, 849–855 (2010)
Pfeffermann, N.: Innovation communication as a cross-functional dynamic capability: strategies for organizations and networks. In: Hülsmann, M., Pfeffermann, N. (eds.) Strategies and Communications for Innovations, pp. 257–289. Springer, Berlin (2011)
Pfeffermann, N.: Innovation communication as an integrative management capability in digital innovation ecosystems. In: Pfeffermann, N., Minshall, T., Mortara, L. (eds.) Strategy and Communication for Innovation, pp. 241–270. Springer, Heidelberg (2014)
Pfeffermann, N.: The role of communication as a dynamic capability in business model innovation. In: Brem, A., Viardot, E. (eds.) Revolution of Innovation Management The Digital Breakthrough Volume 1, Palgrave Macmillan (2016/17)

Rodriguey-Montemayor, E.: Communicating for innovation: the "Social" enterprise and the translation of novel ideas. In: Pfeffermann, N., Minshall, T., Mortara, L. (eds.) Strategy and Communication for Innovation, pp. 329–346. Springer, Heidelberg (2014)
Rogers, E.M.: Diffusion of Innovations, 5th edn. The Free Press, New York (2003)
Rohlfs, J.H.: Bandwagon Effects in High-technology Industries. MIT Press, Cambridge, Mass (2001)
Sandberg, B.: Managing and Marketing Radical Innovations. Routledge, London (2008)
Sowter, C.V.: Marketing High Technology Services. Gower, Aldershot (2000)
Steinhoff, F., Trommsdorff, V.: Innovation marketing—an introduction. In: Hülsmann, M., Pfeffermann, N. (eds.) Strategies and Communications for Innovations, pp. 105–116. Springer, Berlin, Germany (2011)
Talke, K.: Einführung von Innovationen. Marktorientierte strategische und operative Aktivitäten als kritische Erfolgsfaktoren. Deut. Univ.-Verlag, Wiesbaden, Germany (2005)
Teece, D.J.: The foundation of enterprise performance: dynamic and ordinary capabilities in an (economic) theory of firms. Acad. Manage. Perspect. **28**, 328–352 (2014)
Tomochi, M., Murata, H., Kono, M.: A consumer-based model of competitive diffusion: the multiplicative effects of global and local network externalities. J. Evol. Econ. **15**, 273–295 (2005)
Trommsdorff, V., Steinhoff, F.: Innovations Marketing. Franz Vahlen, München (2007)
van den Bulte, C., Stremersch, S.: Social contagion and income heterogeneity in new product diffusion: a meta-analytic test. Mark. Sci. **23**(4), 530–544 (2004)
Weiblen, T., Chesbrough, H.W.: Engaging with startups to enhance corporate innovation. Calif. Manage. Rev. **57**, 66–90 (2015)
Zerfaß, A.: Kommunikation als konstitutives Element im Innovationsmanagement - Soziologische und kommunikationswissenschaftliche Grundlagen der Open Innovation. In: Zerfaß, A., Möslein, K.M. (eds.) Kommunikation als Erfolgsfaktor im Innovationsmanagement: Strategien im Zeitalter der Open Innovation, pp. 23–56. Gabler, Wiesbaden, Germany (2009)
Zerfaß, A., Ernst, N.: Kommunikation als Erfolgsfaktor im Innovationsmanagement. Ergebnisse einer Studie in deutschen Zukunftstechnologie-Branchen. Universität Leipzig, Leipzig (2008)
Zerfaß, A., Möslein, K.M.: Kommunikation als Erfolgsfaktor im Innovationsmanagement: Strategien im Zeitalter der Open Innovation. Gabler, Wiesbaden, Germany (2009)
Zerfaß, A., Sandhu, S., Huck, S.: Innovationskommunikation - Strategisches Handlungsfeld für Corporate Communications. In: Bentele, G., Piwinger, M., Schönborn, G. (eds.) Kommunikationsmanagement. Strategien, Wissen, Lösungen (August 2006: 1.24, 1–30). München: Luchterhand (2004)

Chapter 19
Audience-Centered Approaches to Strategic Planning: Accessing Social Capital Through Sharing Platforms on Social Media

Sherry Devereaux Ferguson

Abstract This chapter responds to the question "What are the strategic considerations in using social media platforms and open source practices such as crowdsourcing as tools in innovating organizations?" The following arguments are put forward in response to this question: (1) Organizations have moved from a learning to a sharing paradigm. (2) The most valuable organizational asset is social capital, accessible through social media and open source practices such as crowdsourcing. (3) Communicators have a role to play in accessing this social capital for purposes of innovation. (4) Changing conceptions of audiences underlie strategic communication planning. (5) Strategic planning for innovation must reflect the character of audiences fashioned by social media. In responding to this last question, the chapter explores seven characteristics of audiences that should be taken into account in planning for innovation and suggests theories that support a user orientation.

19.1 Introduction

Mired in an economic recession of yet unknown parameters, organizations face an unpredictable future. In an environment where the old paradigms have failed, innovation acquires a high value and the communication technologies undergirding innovation become critical organizational resources. In this environment, the users of Facebook, Twitter, Skype, blogs, wikis, Second Life, YouTube, Flickr, mobile technologies, LinkedIn, and other sharing platforms constitute the social capital of an information society (Bourdieu 1986; Coleman 1988; Frank et al. 2004); and practices such as crowdsourcing enable innovating organizations to access this social capital.

This chapter responds to the question "What are the strategic considerations in using social media platforms and open source practices such as crowdsourcing as

S.D. Ferguson (✉)
University of Ottawa, Ottawa, Canada

N. Pfeffermann and J. Gould (eds.), *Strategy and Communication for Innovation*,
DOI 10.1007/978-3-319-49542-2_19

tools in innovating organizations?" The following arguments are put forward in response to this question: (1) Organizations have moved from a learning to a sharing paradigm. (2) The most valuable organizational asset is social capital, accessible through social media and open source practices such as crowdsourcing. (3) Communicators have a role to play in accessing this social capital for purposes of innovation. (4) Changing conceptions of audiences underlie strategic communication planning. (5) Strategic planning for innovation must reflect the character of audiences fashioned by social media. In responding to this last question, the chapter explores seven characteristics of audiences that should be taken into account in planning for innovation and suggests theories that support a user orientation.

19.2 From Learning to Sharing Organizations

Fostering and sharing creative insights through group interactions or the idea of co-creation has a long history in the social sciences in the form of activities such as brainstorming and brainsketching, synectics, lateral thinking, fantasy chaining, and mind mapping. In the same way, businesses have long relied on Delphi and nominal group techniques for extracting knowledge from expert or other populations (Ferguson and Ferguson 1988). In the late 1980s, former General Electric President Jack Welch instituted "work-outs" (akin to New England town hall meetings) with employees, designed to elicit solutions to organizational problems (Krames 2002). Soon after, the innovation literature became peppered with talk of "boundaryless" and "learning" organizations—organizations with no clearly defined boundaries that engage in an ongoing quest for knowledge, value experimentation and improvisation, encourage critical thinking and risk-taking, tolerate mistakes, and value impermanence.

According to Redding and Catalanello (1994), the above characteristics enable the organization to innovate sufficiently fast to survive and prosper in a rapidly changing environment. The concept of the learning organization can be traced back to the double-loop learning advocated by Argyris and Schön (1974). So the movement in the direction of open source practices did not come from a conceptual vacuum. In fact, their main forerunner was open systems theory, which continues to influence scholarship across the spectrum of the social and physical sciences. Open innovation and open source approaches confirm the viability of the open systems model and the "boundaryless" organization, which is said to be characterized by speed, flexibility, and innovation.

Social media have further collapsed the boundaries between organizations and their publics and between content and technology. When we draw upon the collective intelligence through social media platforms and sharing practices, we are accessing what has come to be known as *social capital*. By *social capital*, we mean the value (economic or otherwise) that resides in social relationships and networks (Putnam 2002; Bourdieu 1986). Antikainen and Väätäjä (2010) agree that the new age capital resides in people, not in material goods. Von Hippel (1994) explained the importance of social capital to innovating organizations in the following way:

An inherent tension that plagues knowledge utilization research is the fuzzy, informal, and context-dependent nature of much of the knowledge associated with organizational innovations. This knowledge is not easily transferable because it is often embodied as know-how or practical wisdom in the person or organization that has it (a phenomenon known as *stickiness.*) (cited in Greenhalgh et al. 2005, p. 426).

Rass et al. (2013) argue that open innovation practices generate not only new ideas and concepts for products and services, but also allow the organization to accrue social capital upon which it can draw when needed. Olson and Trimi (2012) attach the term *co-innovation* to this process of value creation through convergence, collaboration, and co-creation with stakeholders.

19.3 Accessing Social Capital Through Social Media

This new age capital can be most easily located in the rapidly developing Web 2.0 phenomenon, first named by Tim Reilly in 2005 (Everitt and Mills 2009). Between 2005 and 2012, the percentage of Internet users with a social media profile catapulted from less than 8–72%—a ninefold increase (Pew Internet 2013). According to the Paris-based analytics firm Semiocast (2013), Twitter had over 500 million members by 2012, up from 27 million in 2009 (Patton 2009). About three million were active tweeters in 2009 (Rose 2009); by 2013, 170 million were active tweeters (Semiocast 2013). Pew Internet (2013) reported that 18% of all Internet users were tweeting by 2012. Skype ended 2010 with 663 million users. BBC statistics (2010) suggested that 450,000 new blogs appear each day.

More than one billion unique users sign into YouTube each month (2013). As of January 9, 2013, LinkedIn counted more than 200 million users (Nishar 2013), up from 55 million in 2010 (Baker 2010). In 2013, Facebook reported 1.11 billion active users, up from one million in 2004. MySpace registered over 33 million unique visitors in the first 6 months of 2013 (Weismann 2013). Constantly in flux, these statistics change minute by minute in an upward direction. The likelihood that social media will further extend their influence into all areas of our lives (business, interpersonal, health, and other) drives the present need to understand the potential contributions of the new communication technologies.

Social media differ from earlier efforts at gathering collective intelligence in terms of quantity of contributions, the uncontrolled nature of the input, and the often anonymous and voluntary nature of the sources—the essence of a phenomenon called *crowdsourcing* (Hudson-Smith et al. 2009). The term *crowdsourcing* (coined in 2006 by *Wired* magazine contributing editor Howe 2008) refers to open source methods of data creation where large groups of users generate content that is shared. The organization makes a deliberate effort through an open call to outsource a task to a community or group (Ekins and Williams 2010;

Tapscott and Williams 2006; Seltzer and Mahmoudi 2012). Others, however, view crowdsourcing as a less centralized happening, where the content appears spontaneously in the form of videos, blogs, wikis, or other sharing platforms. In the context of innovation, this mega trend implies "opening the door to allow more people—your customers, your employees or the public at large—into your innovation process to help improve your products, services, Web site or marketing efforts with the idea that two heads—or 2000 or 20,000—are better than one" (Sullivan 2010).

Applying the potential in crowdsourcing to innovation, Chesbrough and Appleyard (2007) summarized four requirements for using open source practices such as crowdsourcing: the need to engage a broad range of stakeholders over a sustained period of time, to compete effectively for these limited resources, to provide leadership and agendas capable of setting the tone and establishing expectations for meaningful participation, and to identify ways to profit from these policies. Despite its potential for garnering interest and ideas, scholars such as Brabham (2013) urge that organizations should consider open source practices such as crowdsourcing as one (not the only) means to engage or gather ideas for innovation.

In discussing the role of open source practices in innovating organizations, is useful to recognize that *innovation* as a term comes with different definitions in different disciplines and different contexts—sometimes implying products, at other times processes. Sometimes the term suggests recent developments; at other times, it implies new awareness of existing developments. In the same way, adoption of an innovation has a range of meanings, which can relate to individuals or organizations. In terms of corporate or business entities, the concept of adoption can imply full-scale adoption, contracting out the development of an innovation, or purchasing another company with the required innovative skills (Rye and Kimberly 2007). Kastelle and Steen (2011) argue that innovation is more about the managing than the creation of ideas. In other words, any discussion of communication of innovations can have a range of interpretations and implications.

19.4 Role of Communicators in Strategic Planning for Innovation

Both academics and practitioners agree that strategic planning is necessary for the successful integration of new technologies into a corporate vision (Nambisan and Sawhney 2010; Barnes 2010; Sullivan 2010). They also agree communicators have a significant role to play in these strategic processes. Seltzer and Mahmoudi (2012), for example, reference the collaborative planning literature in claiming that "the most active territory for planning theorizing today is 'communicative planning'" (p. 4).

How then can communicators contribute to these strategic planning processes? The communicator looks for ways to support the corporate mission, mandate, and objectives through the framing of communication goals, messages, strategies, and tactics. Without reference to the larger strategic plan, communication planning loses focus; and without a supporting communication plan, the organization has no coordinated way to convey its mission, mandate, or vision or to promote its programs, products, and services.

Despite the importance of including a communication element in strategic planning for innovation, a survey involving 1087 PR practitioners in 22 European countries found that communication personnel are rarely involved in planning for innovation within their organizations. Only one out of every three PR professionals, for example, has any involvement with innovation in their companies; and only one in five communication managers considers innovation to be a strategic issue for communicators (Zerfass et al. 2007).

Cook (2008) proposed a communication model with applications specific to social media and also applicable to innovating organizations. He said that social media perform four functions of relevance to organizations: communication, cooperation, collaboration, and connection. Some argue the need to hire a social media administrator to coordinate these functions (Bradley 2008).

19.5 Changing Conceptions of Audiences

The starting point for any strategic communication plan must be the analysis of audience needs and expectations—a dedicated area of research in communication studies. In the context of this discussion, the term *audiences* will refer to employees, as well as external publics, as social capital resides in both groups.

Following World War II, four major changes occurred in how psychologists, political scientists, and communicators viewed audiences. First, communication studies moved from an emphasis on audiences as passive recipients of information to audiences as active processors of information. The limited effects and two flow models of opinion leadership, which stressed human agency, replaced the hypodermic needle model, which saw audiences as passive and highly susceptible to persuasion (Lazarsfeld et al. 1968).

Much like the early communication models, one of the most popular early innovation models—the Innovation-Decision Process, for example—portrayed "adopters" as passive recipients, who could choose to act or not act on information. This classic model involved five steps: knowledge, persuasion, decision, implementation, and confirmation (Rogers 1995)—none of which required an active contribution to the direction of change. According to Haider and Kreps (2004), over 5000 articles focusing on the distribution process had been published by the 40th anniversary of diffusion research. Nonetheless, as happened in the field of communication, innovation studies have shifted over time to a new and more participatory view of consumers and contributors.

Second, communication theorists began to see senders and receivers as constantly exchanging roles; and the Aristotelian model, which depicted communication flow as one-way and linear, fell into disuse. One of the most popular replacements was the transactional model, which sees communication as a dynamic process, involving continuous changes in the communicators and in the environment in which they operate (Barnlund 1970).

Third, the new models saw audiences as culturally diverse, active, and individualistic in their responses. Talk of *the public* yielded to discussion of *publics*. Uses and gratifications theory (Katz et al. 1974; McQuail 1983) attributed even higher levels of initiative to audiences. This theory argues that audiences actively select media that meet their need for information, entertainment, social interaction, or recognition, among others. Applying these ideas to open innovation, Antikainen et al. (2010) add categories such as personal learning, knowledge exchange, social capital, and enhancement of professional status.

Finally, post-war models moved from views of communicators as conveyers of information (e.g., Hovland et al. 1953) to communicators as builders of social relationships (e.g., Grunig 1992). In support of this view, Paulini et al. (2011) note that social communication increases credibility for organizations when it shows sensitivity to user needs. This idea of building collaborative relationships becomes extremely important when we move into the area of innovation, where motivation to participate becomes extremely important (Antikainen et al. 2010).

19.6 Seven Trends with Implications for Communication Planning for Innovation

The development of social media has further changed and elevated the status of audiences. Thus, this final section of the chapter seeks to identify how social media have influenced the character of twenty-first century audiences and established their status as significant sources of social capital in an information society (see Mandarano et al. 2010). More specifically, I will identify seven trends with the potential to impact upon strategic planning for innovation.

First, the dominant characteristic of all social media is their potential for—and encouragement of—audience participation. An audience member climbs onstage at a Bourbon Street establishment to become a part of the entertainment. Contestants on *American Idol* and *Dancing with the Stars* plead for audience votes that will enable them to continue in the competition. CNN and Deutsche Well invite and publish feedback on online news articles generated by staff members. Artists gain acclaim on the basis of number of YouTube views. Court TV shows and crime stopper infomercials invite questions and feedback from viewers. Citizen journalists and I-reporters publish photographs of tsunamis, tornados, and volcano eruptions; and best-selling author James Patterson invited fledgling writers to pen most of the chapters for his book *Airborne*.

In short, the boundaries between senders and receivers of messages and content and technology have becoming increasingly blurred as audiences demand an active, participative role in the communication process. Recognizing the new user-generated and reflexive technoculture (Han 2010), *Time* magazine named "You" the "Person of the Year" in 2006. Citing Lev Grossman, author of the article accompanying the *Time* cover, Han explains that "the Internet that has allowed 'You' to win the recognition … does not resemble the Internet of the 1990s dot-com boom nor the ARPANET developed by the U.S. Department of Defence 20 years before that" (pp. 200–201). In other words, Web 2.0 is a radical innovation in itself, leaving disruptive change in its wake but creating an environment for "radical inclusion" (Han 2010, p. 201).

Second, social media have encouraged audiences to become active seekers of information. An orthopedic patient arrives at the surgeon's office, armed with information on the latest procedure for resurfacing the hip joint. Potential buyers turn to online reviews in researching the latest innovations in hybrid cars. Interested individuals go to *Britannica Online* to learn more about recent developments in DNA research. The increasing fragmentation and difficulty of using mass media to reach twenty-first century audiences is a well-established finding in communication research (e.g., Webster 2006). For that reason, organizations should not ignore the potential in open source platforms, which allow audiences to seek out the organization.

Third, social media have encouraged a critical mindset in audiences. Users have come to expect a feedback option with every communication. So the possibility to provide critiques of people, organizations, and ideas appears across the spectrum—in online journalism, Twitter, blogs, TV news and entertainment features, and print media.

Whereas the top-down flow of information, dominating the years preceding the development of satellite TV, nurtured a mindset that did not encourage criticism of authority figures in organizations or government, the current flow of information in every direction (upward, sideways, and downward) encourages people to express their points of view and to challenge authority. Even a cursory look at feedback links confirms the critical and cynical nature of much of this feedback (Rice 2010). As a consequence, many organizations have instituted a policy of pulling objectionable comments from the dialog; and some kinds of software allow users to bring unacceptable responses to the attention of the host organization. The struggle of countries such as China to maintain control over social media has led to even stronger policies and practices, such as the demand to censor access points on foreign search engines—an action that caused Google to withdraw services from that country. As illustrated above, the censorship may be initiated at the point of the user, the host organization, or even a national entity.

Fourth, social media draw audiences who seek attention and recognition. As one blogger noted, "There's not a lot I won't put on there" because "I love to be the center of attention" (Miller and Shepherd n.d.). Some studies have demonstrated that audiences stop using sites that fail to acknowledge their presence (Huberman et al. 2009). For that reason, organizations offer a variety of monetary and non-monetary rewards to motivate users to participate in open innovation communities. Common non-monetary techniques include allotting points for valued contributions, listing top innovators on the websites, acknowledging the most active members, and introducing active community members (Antikainen and Väätäjä 2010). Other websites offer financial compensation for ideas; however, people are often willing to forego financial gain to obtain notice from peers or a firm (Huberman et al. 2009).

Fifth, social media encourage audiences to disclose freely, and audiences expect similar levels of openness and transparency in others. High levels of personal disclosure on Facebook, blogs, and websites such as Postsecret.com have created a generation of consumers who expect the same high levels of disclosure from others, including celebrities, politicians, and corporate leaders (Miller and Shepherd n.d.). In the last several years, a number of American and Canadian politicians and generals have resigned from public office after having affairs exposed in the national media and widely discussed on social media. Those facing public demands for accountability not infrequently go on national television to apologize to family, supporters, and a largely anonymous public.

The new level of interconnectedness, offered by social media, has nurtured a culture of voyeurism and incursions into the lives of others. More importantly for organizations, however, the connections do not stop with the personal. Publics expect corporate entities and their leaders to share knowledge and information, including the negative, and to conduct business in the most transparent fashion. In other words, they demand reciprocity: we will share with you, but you must also share with us. As Crescenzo (2010) observed, "Corporate communication—that whitewashed, sterilized, sanitized form of communicating that so many organizations rely on—doesn't really work in the SM space" (p. 11).

Like many other ideas, the recognition of the importance of transparency in communication is not a novel concept. Cleveland wrote an article in 1985 titled "The Twilight of Hierarchy: Speculations on the Global Information Society," in which he discussed the leakiness of information and its impact on hierarchy. In 1988, Ferguson and Ferguson discussed the futility of talking about organizational boundaries and introduced the simultaneous access model as a replacement for the top-down communication model; and in 2001, former GE President Welch (2001) observed:

> Hierarchy is dead. The organization of the future will be virtually layerless and increasingly boundaryless, a series of information networks in which more electrons and fewer people will manage processes. Information will become transparent. No leader will be able to hoard the facts that once made the corner office so powerful. (p. 433)

Sixth, social media have created audiences who expect responses in real time. Instant conversations and instant updates typify interactions on social media. Whereas consumers used to be satisfied with a letter received three or four weeks after an inquiry, they now expect a response within 24 h of receipt of an email. No place or time is sacred space, and meeting the needs of contemporary audiences means accepting their terms of engagement. Yet few organizations are equipped to handle the demands:

> As traditional business intelligence systems and technology intersect with new systems such as Facebook, Twitter, and Google Wave, a conflict arises between traditional information retrieval and discovery of new information available via newsfeeds, blog articles, short text messages from Twitter users, and user-generated videos posted to sites such as YouTube and Vimeo. Most business intelligence systems are not well-equipped to handle real-time information. The future of real time lies in creating applications that require no searching. (Arnold 2009, p. 40)

For governments and organizations that require multiple levels of approval for responses or revelation of information, the problem is serious—and still further aggravated in countries like Canada with requirements for bilingual communications. In speaking of organizational uses of ICTs, Sørnes et al. (2005) note "Given the apparent significance of time in structuring organizational reality, future research should examine more thoroughly the temporal elements that affect members' sensemaking, their communication with one another" (p. 137).

Seventh, social media require a mix of language competencies in audiences, as well as in those who seek to interact with them. Transliteracy is "the ability to read, write and interact across a range of platforms, tools and media from signing and orality through handwriting, print, TV, radio and film, to digital social networks" (Thomas et al. 2007). In response to Twitter's demand for 140 word messages, microbloggers employ a vocabulary of acronyms, abbreviations, and icons to offer brief and to-the-point information to their audiences (DeFebbo et al. 2009). The website blog, on the other hand, encourages a different form of literacy, more akin to the traditional essay or diary. On Instant Messenger, the conversations proceed through the use of multiple and often discontinuous threads. The Social Media Release (SMR), a new public relations tool, provides content to bloggers and other social media users, who may or may not publish or transmit the information to their personal network of friends and acquaintances (Steyn et al. 2010). Even if bloggers choose to share the SMR, they may repackage it or add comments or links to other sources (Bradley 2008).

19.7 Toward a New Paradigm: Connecting Audiences, Social Media, Open Source Policies, and Innovation

The proliferation of social media in the new century has fueled the need for a new paradigm to guide innovation studies and practices—one that sees audiences as participative, active, critical, open, attention-seeking and self-aware, time-sensitive, and transliterate. Some of the theories and concepts relevant to an audience or user orientation include open innovation, symbiosis, social constructionism (also social constructivism), sensemaking, and reflexive modernity. Others (already mentioned) include open systems theory, uses and gratifications, and social influence models such as opinion leadership.

The research into opinion leadership may have new applications in a Web 2.0 world. Jeppesen and Laursen (2009) found that "lead users" (the most active contributors) in online communities possess more relevant solution knowledge than others; and Parvanta et al. (2013) warn that organizations need to identify the most motivated, expert, and creative users to obtain full benefits from crowdsourcing. They claim that only 9% of contributors have the motivation and experience and only 1% the creativity to make a meaningful contribution. Saxton et al. (2013) also point to the importance of identifying "wise" sources.

The extent to which organizations can engage these more sophisticated and "lead users" could also influence commitment to diffuse the innovations.

Christensen (1997) described the reasons that "great firms" fail when faced with disruptive technologies. Unlike sustaining technologies, which involve incremental improvement of established technologies, disruptive innovations typically call for new ways of thinking about products, services, and markets. In these circumstances, large firms rarely cope well, as illustrated by the case of social media:

> The socially transformative innovations in information technology such as email, the World-wide Web, Google, e-commerce, Linux, and eBay have emerged not from the traditional powerhouses of IT innovation such as IBM, Intel, Bell Labs, or Microsoft, but from users of their technologies—business innovators, user groups, and communities of practice outside of the original centers of innovation. (Bers 2005, p. 3)

Accepting that knowledge no longer resides in a few large organizations, Chesbrough (2003) introduced the term *open innovation*, which stresses the importance of going outside the boundaries of the organization to harvest and—and in some cases—develop or out-license innovative ideas and intellectual property. Open innovation theory assumes that knowledge no longer resides in a few large organizations. According to Christensen and Overdorf (2000), viable options for improving the coping potential of larger firms include creating new structures within the corporation, birthing an independent organization that comes from the parent, or acquiring a new company whose processes and values mesh with the demands of the new task. Symbiotic models build on the open innovation concept (Yang and Shyu 2009; Castiaux 2007).

Social constructionist and social constructivist theories also offer user-oriented ways of thinking about social media and innovation (Berger and Luckmann 1966; Bers 2005). In the spirit of postmodernism, social constructionists argue that media technologies have created the reality in which contemporary society moves; however, these scholars do not distinguish between developers and users of the technologies. As in the case of the Linux open source movement, the users are also the developers of the technology, and no one person or organization holds the rights to Linux. In this sense, individuals and groups participate in co-creation of their perceived social realities. Social constructionists such as Bers (2005) argue that recombining and identifying new social ends for existing products and services should be the emphasis of open source innovation research. Even if not applied to every organization, this approach would seem to fit well with large companies that experience difficulty in coping with disruptive innovations.

A psychologically based variation of *social constructionism*, *social constructivism* asserts that we create our own social reality through interaction with the media. Similarly, sensemaking models (Weick et al. 2005; Dervin 1992) are concerned with how we reduce uncertainty and make sense out of our experiences. Moved into the organizational sphere, sensemaking models incorporate concepts related to attribution of meaning in shared and collaborative contexts and help us to understand what motivates people. Theories of reflexive modernity propose that, over time, people become more self-aware and reflective. The focus on "YOU" in modern society would seem to validate the presence of reflexivity in contemporary society, along with its relevance for innovating organizations. Citing Lane (2005), Seltzer and Mahmoudi (2012) assert that all modern schools of thought about planning for innovation regard stakeholder engagement as "a fundamental characteristic of the planning process, not just an adjunct to decision making" (p. 4).

19.8 Conclusion

All of the above models and theories place an important emphasis on audiences—their needs, expectations, and potential to contribute to the collective intelligence through crowdsourcing. Key words in any formula for success will be trust, respect, transparency, openness, sharing, recognition, and timeliness. With the proliferation of related practices such as crowdfunding of business enterprises and even health care (see organizations such as WATSI), little doubt remains that open source platforms and practices will characterize the operations of many different organizations in the coming years; and organizations (innovating or otherwise) risk joining the ranks of endangered species if they do not compete for the new age social capital. In brief, strategic planning for communication of innovations must build on existing knowledge of audiences, social media, crowdsourcing, communication theories, and innovation theories.

Bibliography

Antikainen, M.J., Väätäjä, H.K.: Rewarding in open innovation communities—how to motivate members? Int. J. Entrepreneurship Innov. Manage. **11**, 440–456 (2010)

Antikainen, M., Mäkipää, M., Ahonen, M.: Motivating and supporting collaboration in open innovation. Eur. J. Innov. Manage. **13**(1), 100–119 (2010)

Argyris, C., Schön, D.: Theory in Practice: Increasing Professional Effectiveness. Jossey-Bass, San Francisco, CA (1974)

Arnold, S.E.: Real-time search: where retrieval and discovery collide. Online **33**, 40–41 (2009)

Baker, G.: It's networking, but not as you knew it. NZ Bus. **24**, 40–41 (2010)

Barnes, N.G.: Tweeting and blogging to the top. Mark. Res. **22**, 8–13 (2010)

Barnlund, D.C.: Transactional model of communication. In: Sereno, K.K., Mortensen, C.D. (eds.) Foundations of Communication Theory. Harper and Row, New York, NY (1970)

Bers, J.A.: Contributions of social constructionism to a theory of accelerated radical innovation. Retrieved from http://ari-institute.com/_pdfs/publications/ (2005)

Bourdieu, P.: The forms of capital. In: Richardson, J.G. (ed.) Handbook of Theory and Research for the Sociology of Education. Greenwood Press, New York, NY (1986)

Brabham, D.C.: Crowdsourcing: a model for leveraging online communities. (2013)

Bradley, P.: A selection of social media search engines. Ariadne. **57**, 40. Retrieved 15 May 2010 from http://www.ariadne.ac.uk/issue57/search-engines/ (2008)

Brandel, M.: Baited and duped on facebook: how smart companies are protecting employees from scammers and creating usage policies that work. Computerworld **43**, 28–33 (2009)

Brenner, J.: Pew Internet: social networking (full detail). Retrieved at http://pewinternet.org/Commentary/2012/March/Pew-Internet-Social-Networking-full-detail.aspx (2013), on 10 Aug 2013

Burt, R.S.: The social capital of opinion leaders. Ann. Am. Acad. Polit. Soc. Sci. **566**, 37–54 (1999)

Castiaux, A.: Radical innovation in established organizations: being a knowledge predator. J. Eng. Tech. Manage. **24**, 36–52 (2007)

Chesbrough, H.W.: Open Innovation: The New Imperative for Creating and Profiting from Technology. Harvard University Press, Boston, MA (2003)

Chesbrough, H.W.: Open innovation: a new paradigm for understanding industrial innovation. In: Chesbrough, H.W., Vanhanverbeke, W., West, J. (eds.) Open Innovation: Researching a New Paradigm, pp. 1–12. Oxford University Press, Oxford (2006)

Christensen, C.M.: The Innovator's Dilemma: When New Technologies Cause Great Firms to Fail. Harvard Business School Press, Cambridge, MA (1997)

Christensen, C.M., Overdorf, M.: Meeting the challenge of disruptive change. Harvard Bus. Rev. **78**, 66–76 (2000)

Cleveland, H.: The twilight of hierarchy: speculations on the global information society symposium presentation. Public Adm. Rev. **45**, 185–195 (1985)

Coleman, J.S.: Social capital in the creation of human capital. Am. J. Sociol. **94**, 95–120 (1988)

Crescenzo, S.: Are you a "social media evangelist" or a communicator? Commun. World. **27**, 10–11 (2010)

DeFebbo, D.M., Mihlrad, L., Strong, M.A.: Microblogging for medical libraries and librarians. J. Electron. Res. Med. Libr. **6**, 211–223 (2009)

Delwiche, A., Henderson, J. (eds.): The Participatory Cultures, pp. 120–129. Routledge, New York (2012)

Dervin, B.: From the mind's eye of the user: the sense-making qualitative-quantitative methodology. In: Glazier, J., Powell, R. (eds.) Qualitative Research in Information Management, pp. 61–84. Libraries Unlimited, Englewood, CO (1992)

Ekins, S., Williams, A.J.: Reaching out to collaborators: crowdsourcing for pharmaceutical research. Pharm. Res. **27**, 393–395 (2010)

Everitt, D., Mills, S.: Cultural anxiety 2.0. Media Cult. Soc. **31**, 749–768 (2009)

Facebook: Number of active users at facebook over the years. AP news release, 1 May 2013. Accessed at Yahoo news, http://news.yahoo.com/number-active-users-facebook-over-230449748.html (2013), on 10 Aug 2013

Ferguson, S.D., Ferguson, S.: Organizational Communication, 2nd edn. Transaction Publishers, New Brunswick, NJ (1988)

Frank, K.A., Zhao, Y., Borman, K.: Social capital and the diffusion of innovations within organizations: the case of computer technology in schools. Sociol. Educ. **77**, 148–171 (2004)

Greenhalgh, T., Robert, G., Macfarlane, F., Bate, P., Kyriakidou, O., Peacock, R.: Storylines of research in diffusion of innovation: a meta-narrative approach to systematic review. Soc. Sci. Med. **61**, 417–430 (2005)

Grunig, J.E. (ed.): Excellence in Public Relations and Communication Management. Lawrence Erlbaum Associates, Hillsdale, NJ (1992)

Haider, M., Kreps, G.L.: Forty years of diffusion of innovations: utility and value in public health. J. Health Commun. **9**, 3–11 (2004)
Han, S.: Theorizing new media: reflexivity, knowledge, and the Web 2.0. Sociol. Inq. **80**, 200–213 (2010)
Howe, J.: Crowdsourcing: Why the Power of the Crowd Is Driving the Future of Business. Crown Business, New York, NY (2008)
Huberman, B.A., Romero, D.M., Wu, F.: Crowdsourcing, attention and productivity. J. Inf. Sci. **35**, 758–765 (2009)
Hudson-Smith, A., Batty, M., Crooks, A., Milton, R.: Mapping for the masses: accessing Web 2.0 through crowdsourcing. Soc. Sci. Comput. Rev. **27**, 524–538 (2009)
Jeppesen, L.B., Laursen, K.: The role of lead users in knowledge sharing. Res. Policy. **38**, 1582–1589 (2009)
Kastelle, T., Steen, J.: Proposition: ideas are not innovation. Prometheus. **29**(2), 199–205 (2011)
Katz, E., Lazarsfeld, P.F.: Personal Influence: The Part Played by People in the Flow of Mass Communication. Free Press, Glencoe, IL (1955)
Katz, E., Blumler, J., Gurevitch, M.: Utilization of mass communication by the individual. In: Blumler, J.G., Katz, E. (eds.) The Uses of Mass Communications: Current Perspectives on Gratifications Research. Sage, Beverly Hills, CA (1974)
Krames, J.A.: The Jack Welch Lexicon of Leadership. McGraw-Hill, New York, NY (2002)
Lane, M.B.: Public participation in planning: an intellectual history. Aust. Geogr. **36**, 283–299 (2005)
Lazarsfeld, P., Berelson, B., Gaudet, H.: The People's Choice. Columbia University Press, New York (1968)
Lo, B., Parham, L.: The impact of Web 2.0 on the doctor-patient relationship. J. Law Med. Ethics. **38**, 17–26 (2010)
Mandarano, L., Meenar, M., Steins, C.: Building social capital in the digital age of civic engagement. J. Plann. Lit. **25**, 123–135 (2010)
McQuail, D.: Mass Communication Theory. Sage, London (1983)
Nambisan, S., Sawhney, S.: The Global Brain. Wharton School Publishing, Upper Saddle River, NJ (2010)
Nishar, D.: 200 million members. Linkin Blog. Accessed at http://blog.linkedin.com/2013/01/09/linkedin-200-million/ (2013), on 10 Aug 2013
Parvanta, C., Roth, Y., Keller, H.: Crowdsourcing 101: a few basics to make you the leader of the pack. Health Promot. Pract. **14**(2), 163–167 (2013)
Patton, C.: The social networking express. Prof. Saf. **54**, 5 (2009)
Paulini, M., Maher, M.L., Murty, P.: The role of collective intelligence in design: a protocol study of online design communication. In: Herr, C.M., Gu, N., Roudavsky, S., Schnabel, M.A. (eds.) Circuit Bending, Breaking, and Mending: Proceedings of the 16th International Conference on Computer-Aided Architectural Design Research in Asia, pp. 687–696. Retrieved at http://www.academia.edu/1055048/The_Role_of_Collective_Intelligence_in_Design (2011)
Putnam, R.D. (ed.): Democracies in Flux: The Evolution of Social Capital. Oxford University Press, New York, NY (2002)
Ranger, J.: Exploring the potential of social media as organizational communication tools. Unpublished M.A. thesis, University of Ottawa, (2009)
Rass, M., Dumbach, M., Danzinger, F., Bullinger, A.C., Moeslein, K.M.: Open innovation and firm performance: the mediating role of social capital. Creativity Innov. Manage. **22**(2), 177–194 (2013)
Redding, J.C., Catalanello, R.F.: Strategic Readiness. Jossey-Bass Inc, San Francisco, CA (1994)
Rice, D.: Managing your reputation in a viral world. SDM **40**, 93–94 (2010)
Rogers, E.: Diffusion of Innovations, 4th edn. Free Press, New York, NY (1995)
Rose, B.: Much chatter about twitter. Am. Bar Assoc. J. **95**, 25–26 (2009)
Rye, C.B., Kimberly, J.R.: The adoption of innovations by provider organizations in health care. Med. Care Res. Rev. **64**, 235–278 (2007)

Saxton, G.D., Oh, O., Kishore, R.: Rules of crowdsourcing: models, issues, and systems of control. Inf. Syst. Manage. **30**, 2–20 (2013)
Seltzer, E., Mahmoudi, D.: Citizen participation, open innovation, and crowdsourcing: challenges and opportunities for planning. J. Plann. Lit. **28**(1), 3–18 (2012)
Semiocast: Twitter reaches half a billion accounts. Accessed at http://semiocast.com/en/publications/2012_07_30_Twitter_reaches_half_a_billion_accounts_140m_in_the_US (2013), on 10 Aug 2013
Skype grows FY revenues 20%, reaches 663 mln users: Telecom Paper, posted on 8 Mar 2011 and retrieved at http://www.telecompaper.com/news/skype-grows-fy-revenues-20-reaches-663-mln-users–790254 on 10 Aug 2013
Sørnes, J.-O., Stephens, K.K., Browning, L.D., Sætre, A.S.: A reflexive model of ICT practices in organizations. Inf. Sci. J. **8**, 123–142 (2005)
Steyn, P., Salehi-Sangari, E., Pitt, L., Parent, M., Berthon, P.: The social media release as a public relations tool: intentions to use among B2B bloggers. Public Relat. Rev. **36**, 87–89 (2010)
Sullivan, E.A.: A group effort: more companies are turning to the wisdom of the crowd to find ways to innovate. Mark. News **44**, 23–30 (2010)
Tapscott, D., Williams, A.J.: Wikinomics: How Mass Collaboration Changes Everything. New York, NY (2006)
Thomas, S., Joseph, C., Laccetti, J., Mason, B., Mills, S., Perril, S., Pullinger, K.: Transliteracy: crossing divides. First Monday. **12**. Retrieved 25 May 2010 from http://www.uic.edu/htbin/cgiwrap/bin/ojs/index.php/fm/article/view/2060/1908 (2007)
von Hippel, E.: "Sticky information" and the locus for problem-solving: implications for innovation. Manage. Sci. **40**, 429–439 (1994)
Webster, J.G.: Beneath the veneer of fragmentation: television audience polarization in a multichannel world. J. Commun. **55**, 366–382 (2006)
Weick, K.: Sensemaking in Organizations. Sage, Thousand Oaks, CA (1995)
Weick, K.E., Sutcliffe, K.M., Obstfeld, D.: Organizing and the process of sensemaking. Organ. Sci. **16**, 409–421 (2005)
Weismann, S.: Myspace's pitch to brands. Posted on 8 May 2013 at CraveOnline. Accessed at http://digiday.com/brands/myspaces-pitch-to-brands/ (2013), on 10 Aug 2013
Welch, J.: Jack: Straight from the Gut. New York: NY: Warner Books (2001)
Williams, R., Williams, T.: Adopting social media: are we leaders, managers or followers? Commun. World. 34–38 (July–Aug 2008)
Miller, C.R., Shepherd, D.: Blogging as social action: a genre analysis of the webblog. Into the Blogosphere, online publication of the University of Minnesota. Retrieved from http://blog.lib.umn.edu/blogosphere/blogging_as_social_action.html (n.d.)
Yang, C., Shyu, J.Z.: A symbiosis dynamic analysis for collaborative R&D in open innovation. Int. J. Comput. Sci. Eng. **5**, 74–84 (2009)
YouTube: Statistics. Accessed at http://www.youtube.com/yt/press/statistics.html (2013), on 10 Aug 2013
Zerfass, A., Sandhu, S., Young, P.: Euroblog perspectives on social software in communication management: results and implications. Leipzig/Luzern/Sunderland. Retrieved from www.euroblog2007.org/euroblog2007–results.pdf (2007)

Part IV
Case Examples

Chapter 20
The Innovation Engine: A Framework for Overcoming Cultural and Organizational Impediments to Innovation at Scale

Andrew Breen

Abstract Large, established organizations fear disruption from large technology companies and startups alike. In trying to thwart this, they have explored several innovation approaches such as labs, acquisitions, and spinouts. Most have not succeeded often due to the impediments that traditional twentieth century corporate culture and organizational design bring. The Innovation Engine is a framework developed to overcome organizational impediments to innovation at scale. The framework has been derived from taking organizational design and processes from successful technology growth and mature companies and the learnings from their application in more traditional companies outside the technology industry.

20.1 Cultural Impediments to Innovation

The rapid disruption of many traditional industries such as media and publishing, advertising and retailing by new, rapid growth firms like Google, Amazon, and Facebook has mostly been attributed to the advent of new technologies, principally the Internet, enabling shifts in consumer behavior and commerce patterns. However, an argument can be made that part of the success stems from their unorthodox approach to product development. This can be attributed to organizational and process design which broke from the traditions of twentieth century management canon. This "new norm" now threatens established industries and companies due to intertwining a new generation of organizational approaches with technology, whereby innovation at scale is now feasible. Even companies in capital intense and highly regulated industries like finance and healthcare face existential threats from the technology industry unless they adapt and discover sustainable innovation at scale.

A. Breen (✉)
Leonard N. Stern School of Business, Tisch Hall, 40 West Fourth Street, 900, New York, NY 10012, USA
e-mail: abreen@stern.nyu.edu

N. Pfeffermann and J. Gould (eds.), *Strategy and Communication for Innovation*,
DOI 10.1007/978-3-319-49542-2_20

The first attempts at innovation within large, traditional organizations was as an appendage by starting a "lab," spinning out innovations and investing in or acquiring startup companies. With few exceptions, these approaches have not materialized primarily due to the lack of integration with the core business (Blank and Simoudis 2015). In some cases, the business model was disruptive to the core business threatening existing cash flow. In others, a modern technology research and development culture could not fit with an operational organization focused on optimization of a mature product line. For example, MySpace was the largest social network when News Corp acquired it in mid 2005. Strategically, extending News Corp's traditional business into the media-oriented social network made sense as it was clear younger users were consuming via this new medium. However, cultural differences and the disruptive business model of MySpace clashed with the more traditional News Corp. Revenue pressure arrived from News Corp requiring more advertising per page view hindering the user experience (Gillette 2011). In addition, experimentation by the team ceased as maintaining their incumbent position required most of the team's product development cycles. The upstart Facebook was not only fueled by founder Mark Zuckerberg's "move fast and break things" (Henry 2014) culture of experimentation but also an embrace of external developers to drive disruptive innovation beyond what its own team could develop. This continued until the emergent Facebook overtook MySpace in unique users by 2008. MySpace's decline continued steadily and was written down and eventually sold by News Corp in 2011 (Albanesius 2009).

Another familiar strategy to invigorate innovation in a staid business is the "acqui-hire" of small companies of engineering talent. In this model, the established acquiring company has little interest in the startup's products typically discontinuing them soon after acquisition. The core thesis is that an infusion of engineering talent and disruptive culture can stimulate innovation in the acquiring company. One case of this was Yahoo's numerous acquisitions after Marissa Mayer became CEO in 2012. She arrived after spending most of her career at Google which was famous for its academically freewheeling engineering driven innovation culture instilled by its founders Sergy Brin and Larry Page. Ms. Mayer arrived with an agenda to replicate that culture at Yahoo (Tkaczyk 2013) who, ironically as one of the Internet pioneers, had, by the time of her arrival, taken on a more traditional management culture after being driven toward a traditional media business by leaders coming from traditional media (Blodget 2008). One of the original pioneering startups of the Internet age was now a symbol of a technology company being disrupted by other technology companies primarily due to its adoption of a traditional business model and culture.

To quickly stimulate these changes, Ms. Mayer acquired 20 startups companies in her first 13 months as CEO (Indvik 2013) primarily for their engineering talent and infusion of innovation into the culture. However, this approach did not provide the innovation she envisioned as the fortunes of Yahoo had not changed by October 2015 with declining revenue and significant organizational departures (Lee 2015). One of the subtle aspects missed by Ms. Mayer and by other approaches to infuse an engineering-driven R&D culture is that most of the benefits to the company come from business model, product, or marketing innovation enabled by technology but not the technology itself. As the annual studies by Booz & Co. have shown, there is no correlation between corporate research and development spending and financial performance (Jaruzelski and Katzenbach 2012). As has been identified in famous labs like Bell Labs and Xerox Parc, even a massive budget with large groups of sophisticated scientists and engineers did not lead to successful commercialization of the innovations due to lack of focus not only on the business model but in how the operation could absorb them without disrupting the existing business (Blank and Simoudis 2015). Thereby, one can assert that technology itself does not represent innovation. A culture and process integrated into the existing organizational fabric begets innovation.

20.2 Organizational Challenges in Cultivating Innovation

Traditional organizations are designed to manage away risk optimizing existing products in existing markets to ensure the highest margins. On the occasion of market expansion, an existing product, and its associated business model, is brought to new geographies or adjacent segments with refinement of its marketing messaging to communicate the value proposition to the new market. In these moves, there is no innovation around the product and only minor evolution in the product promise through marketing. It is an extension and optimization of the product. Historically, when a company wanted to develop a new product for an existing customer base they typically did that through careful exploration in focus groups and market tests. The new products were additive to the product mix but not typically threatening their existing market accepted products nor business model. By this definition, innovation constitutes an attempt to change the business model on an existing solution for an existing market need or creating a new market by developing a solution to an unmet yet unknown need displacing the current solutions. For an existing business, both of these are disruptive and risky finding little appetite in the organization.

Exploration of new, disruptive products and models is typically only supported when there is a real or perceived threat to the existing business. In those cases, companies often try and stimulate innovation internally thinking their employees who are closest to their existing products and customers will most easily generate new ideas to defend and extend their market. In some ways, this is a rational behavior but often can generate local maxima due to the existing employee base solely thinking about the company and its traditional business. Even when a new idea might be organically developed by the existing employees, it struggles to be fully realized. Existing employees are in role to optimize the existing business not disrupt it. Even if they were given the freedom to innovate, their skillset does not lend itself to risk taking as they were hired based on their ability to run the existing business. Developing a mindset for experimentation and risk taking is not developed overnight and might even be psychologically distinct (Wolchover 2011).

Thus, new ideas emerge as the currency of the organization once leadership sets innovation as a cultural goal but fails in execution and testing those ideas for market readiness.

Large, multi-product, multi-market organizations typically organize in distinct business units so they can operate more freely. These companies within the larger enterprise can be quite independent and collaboration across them is difficult and not incentivized by the business unit's goals leading to gaps and overlap. Ideas which may lead to innovation needing different business units to collaborate require employees to rise above not only their day-to-day responsibilities but the lack of processes to share information and work across business units. Without execution ability, information hoarding becomes common to ensure power strictures amongst the business units. Pursuit of opportunities by each unit arising from partial information leads to overlap and unhealthy competition amongst units and little collaboration stifling innovation.

Trying to solve for this, leadership will instill processes across the organization for information sharing. Process is the lifeblood for any scale organization as their sheer size and inability to communicate uniformly amongst the employee base requires policies and controls. For new ideas to work through these processes, alignment is needed for all decisions to ensure that all affected groups agree with the new approach. The greater the number of operational employees who are invested in the status quo and who are involved in evaluating new products and business models, the more that organizational inertia takes over and objections will be found. Even when leadership dictates some new approach and the lower level operational employees embrace it, operational middle management often passively resists thwarting new initiatives. Middle management as a practice was installed at the beginning of the twentieth century as part of the new class of risk deferment in the modern corporation as embodied in General Motors (Davidson 2014). An unpredictable side effect of building a professional class of risk managers was that the focus of the employees in that class was not on the organization but on oneself. If you failed, you were fired. New ideas might not work, might eliminate jobs or responsibilities or expose the need for new skills not shared in existing managers. In this sense, a twentieth century style organization cannot accept failure as part of its mission.

Most large businesses are now enabled by technology even if they are not in the technology business themselves. However, in not having technology as the product or service, they typically rely on an information technology (IT) organization to develop, manage, and support their technology needs. As the shift toward more technology inside large, traditional businesses accelerated starting in the 1990s, businesses often outsourced development as it was not a core competency where they could justify building a large and expensive staff. The IT organization thus became filled with project managers responsible for the operations of new technology development as well as for the operation of the resulting technology products (e.g., the company website). In this way, they mirror their business counterparts in being focused solely on the operational aspects of technology not research and development. Also, like their operational business counterparts, they have little incentive to disrupt their existing models and were hired for their skillsets as such. The consulting vendors they are dependent upon also have little incentive to disrupt their relationships and the established systems they have developed for their client. This self-reinforcing loop creates a high entrenchment culture difficult to displace in trying to stimulate innovation. This is demonstrated in IBM and Hewlett-Packard's shift from high margin yet risky technology products to low margin, low risk services over the past 20 years (Balakrishnan 2015). The steady revenue born from these parasitical relationships became favored over the risk of trying to innovate on new technology as well as convincing clients to reinvest and adopt (i.e., take a risk with you).

IT organizations also have one additional constraint limiting their interest in innovation processes. A big regression risk is introduced for every new technology solution they try to deploy having them take a conservative approach. This is why mainframes are still the backbone of many data processing intensive businesses like finance and healthcare even though their favor as technology solutions dwindled in the 1990s. They work and thus what incentive does a technology manager have to replace them even when faced with the opportunity to increase performance, add new features or decrease costs. Like all managers, technology managers are hired to reduce risk. All new technology—but especially experimental products—must be analyzed and rigorously tested so as to not negatively impact the technical operations of the existing business. Given that many experiments will be small scale and never fully realized, the cost–benefit analysis for the testing and risk exposure alone often defeats these innovation initiatives. Startup companies are not burdened by trying to maintain existing systems and have a greater risk profile. Hiring and retaining those who have experienced the freedoms of growth stage companies into large organizations poses a notable challenge. Those employees often become frustrated with the real and perceived bureaucracy in large technology organizations they are not familiar with in growth stage companies and academia.

In trying to break the risk averse IT culture and stimulate a more nimble environment familiar to a generation of startup product developers, adoption of Agile methodologies is now common in large IT organizations. While Agile has generally been accepted as a better approach than traditional waterfall, IT leaders need to understand the cultural impact on the existing operational employee base that comes with this more decentralized, less structured approach. IT middle managers are often less comfortable with the lack of planning and predictability. Waterfall was born of traditional engineering methodologies, such as bridge building, which logically said that progressing to the next stage could not occur until the prior one was fully completed and signed off on by all stakeholders. Sensibly one cannot order the bricks, steel, and labor to build a bridge until the design is completed and approved by the local authority or else face a huge regression expense if something changes or fails. Early software engineering practices followed a similar model with heavy analysis and design phases before any engineering work took place. As technology become more deeply rooted into business and consumer adoption accelerated, it created an environment whereby, even if all your analysis and design was accurate (and that was rare), the technology capabilities and user expectations had changed while you were developing your solution. It was the equivalent of having the river widen and people shifting from horse and buggy to cars while you were building your bridge. You could never anticipate all factors given the speed of technological change. Agile was a response which said that it was better to proceed with unknowns rapidly iterating the software development process favoring releasing the resulting technology after short iterations and learning.[1] The Lean Startup (Lean) methodology pioneered by Steve Blank in his book "Four Steps to the Epiphany" and written about by Eric Ries in "The Lean Startup" has extended the philosophy to cover the whole product and customer development process based in the scientific method. By embedding innovation around a repeatable process, Lean seeks to ensure that experimentation and customer testing is at the root of all new product development.

While Lean and Agile are now fashionable for large company leadership to show they are innovating, the cultural impacts cannot be overlooked. By focusing on user needs and breaking the problems down into small parts, this new approach requires less of project management skillset and one more of product management, design and having engineering skills more deeply involved in the business problem early in the process. All of these are not typically the skillsets nor processes of employees who were hired to manage IT operations. The business and IT do not share a common language even though they are more dependent on one another in a digital world (Cigaina and Riss 2016). Re-skilling is a common approach but most have found that the skillset and mindset is so notably different that wholesale organizational shifts are required. When faced with a disruptive threat, business leadership has the decision to significantly expand and upskill its employee base while also protecting its existing business. This is not an easy sell especially to bottom-line focused public market investors. Consultants can fill specific skill gaps for a time but their mercenary engagement tends not to leave behind the lasting cultural impact needed for sustainable innovation.

The lack of comfort in rooting your technology approach in the unpredictable nature of the scientific method also extends to some sales and marketing teams. Innovative, Lean and Agile product development teams carrying out a process of test and learn often face resistance from the groups that control access to the customer. The notion of engaging a subset of customers in a qualitative survey let alone thousands to test scale engagement with a speculative new product is foreign and threatening. This most vital piece to the Lean methodology is ironically rooted in marketing focus groups and market testing but the lack of control by those market facing teams leads to conflict in organizations around customer touch.

All of these impediments to innovation can most obviously be seen in performance review systems within large organizations. The best are derived from the company's mission and values which are communicated, to varying degrees, to

[1]Beck, Kent et al. The Agile Manifesto. agilemanifesto.org.

form the organizational culture. Humans respond to incentives which, in most commercial organizations, materialize in compensation and promotion rewards. Those incentives are realized through attainment of goals. Goals are rooted in the near term, typically annual, objectives of the organization as well as the mission and values which, if done well, layout the leadership qualities for the successful individual in the organization. If those values do not include innovation, experimentation, and calculated risk taking then no such similar behaviors will flow through to the performance review system. With behaviors not incentivized toward innovation but more toward operational efficiency and consistency, the organizational culture will not support those who operate in a contrarian risk-seeking manner.

Fundamental new product development especially that which is based upon emerging technology has been shown to require heavy collaboration by a group of diverse yet similarly focused people. As Walter Isaacson explains,

> The lesson of Bell Labs is that most feats of sustained innovation cannot and do not occur in an iconic garage or the workshop of an ingenious inventor. They occur when people of diverse talents, mind-sets and expertise are brought together, preferably in close physical proximity where they can have frequent meetings and serendipitous encounters. (Isaacson 2012)

No one person is able to achieve innovative breakthroughs due to the complex nature and rapid evolution of technology. Performance review systems such as stack ranking disincentivize collaboration since they require judging of employees against one another and are directly tied to compensation and promotion considerations. As was described in a study of Microsoft's stack ranking system, top tier R&D employees would not work together out of fear of losing out on compensation and promotion due to force ranking (Nisen 2013). Thus, the combined effect of not memorializing innovation, risk taking, and experimentation in company values coupled with a lack of group incentives tying compensation solely to individual performance ranking drives employee behavior toward self-interested, conservative behaviors not conducive to an innovation culture. In this setting, experimentation is taken at great personal risk.

Innovators within large organizations face many obstacles, but the most common are inadequate funding and support from leadership, risk avoidance, "siloing," time commitments and incorrect measures for innovation programs (Andrews 2006). Traditional business planning and controls do not work for innovation given a new initiative, be it a startup or emerging inside a larger organization, does not yet have a business model which can easily be forecast and measured (Blank 2013). A new model is required to reset the organizational design in favor of innovation. Having leadership speak of innovation is important but that alone has shown not to be sufficient to produce sustainable innovation which affects company performance. Innovation processes, norms, artifacts, and innovative behaviors are required to reinforce the desires of leadership (Hogan and Coote 2014).

20.3 Responsive Organizations

As innovation driven technology organizations have matured, the fundamental elements that allow innovation to thrive at scale has emerged. One philosophy memorializing this is the responsive organization. In an age of constant disruption, responsive organizations are built to learn and respond rapidly through the open flow of information. Transparency is a core value recognizing that information hiding for power and control are relics of an old operating model. Responsive organizations encourage experimentation and learning in rapid cycles. Along the philosophy of Agile and Lean, responsive organizations encode experimentation processes in the organizational fabric to ensure they are always close to the market. Responsive organizations assemble as a network of employees, customers, and partners motivated by shared purpose.[2] Without organizational silos to bound them, their cross-functional relationships allow them to exploit new opportunities across the organization.

The responsive mindset is a direct affront to the command and control structures common to twentieth century management thinking. Prior to the technology disruption of the past 30 years, large organizations scaled and managed their operations through hierarchies not unlike the government. When the method to scale a business was adding low-level laborers and, with the twentieth century practice of distributing risk across a professional class of middle managers, a sizeable hierarchy was inevitable. Business evolved slowly and was predictable primarily because large entrenched players would invent a new product and then use their marketing scale to tell customers its benefits. As Amazon CEO Jeff Bezos says:

> In the old world, you devoted 30% of your time to building a great service and 70% of your time to shouting about it. In the new world, that inverts. (Anders 2012)

In the digital era, challenges are less predictable and controllable. Consumers are empowered. Most traditional marketing is thwarted by online consumer review sites like Amazon, Yelp, and social media. Any marketing claim made can be substantiated and refuted in seconds by thousands of people online. Consumer behavior and expectations have fundamentally changed. Information has been democratized and made transparent. Communication is instantaneous and ubiquitous. The only constant is change.

Many of our large organizations are vestiges of twentieth century management thinking. twenty-first century responsive organizations are designed not only to survive but thrive in less predictable environments (Table 20.1).

Most successful scale innovation companies demonstrate these attributes. From Google's "Don't be evil" moniker to its "20% time" allowing workers to explore projects of interest one day a week outside of their day-to-day responsibilities, the employees are engaged to set both the company's business and cultural direction.

---

[2]Responsive.org.

Table 20.1 Attributes of responsive versus traditional organizations

More predictable	Less predictable
Profit	Purpose
Hierarchies	Networks
Controlling	Empowering
Planning	Experimentation
Privacy	Transparency (see Footnote 2)

20.4 The Innovation Engine

With this backdrop of challenges faced by traditional organizations and lessons from scale innovation companies, I have developed a framework called the Innovation Engine. It provides a set of guardrails with a new view on how to structure an organization for sustainable and disruptive innovation while operating its existing business (Fig. 20.1).

The framework bifurcates the product development organization into two segments: the optimization engine and the innovation engine. The optimization engine can most simply be thought of as the existing product teams maximizing the value of mature or declining products. They operate to ensure they continue to meet known customer needs in a low risk environment improving and enhancing as necessary. Well run traditional organizations should have some semblance of this existing today. Sixty percent of the product development team's focus—and likely

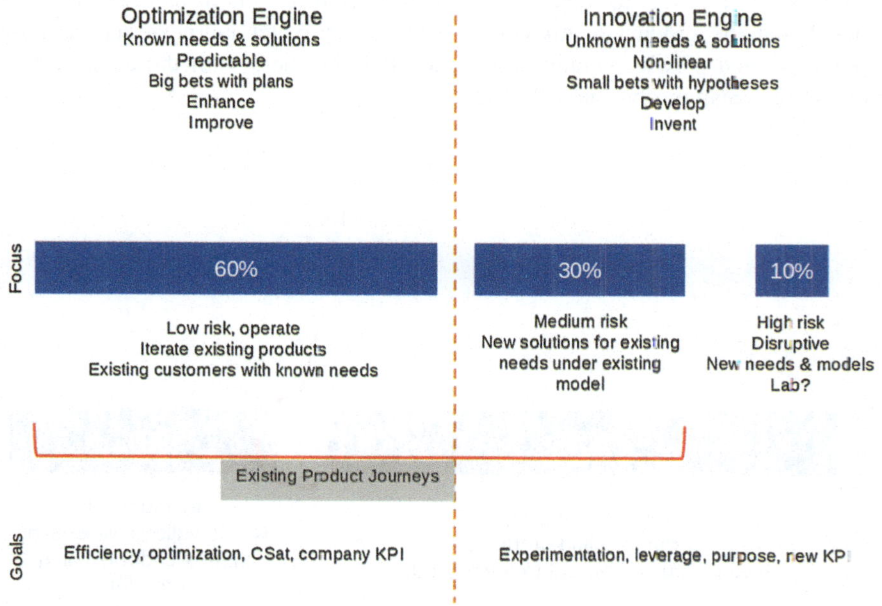

Fig. 20.1 The innovation engine framework

the corresponding investment—should be dedicated to these efforts. This might scale up or down depending on the intensity of disruptive threats but it's important that it remain the majority of the effort assuming the products are not in rapid decline.

The second segment is the innovation engine. This team is focused on new products for user needs that are not yet well understood. The needs and the solutions for those needs are realized through constant experimentation which might take them on a nonlinear path including one where hypotheses are refuted and ideas killed.

The innovation engine is further split into two parts. The thirty percent segment focuses on medium risk new solutions for existing or emerging needs under the existing business model. In essence, is there a better or more efficient way to operate and deliver against known user needs. Taking an existing product into a new market that required more than just adjusting marketing messages would fall here. On the other end of its responsibilities would be more radical solutions such as when Netflix embraced online video streaming at the possible threat to its existing DVD business. They recognized that there were great advantages in shifting user behavior to online video in the reduction in operational costs and inventory risk. However, one click access could facilitate increased viewing raising already high content costs. They experimented their way transitioning customers slowly over a number of years as consumer technology more readily supported this behavior. However, it was not the technology itself that smoothed the transition but that Netflix found a way to support streaming under their existing monthly subscription plan. Their innovation was in the business model or lack of changing their innovative business model to radically shift consumer behavior. In this sense, it was critical that the product teams coordinated their product journeys across those optimizing and slowly winding down the DVD business with those innovating around the streaming business (Fig. 20.2).

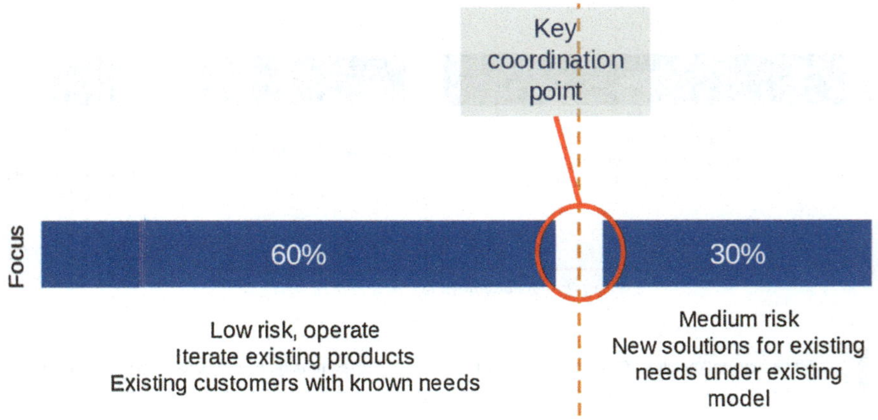

Fig. 20.2 Key coordination point

As a guideline, thirty percent of a company's focus should be in this area of sustainable innovation around existing user needs and business models to smooth the transition to new customer relationships.

The last responsibility of the innovation engine is a ten percent focus on disruptive innovation. These are high risk experiments which explore new user needs and/or business models which do not easily align with the core business. This small but impactful focus has a high failure rate but ensures that companies are playing defense by being on offense exploring what disruptive innovators might be developing in their current and adjacent markets which might, at some time, threaten the core business. Google's "X" business unit is dedicated to this mission operating outside its mainline advertising business seeking new opportunities that are 10 or more years away (if ever). They are pursuing self-driving cars, wearable biometric sensors and wearable computers none of which have an obvious market or business model today. These disruptive explorations have little chance of being absorbed into the main business even after validation and scaling. Google stated mission is to organize the world's information and business is monetizing that information. In the future, they realize that information will be both produced and consumed in different ways and thus do not want to be pushed aside by disruptive startups who define this future.

Within the framework, only the ten percent disruptive innovation is suitable for a lab. The thirty percent sustainable innovation needs to stay close to existing customers and products easing the transition. In revisiting the failure of labs noted earlier, historically there are two main problems:

1. The lab is solely focused on technology innovation and not product or business model innovation meaning technology tends to emerge looking for problems. Labs must be rooted first in seeking solutions to user needs including those where the users may not yet know they have them.
2. The lab focuses on sustainable innovation around its existing products and users. Any innovation that emerged was not easily absorbed back into the main operational organization either due to lack of understanding of the current operating environment or resistance from existing product teams of anything not invented by them.

Again, understanding and getting cultural elements right prove as, if not more, important that the technical innovations themselves.

20.5 Organizing for the Innovation Engine

When the optimization and innovation engines are not clearly delineated with well understood goals and operating environments, the ambiguity can create resentment. One reason operating teams are often unwilling to take innovations from other teams for their products—be it from a lab or not—is due to resentment. Resentment is created when innovation teams put up walls, believe they are the idea people and stop listening to external input. Ironically, too often this includes not listening to users who are the most important source of insights leading to ideas. Innovation has become a sought after title within traditional organizations. Besides being a boon to one's resume, the perception of the perks that come along with it—foosball tables, a looser dress code and freedom—are undeniable. In some cases, perception is reality but it's important that any employee wanting to join an innovation team also understands the different mindset, skillset and responsibilities that come with it.

First, they are taking on increased risk. By definition, the innovation teams are pursuing riskier bets meaning a large portion will fail. One study of companies following Lean, hypothesis driven experimentation found that over 2/3rds of hypotheses are refuted.[3] Many people are not effective when faced with risk and do not thrive in being found wrong so often. Second, there is little need for traditional management and strategy roles in innovation teams. The groups should be organized into small pods or squads of cross functional teams with the necessary skills to build potential solutions addressing the user need. These teams are often described as "one or two pizza" meaning small enough to feed with one or two pies. With this small size, there are fewer people to update and communication flows freely between them given their commitment to transparency. More rigid process tends to only be needed once a team grows and all members cannot be intimately aware of everyone else's day to day work and findings. This is why startup companies often struggle at their first major growth stage from 10 to 50 people. With small squads, the company insulates itself from this problem by scaling horizontal and in parallel via adding additional independent squads (Fig. 20.3).

Importantly, the skills that the team does need are builder or "maker" where they are contributing hands-on, day-to-day as analytical problem solvers. In technology companies, this spectrum revolves around the three corners of the product development triangle: product management, design and engineering. The more each of those disciplines understands about the adjacent disciplines—often called T-Shaped people (Hansen 2010)—the more effectively they collaborate. With both optimization and innovation teams following strong test and learn models sourcing ideas from and validating with their customers, resentment lessens. Better understanding of their common interests and fewer chances one is perceived as having a more valuable or impactful job than the other breeds collaboration.

[3]Launchpadcentral.com.

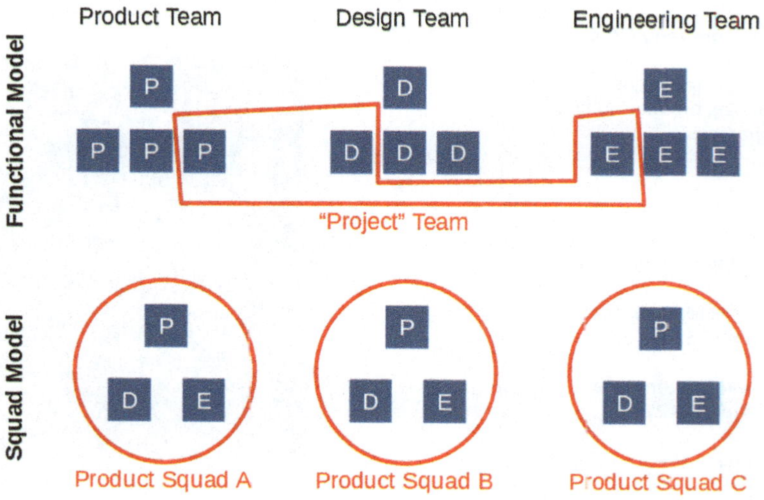

Fig. 20.3 Squad versus functional models

20.6 Performance Reviews and Compensation for the Innovation Engine

As noted, most corporate review systems are designed as annual reviews of individual performance based on well defined goals. For optimization teams, this might not be far off what they need. However, a system of goal setting called Objectives and Key Results (OKRs) (Klau 2012) has emerged in technology firms as an alternative way to tie company, product and individual goals together. In addition, the most important innovation might be in the breaking of goal setting and measurement from compensation and promotion considerations. In this way, OKRs encourage risk taking by asking employees to set stretch goals they are not sure they can achieve in the defined period.

If adopted by the full company—but full company adoption notwithstanding—OKRs flow down from overall company goals and goals of leadership. Any individual on a product development team should form their OKRs based on a combination of company goals and the Key Performance Indicators (KPI) for their products. The objective is a top level goal. For each objective, there are one or more *measurable* key results such as (Fig. 20.4):

At the end of the defined cycle, OKRs are measured on a zero to one scale not simply as a binary yes or no as to whether they were achieved. The measurement removes subjective input from the leader and the scale allows for partial credit. In fact, OKRs are considered successful if, in aggregate, the employee achieves a score in the 0.6–0.7 range. If the score is too high closer to 1 then the employee likely did not set aggressive enough stretch goals. The system also allows for the dynamic nature of rapidly changing business environments where innovation teams

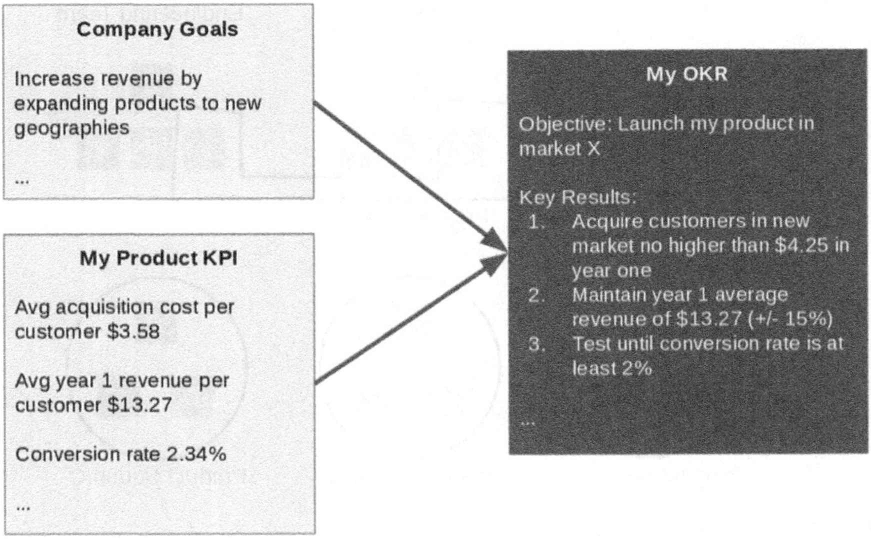

Fig. 20.4 OKR example

cannot foresee far into the future. They also focus the teams on a product KPI mindset where success is measured by achieving the product goals and not simply getting credit for delivering the product—a historical project mindset that achieved no tangible outcome for the business. From there, compensation and promotion considerations are taken on a different cycle. OKRs obviously will have some influence but by breaking the two apart, employees no longer optimize and conform their work behavior solely to their incentive system.

20.7 Conclusions

The Innovation Engine framework provides a working model that can be morphed for many different organizations in different cycles of product development threatened by different disruptive elements. One constant that has emerged is the need for executive sponsorship, interest, and involvement in what the innovation teams are working on. Creating innovation labs as executive vanity plays dooms them to failure almost from the start and their funding gets pulled as soon as the leadership team changes or a few bad quarters bring investor pressure to rationalize investments.

As noted, organizational inertia can arise from several places. Having clear definitions of roles and goals especially within the IT organization is a must. Constant communication with optimization teams training them on the test and learn process reduces resentment and can create opportunities for partnership helping them with problems where they do not have time nor expertise to delve into. If an immediate opportunity for innovation is not there, finding a challenging product problem the organization has not been able to solve and execute against is another path toward proving the model.

Finally, careful selection of the innovation team is a key. Besides the necessary skills, confirmation that they have the appropriate aptitude and attitude to take on risk, adapt in the face of failure and can work with radical transparency are fundamental. They will face many hurdles beyond just their products and a strong conviction and buy-in on the philosophy will be key as well as their advocation of it to others outside of the innovation team. The team should also constantly reflect upon and adjust the innovation process itself. No single process works for all teams in all situations and learning about and iterating on your own process is just as important as for your product. The product development team are the users of the product, test and learn with them to innovate.

Bibliography

Albanesius, C.: More Americans go to facebook than myspace. PCMag.com (2009)

Anders, G.: Jeff Bezos's Top 10 Leadership Lessons. Forbes.com (2012)

Andrews, P.: Five barriers to innovation: key questions and answers. IBM Executive Technology Report (2006)

Balakrishnan, A.: Why IBM could be the next HP. CNBC.com (2015)

Blank, S., Simoudis, E.: Innovation Outposts and the Evolution of Corporate R&D. steveblank.com (2015)

Blank, S.: Why the lean start-up changes everything. Harvard Bus. Rev (2013)

Blodget, H.: Was Yahoo's Terry Semel the Worst Internet CEO Ever? BusinessInsider.com (2008)

Cigaina, M., Riss, U.: Digital business modeling: a structural approach toward digital transformation. SAP White Paper (2016)

Davidson, A.: Welcome to the Failure Age! NYTimes.com (2014)

Gillette, F.: The Rise and Inglorious Fall of Myspace. Bloomberg.com (2011)

Hansen, M.: IDEO CEO Tim Brown: T-Shaped Stars: The Backbone of IDEO's Collaborative Culture. chiefexecutive.net (2010)

Henry, Z.: Mark Zuckerberg's 10 Best Quotes Ever. Inc.com (2014)

Hogan, S.J., Coote, L.V.: Organizational culture, innovation, and performance: a test of Schein's model. J. Bus. Res (2014)

Indvik, L.: The 20 Startups Marissa Mayer has Acquired at Yahoo. Mashable.com (2013)

Isaacson, W.: The idea factory. New York Times (2012)

Jaruzelski, B., Katzenbach, J.: Building a Culture that Energizes Innovation, Financial Executive (2012)

Klau, R.: How Google sets goals: OKRs. gv.com (2012)

Lee, W.: Executive departures point to deeper trouble at Yahoo. SFChronicle.com (2015)

Nisen, M.: Why Stack Ranking Is a Terrible Way to Motivate Employees. businessinsider.com (2013)

Tkaczyk, C.: Marissa Mayer breaks her silence on Yahoo's telecommuting policy. Fortune.com (2013)

Wolchover, N.: Why Do People Take Risks? LiveScience.com (2011)

Chapter 21
Strike While the Iron Is Hot: User Centricity Adapted to the Agile Innovation Development Process

Jörn Schulz, Fee Steinhoff and Kathrin Jepsen

Abstract The increasing demand for more agility and flexibility in innovation development projects also sets new requirements for market and user experience (UX) researchers. In this chapter, we discuss several challenges and issues inherent in the liaison between agile innovation development and the claim for user centricity. We outline one specific approach to agile market and UX research in the project field User Driven Innovation at the Telekom Innovation Laboratories. By introducing two different types of user researchers, the UX consultant and the UX tester, who are deployed in different stages of an innovation project and who operate in different working modes, we offer a hands-on solution based on experience for the issues described. Two short case studies exemplify more agile and flexible methods that are tailored for fast but yet profound market and UX research. We argue that Rapid Ethnography and the Customer Advisory Board are ideal for the new requirements mentioned. In conclusion, we list critical success criteria for user integration in agile innovation development projects.

21.1 Introduction

The telecommunications and the information technology (IT) industry are characterized by an extremely high pace of innovation and a high risk of market failure for new products and services (e.g., EIT 2015). Especially incumbent operators have to rely on the continuous introduction of convincing innovations—and they have to be fast in doing so because:

> "Every morning in Africa, a gazelle wakes up, it knows it must outrun the fastest lion or it will be killed. Every morning in Africa, a lion wakes up. It knows it must run faster than the slowest gazelle, or it will starve. It doesn't matter whether you're the lion or a gazelle–when the sun comes up, you'd better be running" (McDougall 2009).

J. Schulz (✉) · F. Steinhoff · K. Jepsen
Deutsche Telekom AG, T-Labs (Research & Innovation), Winterfeldtstraße 21, 10781 Berlin, Germany
e-mail: joern.schulz@telekom.de

N. Pfeffermann and J. Gould (eds.), *Strategy and Communication for Innovation*,
DOI 10.1007/978-3-319-49542-2_21

Time as a scarce resource is becoming more and more crucial—not only in nature and other parts of "real life" but also in management and business processes. Time-to-market has become an essential success factor (Cooper and Kleinschmidt 2007). The German engineer and economist Klaus Schwab noted on the Davos World Economic Forum: "We have moved from a world where the big eat the small to a world where the fast eat the slow" (Friedman 1999, p. 206). This statement characterizes the increase of speed in the race for successful innovations: Innovative products and services are introduced in ever shorter intervals into the market. This innovation spiral is based on various factors, one being the generally known phenomenon of increasingly saturated, stagnant, and declining markets. The intensity in competition related with it leads to a shortening of life cycles of many products and services. At the same time, the cost of development increases, especially for technical products, which in turn results in prolonged break-even times. Thus, companies are faced with a dilemma: Due to higher development costs and shorter product life cycles at the same time, there is less and less time for the amortization of the innovation investment. Companies try to handle this so-called "time trap" with ever shorter developmental periods in order to be as early as possible on the market and thereby gain more time for the commercialization of innovations. Some instruments to shorten time-to-market are more agility and flexibility in form of agile development and advanced management approaches, such as Scrum (Schwaber and Beedle 2002), which aim to make development processes more flexible and leaner (Trommsdorff and Steinhoff 2013).

Not only a shorter time-to-market, but also the ability to derive and implement user insights into innovation development is mandatory today. Successful innovators integrate users as codesigners in the creation of value (Cooper and Kleinschmidt 2007; Steinhoff 2006). Until the end of the 1970s, management was dominated by the "manufacturer-active-paradigm" (von Hippel 1978): It was the company's task to identify target groups, to discover user needs, and, building on this, to develop and to implement innovative ideas. The role of the user within this paradigm was purely passive, in the sense of "speaking only when spoken to" (von Hippel 1978). In our days, agile user research is understood as a type of intensive interaction between innovating companies and users, which goes far beyond traditional market research. A balance between technology and user benefits is based on a learning process on both sides. In other words, selected users adopt the role of active co-creators in the innovation process (Greer and Lei 2012).

In the following paragraphs, we focus on challenges and solutions for agile user research. First, we address challenges of user research in agile innovation development. Afterwards, Telekom Innovation Laboratories (T-Labs) are introduced. T-Labs have developed a user-centric approach named User Driven Innovation (UDI) to leverage external knowledge for innovation projects. One essential asset of UDI, the Innovation Forum, is a local user panel in Berlin, designed to support the wide range of T-Labs open innovation (Chesbrough 2003) activities. Qualified user samples can be recruited in a fast and reliable fashion, and the local availability enables live user interactions. UDI secures user centricity in agile innovation development, from strategy over product vision and concept up to productization.

This is done with the help of two distinctive user researcher roles: the UX Strategist and the UX Implementer. Two case studies describe exemplary methods for agile user research: Rapid Ethnography and the Customer Advisory Board (CAB). The paper sums up with the derivation of four distinct success criteria for agile user research.

21.2 Agile Innovation Development and Challenges for User Research

Due to its connection to the information and communication technology (ICT) industry, innovation development at T-Labs is strongly connected with agile software development. Today's telecommunication services and products are driven, controlled, and processed by software and protocols. The migration from the old Public Switched Telephone Network (PSTN) to All IP at Deutsche Telekom is just one indicator that software becomes more and more important for the traditional telecommunication industry. But also other industries increasingly deploy agile principles in their software development projects. The international study by Komus et al. (2014) with more than 600 IT experts from 20 different industries shows almost two-third use agile methods since at least four years. In this way, it can be said that agile innovation and software development has been progressively established as framework for managing complex product and service innovations.

There are numerous processes in agile software development like Scrum (Schwaber and Beedle 2002), Adaptive Software Development (Highsmith 2000) or Extreme Programming (Beck 2000). All these agile software development approaches share as common ground the philosophy and the application of values from the Agile Manifesto. Examples of agile values include the following: (1) Individuals and interactions over processes and tools; (2) Working software over comprehensive documentation; (3) Customer collaboration over contract negotiation; (4) Responding to change over following a plan (Beck et al. 2001).

In contrast to that, traditional software development approaches like the V-Model or Waterfall Model work with sequentially successive phases. During the ramp-up phase, a detailed requirements analysis of all features is conducted to secure a complete feature list that is decisive for the subsequent stages of design, implementation, and quality tests. Thus, traditional models focus less on flexible and lean processes.

The success of agile development approaches can be explained by the following aspects: Due to short planning and development phases, executable software can be coded in a relatively short amount of time. In addition, a clearly structured process with short iterations makes it possible to detect problems early and implement changes without delay. Since agile development approaches encourage a very cooperative, collaborative, and cross-functional approach, an excellent visibility for customers, both of the project's progress and of the product itself can be achieved.

In this context, it should be noted, that the customer is not a person external to the product team who purchases or uses the product, but a role filled by one or more members of the product team (Sy 2007). However, the agile approach leads to various challenges for the collaboration between software development and user research.

1. *Insufficient product vision*: The main objective of agile software development approaches is the optimization of the process. That means how to proceed is most important. By this, the efficiency of the development process is to be increased. However, if neither the product owner nor the customer has sufficiently defined what to code in advance in a holistic way, the advantages of efficient development with the aim to create an executable piece of software quickly reach its limits. Thus, a clear product vision is crucial for successful innovations that meet the needs of customers respectively. the end users and should be created in the beginning of a development process (Schwaber 2004). Pichler (2010) mentions in this context it is an obvious, but yet surprisingly common mistake to start the product development without a product vision.

2. *Compromise between agile and traditional approaches needed*: In traditional approaches to innovation and software development, time blocks of several weeks or even months are reserved to formulate a clear product specification and requirements within the ramp-up phase. In exchange, that specification has to consider all eventualities concerning the major aspects of the product or service. However, in real life all facts about markets, technical conditions, and requirements from users can never be known upfront, but they develop over the time and during the development process. Thus, a balance between the depth of details for the product vision and time restrictions has to be found. On the one hand, the product vision has to be understood as a process of learning and acquiring knowledge and experimenting guided by the principle "less is more" (Pichler 2010). On the other hand, the product vision is supposed to be as precise and focused as possible in order to cover the key user needs, the features resulting from them, the USP, and the target group. Only then relevant user stories can be written, which are the basis for the development phase. Therefore, new working and collaboration modes have to be created to find a compromise between agile and traditional approaches (Sy 2007).

3. *Traditional user research methods reach their limits*: Another challenge for agile user research is the selection and creation of adequate methods to ensure user centricity during the entire development. In order to meet the demand for quick user feedback, the user researcher has to utilize lightweight and lean methods, which fit for the Scrum process and that enable direct interactions with users nevertheless (Evnin and Pries 2008). "Just-enough research" is the name of the game. That means traditional methods have to become slimmer in order to be able to deliver relevant results for the further product development in a given time frame of 2–3 weeks, according to each Scrum sprint. Hence, it is the responsibility of the user researcher to synchronize the user research activities with the pace of the development. This typically covers design reviews as well as iterative user experience (UX) tests. Nevertheless, the user researcher is also in charge when it comes

to asynchronous and more long-term user research activities that cover the 'holistic' view by shaping the overall proposition, e.g., target group analysis the definition of user needs, usage patterns and UX criteria for the specific user stories.

4. *Dedicated human resources and time slots for user research needed*: As an obvious consequence of the issues mentioned above, especially the acceleration of user research, there is the need for more staff. In traditional development processes, the user researcher compiles his results in an insight report and hands it over to his project partners to feed the overall specification. With agile development it is even more the responsibility of the user researcher to secure the implementation of those insights during the development (Fox et al. 2008). Furthermore, agile development often comes with daily meetings and very close collaboration in one place. To be part of this, the user researcher has to become an integral part of the development team (Fox et al. 2008). It is apparent that these requirements can only be met with full-time dedication. Furthermore, the user researcher also has to insist on dedicated time slots for user research activities in each sprint. Only by this, interdisciplinary collaboration, transparency, and user-centric innovations can be implemented.

21.3 Agile User Research at Telekom Innovation Laboratories

21.3.1 User Driven Innovation and the Innovation Forum at Telekom Innovation Laboratories

Telekom Innovation Laboratories (T-Labs) were constituted in 2004 as an affiliated institute of the Technical University Berlin, to drive research, development, and innovation across Deutsche Telekom AG. The mission of T-Labs is directly derived from the vision of Deutsche Telekom, that is: As the leading telecommunication services company, Deutsche Telekom networks society for a better future with top quality, efficiency, and innovation for the benefit of the customers.

All T-Labs projects support the objective of developing new products and services for Deutsche Telekom customers. Scientists and industry experts study and develop technologies and solutions for the communication of tomorrow. They concentrate on the development of applications and services, in such key areas as new media, future communication, and smart data analytics. Furthermore, T-Labs innovations help to secure complex services in the areas of convergent networks and infrastructures, cross-domain middleware, and IT and cloud evolution. The results of the innovation projects are primarily transferred to the group's strategic business units. Topics that do not fit into the core areas of the company or that demonstrate disruptive potential are candidates for start-ups and spin-off organizations.

T-Labs' headquarters is located in Berlin, and there are offices in Darmstadt and Bonn in Germany, in Silicon Valley in the United States, and at the Ben-Gurion

University in Israel. The open innovation model (Chesbrough 2003) is a guiding principle for the work of T-Labs. This encompasses the inclusion of customers, users, and experts in the innovation process, enabled by advanced methodology. The link between Telekom's innovation activities with Technical University Berlin and a variety of other universities, institutes, and industry partners worldwide creates close ties between the worlds of science and business.

The project field User Driven Innovation (UDI) is a multidisciplinary team of business economists, psychologists, social scientists, and designers. UDI supports core topics at T-Labs in securing the implementation of the users' perspective into the entire innovation process. The mission of UDI is the (further) development and implementation of suitable methods of user centricity in research and innovation projects. UDI's daily work is based on user research and user experience tools tailored to three phases or thematic layers: strategy, product vision & concept and productization (see Sect. 21.3.2 for more details).

A major asset of the UDI approach is the "Innovation Forum", a local user panel for T-Labs, which was established in 2009 (see also Breuer et al. 2011). This group of participants is initiated and managed by the corporation. Access panels managed by intermediates like market research institutes allow representative (usually online) surveys and targeted sampling. But access panels cannot fulfill all specific needs of a single company, including recruitment adapted to individual project demands, local availability, and deep industry specific knowledge about the panel members. Its major advantage is that it allows for method, data, and theory triangulation by integrating heterogeneous user perspectives in a very efficient way.

The overall goal of the Innovation Forum is to enable a fast and reliable on-demand-recruiting of participants to flexibly support heterogeneous individual studies. The starting point and one of the most critical issues of user insights generation is the identification and recruitment of suitable participants (Alam 2006). Currently, the Innovation Forum consists of over 1600 members from the Berlin area with different usage behaviors, characteristics, skills, and lifestyles. It includes employed and self-employed persons from over 30 industries. Age groups include teens, students, young professionals, mid-agers, families and best agers, all defined customer segments of Deutsche Telekom. Subpanels include a group of lead, respectively, advanced users (measured through the criteria innovativeness, their technology expertise and their involvement), and other groups according to the specific demands of the project partners (Breuer et al. 2011).

The Innovation Forum members participate in research activities along the entire innovation process—from strategy, over product vision and concept, up to productization. The participants contribute with different roles and competencies: Panelists deliver user insights in ethnographic studies. They engage in ideation workshops or act as co-creator in product and service specification. They rate product ideas and concepts in quantitative surveys, or test the usability and experience of prototypes.

The key strength of the Innovation Forum is its flexibility serving a great variety of user-related questions. The potential of this approach includes flexible recruitment for diverse purposes, highly targeted sampling and local availability for live events and operational efficiency. This enables the UDI team to collect market and development critical information for all T-Labs projects from qualified contributors in a very fast and flexible way.

21.3.2 *Roles for User Researchers in Agile User Centricity: UX Strategist and UX Implementer*

One essential challenge of agile user research is the necessity of asynchronous user research, which means before or besides the actual development process, and synchronous user research, which means aligned with the sprint cycles. Agile innovation development projects need both types of support: On the one hand, a meta-level and holistic perspective, which defines the frame for user centricity; on the other hand, the continuous implementation of user centricity along the entire agile development process (see also Isomursu et al. 2012; Sy 2007).

At UDI, we meet this challenge with two different types of user researchers: the user experience (UX) Strategist and the user experience (UX) Implementer. In each agile innovation project at least two team members are allocated to embody these roles. The UX Strategist focuses on user-centric research questions of effectiveness, which means he asks the question “What is the job to be done?” and secures to “making the right things”. His results set the scope of action and the underlying framework for the mission of the UX implementer. He ensures the implementation of user research insights in the agile development process and asks “How are things done?” and is responsible for “making things right”. While the UX Strategist operates before or besides the agile innovation development, the UX Implementer is deep-seated in the actual development process and e.g. the Scrum team. Thus, both roles stand for different competencies and focal points. The methodical scope of action of both roles is tied to the three thematic layers mentioned above: strategy, product vision and concept and productization.

The key question of the first layer, the strategy layer, is: How does the project topic resonate with requirements of future societies? Finding answers to this question lies—as the name already triggers—in the area of responsibility of the UX Strategist. The challenge is to build and realize future-oriented strategies that address relevant user needs and requirements. This refers to the initial fuzzy front end of an innovation project and refers to proceeding from nothing to something (Breuer et al. 2009). To do so, it is imperative to understand and interpret the implications of social and technology trends and use them as a base for innovation decisions. By using lean tools like scenario and future workshops (Fink et al. 2000), the UX Strategist maps emerging technologies on user-centric key factors and creates scenarios of possible futures.

The second layer, product vision and concept, concentrates on the question: What should my product be and what makes it valuable for users? The main responsibility for this layer also lies in the hands of the UX Strategist, but with an overlap to the UX Implementer. In a first step, ethnographic tools (e.g., Madsbjerk and Rasmussen 2014) aim at a deep and integrated understanding of future users in terms of behavioral pattern, unsolved problems, needs and wants. Desirable results are user insights, vivid descriptions of unmet needs, presented from the user's own perspective. The following idea generation uses creativity and design thinking (Brown 2008), in which external impulses, apart from internal sources, are essential.

User-centric tools like personas, codesign of concepts, and prototypes (Ungar and White 2008) as well as early business modeling (Osterwalder and Pigneur 2010) lead step-by-step to solid and compelling product and service concepts. This approach builds on the lean validation of central assumptions and the application of suitable methods for quick and iterative generation of user feedback. A joint task of the UX Strategist and the UX Implementer is the subsequent translation of derived user insights into user stories and their prioritization for the Scrum backlog.

The third thematic layer, productization, refers to the question: How can my product deliver a superior user experience? Securing a superior UX through the realization of user stories along the Scrum process is the main task of the UX Implementer. His objective is to provide the Scrum team with insights on both, the functional and non-functional level (e.g., novelty, aesthetics), and to assess all of these aspects with regards to user needs. This refers to a holistic perspective and considers the complete experience a user has while interacting with a new product or service. Expert reviews serve to identify major usability problems and potential barriers for the usage. Interactive prototypes are tested with the target user groups by carrying out scenario-based UX walkthroughs to see if the user needs and expectations are met (e.g., Nielsen 1989; Beyer et al. 2004). Minimal viable products (MVP) are put into a real world context by conducting lean field tests to gather input for final adjustments. In sum, the outcomes wanted of the productization layer are resolved usability issues, optimized design, and superior user experience.

Put in a nutshell: Good usability is imperative nowadays—products must be easy to use otherwise they will fail in the market. Nevertheless, usability does not equal desirability. Accordingly, both, the UX Strategist and the UX Implementer strive for products and services that are not only easy but joyful to use and that can evoke meaningful experiences. They work hand in hand to reach this goal.

21.4 Case Studies

21.4.1 Case Study #1: Rapid Ethnography for WiFi Community

The first agile user research method described here and utilized by UDI is Rapid Ethnography. Traditional ethnography is a methodology that helps studying people, communities, and cultures-foreign as well as domestic ones-systematically with participant observation, interviews, and other methods in a specific field (Geertz 1987). Rapid Ethnography, however, is a methodological concept introduced and popularized by Millen (2000) in his paper *Rapid Ethnography: Time Deepening Strategies for HCI Field Research*. In this text, Millen characterizes Rapid Ethnography as a form of ethnography that is more focused and can, therefore, be conducted in a shorter time frame. Key elements of Rapid Ethnography are narrowing down the research field as well as the research question. Furthermore, Millen emphasizes the conduction of interviews with key informants, the parallelized collection of field data by more than one ethnographer and the collaborative analysis together with the project partners.

When it comes to creative naming of this lean qualitative method, several other terms from various authors have been suggested. Rapid Ethnography is also known as Quick Ethnography (Handwerker 2001), Focused Ethnography (Schmid and Kaufmann 2005), Agile Ethnography (Dudek 2012) and Lean Ethnography (Higgins 2013). All variations share, more or less, the view that ethnography is very useful for the exploration of user needs, wishes, and problems. However, in its traditional form ethnography is too time-consuming and laborious for applied contexts like business environments. Hardly any company can afford to send researchers for weeks, months, or even years into a field like classical ethnographers did and still do for their research (e.g., Malinowski 1922; Mead 1929; Myerhoff 1980). Time is precious in the business world, and agile innovation development has intensified this situation even more. Drawn on experiences collected at T-Labs, it can be said that most innovation development teams need explorative insights into the field under examination rather quick in order to continue with their development process and to meet their milestones. Hence, the need for accelerating and boiling down ethnographic user research is obvious. The Senior User Experience Researcher Ellen Isaacs who worked at the Xerox Palo Alto Research Center (PARC) and at AT&T Labs writes:

> Today, ethnographers working in corporate settings have to accept the need for speed and adjust their approach. While surely requiring tradeoffs, rapid ethnography brings the ethnographic perspective into organizations in a way that is culturally palatable and gives ethnographers a seat at the table (Isaacs 2012, p. 92).

To not miss out on valuable, qualitative insights from real user living environments, UDI has adopted Rapid Ethnography as a method for agile innovation development projects and altered it slightly according to T-Labs specific requirements. Hypotheses as a starting point for the exploration are an important ingredient for Rapid Ethnography at UDI. This is seemingly contradictory to the traditional and inductive approach of ethnography and a move to a deductive procedure. However, the more focused and efficient the ethnography has to be, the more trade-offs have to be taken into consideration. In this way, the formulation of hypotheses is just the next step in the adaption of ethnography for the needs and requirements in companies. Anyway, UDI researchers try to stay as open and unbiased as possible in order to explore new perspectives not thought of before.

After all, this exploration of unknown or hidden user practices is one of the strengths of ethnography in business contexts.

Due to the adoption of this lean method, ethnographic insights can now be delivered in around four weeks—including the research preparation, the recruitment of participants, the field phase, the analysis, and the production of an just-enough insight report. For this type of user research, a UDI UX Strategist is deployed to create relevant ethnographic insights and to secure—together with the UX Implementer—the placement of those in the product visions as well as in user stories for the Scrum backlog.

Innovation Project: WiFi Community

In the autumn of 2014, UDI was given the chance to test its own Rapid Ethnography concept. A developer and engineer team of T-Labs, which is engaged in the infrastructural advancement of telecommunication networks, wanted to know how WLAN access has to be designed in order to be user-friendly and convenient. Furthermore, they wanted to know what attitudes and practices concerning WLAN and Internet sharing exist among users, e.g., when sharing an Internet connection with neighbors or in public space. The research questions, which UDI drafted together with the product owner, were the following ones:

- How does an attractive and convenient handling of WLAN connections look like in general and when sharing WLAN with others?
- What does the status quo concerning the handling and sharing of WLAN connections look like today?
- Which concerns, usage barriers, and sensitivities prevail among users concerning the sharing of WLAN and Internet connections?
- Which motives, wishes, and needs do users have relating a shared Internet and WLAN connection?

Goals for the project WiFi Community were, e.g.,

- Revision of five hypotheses related to the handling of WLAN connections and subsequent verification, falsification or specification of those hypotheses
- Identification and explanation of drivers and barriers for the usage of shared Internet connections
- Derivation of technical and emotional user requirements for a convenient WiFi Community
- Formulation of consumer-centric recommendations for a user-oriented product definition for a WiFi Community

Research setting

For the ethnographic user research for the WiFi Community project, UDI recruited ten technology affine members from the Innovation Forum who were using WLAN at home and on mobile devices outside their living spaces. Because we were interested in how users are handling and treating different WLAN networks, the physical research field could not be limited to the private house or apartment, but the field was expanded to one place outside where the participants usually used

WLAN connections of other parties. Thus, we also conducted research at Starbucks, in other cafés, and in one university library. In order to remain rapid in our approach, we limited the time spent at the homes of the participants and outside to a maximum of three hours per user. Regarding the tools, we applied semi-structured interviews and participant observation at home and outside to collect qualitative data on the different usages of WLAN connections.

Outcome
We verified, falsified or specified the five hypotheses we started with on the base of the data collected and also added two new theses, which came out of the research. To translate the insights and to make them more digestible, we derived fourteen recommendations in different dimensions (functional, technical, emotional, and communicative) in order to point out user requirements for a convenient WLAN sharing service. The outcome of the Rapid Ethnography became the base for further agile user research in form of a quick poll to quantify attitudes and practices of users, and also functioned as guard rail for a lean ideation workshop, and the development phase. The derived insights were translated into user stories and were continuously integrated and tested by the UX Implementer in the project.

21.4.2 Case Study #2: Direct Interaction with Users at Eye Level at the T-Labs Customer Advisory Board (CAB)

The second agile user research method described here and adopted by UDI is the Customer Advisory Board (CAB). Originally a method for customer relationship building, Carter (2003, p. xiii) describes the method as "an extraordinary and successful tool that gives organizations the opportunity, in a nonthreatening setting, to get closer to their clients. The chance for decision makers to learn what it takes for a seller to be the kind of firm a customer will buy from is a very compelling, dynamic encounter". Also Lombardo (2004, p. 21) notes that the Customer Advisory Board is a great method for B2B customers in "helping resellers fulfill their clients' wish list". UDI altered the original purpose and adapted it for the innovation context at T-Labs. Roughly speaking, the T-Labs CAB is a format that allows for direct and straightforward interaction with potential users in an informal setting to test ideas, hypotheses, and prototypes.

The CAB is being conducted by UDI since 2014 and has become a regular event for T-Labs innovation projects. Due to its lean and hands-on character concerning the event organization and the analysis of the data collected, it has changed the image of user research from a cumbersome critical testing instance into a continuous and collaborative support that helps to amplify the quality of T-Labs projects. This user feedback format has also led to a more continuous user integration and feedback implementation in the development process and, thus, a cultural change towards user-centric innovations.

Why did UDI introduce the CAB at T-Labs? Firstly, the fast moving innovation development projects of T-Labs demand more flexible as well as quick formats and fast results rather than long and laborious user research methods. This is especially true for minor or low-priority questions, which regularly pop up suddenly in the innovation development process. Often, there is not enough time and urgency to address these questions with more elaborate user research methods, or it would seem like taking a sledgehammer to crack a nut. This situation has intensified since T-Labs innovation projects are more and more conducted in Scrum or Kanban mode. The need for a faster user feedback format was evident.

Second and another circumstance which led to the introduction of the T-Labs CAB format is the often missing integration of the Scrum team into the user research process itself and, thus, the exposure to direct contact with and feedback from potential users. If user feedback is only collected by user researchers and concentrated results are handed over to the project partners later on, the mere presentation of the results does not guarantee that the user perspective will be understood and considered as urgent enough. "The user" often stays an abstract and blurry concept instead of being an integral part of the innovation process.

Third and yet another problem is that in the early and vague stages of innovation development, designers and developers often hold the opinion that there is nothing to test right now or that the existing development stages are too incomplete to expose them to users. But even later on, user tests are often seen as an obstructive control instance that may lead to negative consequences for the project instead of being an opportunity to sharpen the value and the usability of a product or service together with users.

These surrounding conditions led to the necessity for another user feedback format, which is faster, more integrative and less intrusive. With the T-Labs CAB, UDI was able to offer its project partners a rapid user feedback format that tackles all these problems and paves the road for a close cooperation between different agents of the innovation development process, namely, project managers, developers, designers, user researchers and, of course, users.

Research setting

The T-Labs CAB is an interactive, informal, event-like and efficient format, in which all stakeholders of a project are brought together and confronted with potential users and where projects are presented and evaluated in a direct exchange. Scrum teams and product owners at T-Labs get the chance to meet users from their target group on an iterative and bimonthly basis and to discuss prototypes, concepts, ideas or hypotheses from their projects with them and without a deterrent long-term preparation. To ensure capturing relevant insights, a UX Implementer supports each project team in finding one focused research question and in selecting appropriate test methods.

In each CAB event, 3–4 topics of different innovation projects are included and presented to around 20 users from relevant target groups. The circuit testing concept, similar to a circuit training with 3–4 stations, secures that small groups of users get to see all stations and that the UX Implementer gets as much feedback as possible from various users. They provide quick and direct feedback, which indicates potential opportunities, ideas, barriers, weak points, or open questions. Moreover, the confrontation of the entire team with direct user feedback clarifies the importance of the user perspective and strengthens the significance of the user as an integral part within the Scrum team.

To assure a culture of open discussion, each CAB is organized as a casual after work event in a relaxed atmosphere that enables participants and members of the Scrum team to meet at eye level and that encourages users to tell their honest opinion about a certain idea, concept or prototype without reservation.

Afterwards, a standardized, hands-on analysis process guarantees fast feedback without time-consuming evaluation. It includes a rough documentation of key takeaways and some photos about a few days after the event. However, the focus of the CAB lies on the direct interaction between the Scrum team and users.

To give one example of a test station at a CAB event: Together with a T-Labs Scrum team engaged in innovations in the area of customer service, UDI developed a concept that addressed several questions. The first activity in the test station was a small group discussion about online help forums in general. Do users from the target group utilize them? How do they locate help forums? And what do they think of them in general? The second part of the test station consisted in the individual testing and evaluation of a click demonstrator for a new forum and a virtual assistant designed to guide users through different customer service channels. The user feedback helped the UX Implementer and his Scrum colleagues in tweaking their concepts and click demonstrators.

Outcome

The impact of the CAB format for T-Labs projects has been enormous. The integration of the Scrum team as an active part in the feedback sessions has led to a far more reaching integration of user insights in the development process. After all, self-experienced insights are easier to digest than abstract insight reports. This supports the implementation probability of user research insights. The involvement of all stakeholders as an active part also strengthens the shared responsibility and helps to understand and learn from each other's perspectives. Since UDI started the CAB, the bimonthly appointment has evolved to an internal deadline for Scrum teams, which helps them to work even more efficient by focusing on and developing testable ideas, concepts or prototypes iteratively. Now, the teams integrate user tests more proactively in their frameworks because they have experienced the advantages of continuous user collaboration. The barrier for user integration is much lower in contrast to preparing and conducting extensive user research and evaluation methods. The synergistic advantage of the CAB's circuit training setup is: By inviting participants to one CAB, it is possible to collect insights on three to four projects in one evening.

Overall, the CAB at T-Labs supports a shift in working culture—from silo mentality in large cooperation to interconnected and simultaneous collaboration and a continuous user integration and feedback implementation in agile innovation development.

21.5 Summary: Success Criteria for Agile User Research

As elaborated in the previous sections, agile innovation development and the different working modes of software development processes connected with it, require new approaches to user research. Traditional methods and processes quickly reach their limits when trying to integrate, e.g., ethnography or a quantitative online study into a biweekly sprint. Therefore, specific measures have to be taken to secure user centricity in agile contexts nevertheless. The following success criteria have been identified by UDI and base their validity upon the work experience in agile innovation projects at T-Labs.

1. *Adoption of lean and agile methods*: Thorough user research with traditional methods takes time—time that agile innovation projects often cannot grant due to their tightly structured schedules. In order to conduct user research nonetheless, there is two ways of handling this situation. First, classical methods like ethnography and others can be adapted to the new circumstances and boiled down to meet the new requirements. Thus, inductive ethnography becomes Rapid Ethnography with hypotheses to be checked by field research. Second, innovative methods can be created to meet tough goals concerning the time. In this way, the Customer Advisory Board can be turned into a circular testing and discussion event that bundles several topics and research questions. These methodological transformations entail certain benefits like a higher pace in conducting research, but they also incorporate drawbacks like a limited scope and depth. But as the saying goes: it is better to be roughly right than precisely wrong. To compensate for this, an iterative approach to user research with several and continuous activities is highly important.

2. *User researcher as integral part of the Scrum team*: To be able to react quickly to all sorts of new developments within the agile innovation project and to recommend appropriate user research methods for each specific development phase, at least one user researcher has to become a fully integrated team member of the Scrum team. Working side by side with developers, designers, and product managers, the UX Implementer has to sense continuously the current needs of the project for user integration, feedback, and/or testing. On this basis, he is responsible for formulating research questions, for delivering answers to those questions, and for securing the integration of the insights gathered into the innovation project. This round-the-clock user research support comes with the need for more human resources though because one user researcher cannot support several projects at the same time anymore but is integrated into one specific project team.

3. *Fast access user panel and direct dialog with users*: To secure user integration in agile innovation projects, fast and immediate access to a user panel is vital. Thereby, the quick selection and recruitment of relevant user and consumers of project-specific target groups can be ensured. Once recruited, exchange and discussions at eye level with those users is another important aspect in making sure to get honest opinions about innovative idea, concept or prototype without reservation. This is best achieved in an informal setting as in an after work event like the CAB with a casual atmosphere as opposed to the typical testing situation in laboratories.

To harness the full potential of this direct meeting, Scrum teams are best integrated into this dialog in order to co-create new visions and perspectives together with users and to learn from them.

4. *Cultural change in the organization*: Last but not least and not described in details yet, but just as important as the other criteria, the value of user research in the organization has to be reconsidered if not already done so. In some companies and project teams, user research is seen as annoyance due to the often critical view on things. This is understandable because no one likes to be criticized. However, this "grumbling from the user perspective" is supposed to add a different view on the innovation project and to help in making innovations more user-centric and usable. Thus, the attitude that users and their opinions add valuable input to the development of any innovation project has to be established in the organization. Learning from the user should be a guiding principle since it helps all stakeholders to produce innovative and user-friendly products and services. To achieve this paradigm, iterative research as well as continuous testing and interdisciplinary collaboration have to be core elements of innovation projects. This means that time and dedication are required to integrate user research in agile innovation projects. The participation of a developer, a designer or a product owner in user research activities is not just a favor for colleagues but an actual duty and essential for the success of the project. Yet another point, smaller samples of users have to be accepted as immanent in the process and not be used as counter-arguments. Thus, agile user research demands a cultural change in innovation projects as well as organizations by enforcing to rethink traditional habits, working modes and authorities. In the end, the user should be the "approving authority" as he is the one paying for the new product or service.

To wrap it up: There are a number of crucial success criteria for the implementation of user research in agile innovation projects. Nevertheless, agility and flexibility must not turn into new dogmas that dominate everything else. Becoming a slave of agility and omitting common sense is doubtlessly the wrong path. Sometimes, investing time and thorough thinking in important aspects like the definition of a value proposition or the specific target group are just as important as speed in the development process. "Quick and dirty" is apt for some phases, at other times in-depth research and profound mental work is needed. Thus, the UX Strategist and the UX Implementer roles at UDI offer different methods and working modes for synchronous as well as for asynchronous user research in agile innovation development.

Bibliography

Alam, I.: Removing the fuzziness from the fuzzy front-end of service innovations through customer interactions. Ind. Mark. Manage. **35**(4), 468–480 (2006)

Beck, K.: Extreme Programming Explained: Embrace Change. Addison-Wesley Professional, Boston (2000)

Beck, K., et al.: Manifesto for Agile Software Development. Retrieved from http://agilemanifesto.org (2001)
Beyer, H., Holtzblatt, K., Baker, L.: An Agile Customer-Centered Method: Rapid Contextual Design. Extreme Programming and Agile Methods—XP/Agile Universe, pp. 527–554 (2004)
Breuer, H., Hewing, M., Steinhoff, F.: Divergent innovation-fostering and managing the fuzzy front end of innovation. In: Proceedings of 9th Portland International Center for Management of Engineering and Technology Conference (PICMET Conference), pp. 1–8, 2009
Breuer, H., Wogatzky, M., Steinhoff, F.: On the unique potential of local innovation panels. Paper presented at international product development management conference (IPDMC), Delft, 2011
Brown, T.: Design thinking. Harvard Bus. Rev. 85–92 (2008)
Carter, T.: Customer Advisory Boards: A Strategic Tool for Customer Relationship Building. Routledge, London and New York (2003)
Chesbrough, H.W.: The era of open innovation. MIT Sloan Manag. Rev. 35–41 (2003)
Cooper, R.G., Kleinschmidt, E.J.: Winning businesses in product development: the critical success factors. Res. Technol. Manag. 18–29 (2007)
Dudek, A.: Agile ethnography: a proposition. In: Thinking About the Future & Ethnography, Alicia Dudek's Blog. Published on 23 July 2012. Retrieved from https://aliciadudek.wordpress.com/2012/07/23/agile-ethnography-a-proposition (2012)
EIT Digital IVZW: Innovation Radar 2015-Annual Trend Report EIT Digital, Brussels (2015)
Evnin, J., Pries, M.: Are you sure? Really? A contextual approach to agile user research. AGILE-IEEE Comput. Soc. 537–542 (2008)
Fink, A., Schlake, O., Siebe, A.: Wie Sie mit Szenarien die Zukunft voraus denken. Was Szenarien für die Früherkennung leisten und wie sie konkrete Entscheidungen unterstützen. Harvard Bus. Manag. **8**(2), 34–37 (2000)
Fox, D., Sillito, J., Maurer, F.: Agile methods and user-centered design: how these two methodologies are being successfully integrated in industry. In: Proceedings of the Agile, pp. 63–72 (2008)
Friedman, T.L.: The Lexus and the Olive Tree: Understanding Globalization. Farrar Straus Giroux, New York (1999)
Geertz, C.: Dichte Beschreibung. Beiträge zum Verstehen kultureller Systeme. Suhrkamp Taschenbuch Wissenschaft: Frankfurt am Main (1987)
Greer, C.R., Lei, D.: Collaborative innovation with customers: a review of the literature and suggestions for future research. Int. J. Manag. Rev. **14**, 63–84 (2012)
Handwerker, W.P.: Quick Ethnography: A Guide to Rapid Multi-Method Research. AltaMira Press, Lanham (2001)
Higgins, K.: Lean Ethnography: Sketchnotes from UXLX. In Krystal Higgins' blog. Published on 15 May 2013. Retrieved from http://www.kryshiggins.com/lean-ethnography-sketchnotes-from-uxlx (2013)
Highsmith, J.A.: Adaptive Software Development: A Collaborative Approach to Managing Complex Systems. Dorset House Publishing, New York, NY (2000)
Isaacs, E.: The value of rapid ethnography: three case studies. In: Jordan, Brigitte (ed.) Advancing Ethnography in Corporate Environments: Challenges and Emerging Opportunities, pp. 92–107. Left Coast Press, Walnut Creek, CA (2012)
Isomursu, M., Sirotkin, A., Voltti, P., Halonen, M.: User experience design goes Agile in Lean transformation—a case study. In Agile Conference (AGILE), pp. 1–10, 2012
Komus, A., Försterling, R., Schuh, M.: Status Quo Agile Verbreitung und Nutzen agiler Methoden. Eine Studie des BPM-Labors der Hochschule Koblenz, Prof. Dr. Ayelt Komus. Retrieved from www.status-quo-agile.de (2014)
Lombardo, C.: Customer advisory boards: honoring your customers. Acc. Technol. **20**(4), 21–26 (2004)
Madsbjerk, C., Rasmussen, M.B.: An Anthropologist walks into a bar… To understand what makes your customers tick you have to observe them in their natural habitats. Harvard Bus. Rev. 1–10 (2014)

Malinowski, B.: Argonauten des westlichen Pazifik. Ein Bericht über Unternehmungen und Abenteuer der Eingeborenen in den Inselwelten von Melanesisch-Neuguinea. Frankfurt/Main, Klotz (1922)

McDougall, C.: Born to Run: A Hidden Tribe, Superathletes, and the Greatest Race the World Has Never Seen (2009)

Mead, M.: Coming of Age in Samoa: a psychological study of primitive youth of western civilisation. Cape, London (1929)

Millen, D.R.: Rapid Ethnography: Time Deepening Strategies for HCI Field Research. In: DIS '00 Proceedings of the 3rd Conference on Designing Interactive Systems: Processes, Practices, Methods, and Techniques, pp. 280–286 (2000)

Myerhoff, B.: Number our days. Simon & Schuster, New York (1980)

Nielsen, J.: Usability engineering at a discount. In: Salvendy, G., Smith, M.J. (eds.) Designing and Using Human-Computer Interfaces and Knowledge BasedSystems, pp. 394–401. Elsevier Science Publishers, Amsterdam (1989)

Osterwalder, A., Pigneur, Y.: Business Model Generation. Wiley, Hoboken (2010)

Pichler, R.: Agile Management with Scrum. Creating Products that Customers Love. Boston, Addison-Wesley (2010)

Schmid, S., Kaufmann, R.: Fokussierte Ethnographie. Der neue Königsweg in der qualitativen Marktforschung? *Planung & Analyse. Zeitschrift für Marktforschung und Marketing* **6**, 32–34 (2005)

Schwaber, K.: Agile Project Management with Scrum. Microsoft Press, Remscheid (2004)

Schwaber, K., Beedle, M.: Agile Software Development with Scrum. Prentice Hall, Upper Saddle River, NJ (2002)

Steinhoff, F.: Kundenorientierung bei hochgradigen Innovationen-Konzeptualisierung, empirische Bestandsaufnahme und Erfolgsbetrachtung. Deutscher Universitäts-Verlag, Wiesbaden (2006)

Sy, D.: Adapting usability investigations for agile user-centered design. J. Usability Stud. **2**(3), 112–132 (2007)

Trommsdorff, V., Steinhoff, F.: Innovationsmarketing, 2nd edn. Franz Vahlen Verlag, Munich (2013)

Ungar, J.M., White, J.A.: Agile User Centered Design: Enter the Design Studio-A Case Study. In: CHI 2008 Proceedings, pp. 2167–2177 (2008)

von Hippel, E.: A customer-active paradigm for industrial product idea generation. Res. Policy **7**, 240–266 (1978)

Chapter 22
Disrupting Communication: Innovation Communications in the Digital Age

Johannes von Karczewski and Sandra Zistl

Abstract For nearly 170 years, the foundation of Siemens' success has been innovations that do not simply offer new ideas but instead disrupt the market in the form of new products, technologies, or services. Innovations ensure growth and competitiveness. The underlying ideas begin in the minds of individuals and are then advanced within Siemens in open development processes that follow the "Open Innovation" paradigm. Development is not everything, however: These ideas also have to be communicated. Therefore, a targeted communications strategy is crucial for success. The case study in this chapter shows how Siemens' newly aligned Innovation Communications meets this challenge. The chapter follows the path from communications concept and its interdepartmental implementation to subsequent evaluation of the actions performed.

22.1 Importance of Innovation Communications for a Company's Success

The formula for success in corporate environments has been fundamentally transformed. Established companies used to be able to defend the market position they had achieved by constantly improving their existing products, processes, and services and adapting them to market trends. But today, that is no longer enough. A company that wants to compete successfully in today's world must move beyond incremental improvements and become proactive: looking beyond current trends and setting standards for the future with new ideas and trailblazing innovations, both large and small. The triumph of the start-ups—small, young companies that act quickly, flexibly, and often unconventionally and have significantly shortened innovation cycles, which large companies must now emulate—clearly shows that this is the future that companies are facing. Global players will also have to do more to come up with new

J. von Karczewski · S. Zistl (✉)
Siemens AG, CC CT 51.441, Sandra Maria Zistl, Otto-Hahn-Ring 6, 81739 Munich, Germany
e-mail: Santra.zistl.ext@siemens.com

N. Pfeffermann and J. Gould (eds.), *Strategy and Communication for Innovation*, DOI 10.1007/978-3-319-49542-2_22

inventions, and be faster at it. The innovation landscape has fundamentally changed, and so has the environment in which companies can succeed.

For today's companies in all sectors, the motto must be "Unconventional ideas allowed"—or even "Desperately seeking ideas." Siemens has recognized the competitive advantage offered by these assumptions, which were previously the domain of the start-ups. With its concept of "Open Innovation", Siemens had begun adopting a more open attitude to start-ups 20 years ago, among other things. When paired with the infrastructure of a multinational company and the experience stored in the minds of its employees, this opens up untold potential for innovations. Of course, it is not enough to just look at the start-up scene. A great deal of momentum comes from in-house competitions for new ideas and collaborations between Siemens and top international universities and non-university research institutions.

The numbers also show that innovation enjoys a high status at Siemens: The company is projected to increase its spending on research and development (R&D) in fiscal year 2016 to €4.8 billion, up some 20% from 2014. In fiscal 2015, 32,100 employees all over the world were working in R&D at Siemens, producing 7650 inventions that year alone—35 for every working day. During the same time period, Siemens employees filed about 3700 patent applications, about 17 per day. Siemens currently holds approximately 56,200 patents all over the world.

This innovative, cosmopolitan attitude in the area of R&D led to a realignment of Innovation Communications at Siemens in early 2015, because talking about inventions is just as important as inventing them. Innovations must be fully supported by professional communications if they are to develop their full potential as guarantors of growth and competitiveness. Innovation Communications makes a decisive contribution to the success of new ideas and therefore to the creation of new products, processes, and technologies.

Networking, thinking beyond stereotypes, and talking to, making suggestions to, and supporting people in other departments is how innovations come about at Siemens—and this also describes the new form of Innovation Communications. Internal and external communications were linked more closely. Content is being disseminated across different channels—online, print, social networks, newsletters, and events—with its form and language tailored to the target groups.

Two highlights of the first year of the realignment of Innovation Communications were Digitalization Day in July 2015 and Innovation Day in December that same year (see Sects. 22.3.2 and 22.4). Those events received a lot of media attention. So the goal of Innovation Communications during the preceding months was to familiarize employees around the world with the new strategy and get them to buy into it. That concept, which addressed multiple target groups, also included transforming the self-image of the communicators, which is most clearly reflected in social networks. Communicators do not just send messages, they also receive them, so they are dialog partners.

The goal of an integrated, open innovation communications is both to disseminate messages to the outside world and to shape their content into a culture that is also experienced internally. In brief, we want to reorient how people talk and write about innovations.

22.1.1 Communications Strategy

Innovation Communications at Siemens has had strong internal and external channels for many years. However, the fundamental importance of this topic for the company—the strategic umbrella covering the many fascinating innovation stories—was not always consistently conveyed. A personnel change in Innovation Communications management in early 2015 presented the opportunity to design an overarching, multi-media communications strategy. The goal of the integrated communications model is to highlight the importance of innovations for individual Siemens businesses and for the company as a whole more deliberately and consistently across the company's media. The development of a higher order communications strategy also includes enhancing the perception of the company for what it is—a global company that has been distinguished by its innovative strength for almost 170 years—in the eyes of its own employees and of players in industry and society, as well as shaping opinions (Fig. 22.1).

The point of departure for this communications objective is the company's strategy with the future-oriented name "Vision 2020," which addresses future issues that will impact Siemens as a global company. Anyone who wishes to orient a company to the future must answer the following questions: "What do we stand for?" "What distinguishes us?" and "How can we be successful over the long term?" Through Vision 2020, the company has discovered the answers and set the stage for a successful future. Based on a strong mission statement that reflects the company's self-image in the expressive phrase "Siemens impresses with innovations that make real what matters," along with a "culture of ownership" that makes each and every employee responsible for the success of Siemens, the corporate strategy sets the stage for Siemens' entrepreneurial success. Accordingly, Vision 2020 and the company's mission statement also form the strategic foundation and framework for all content to be communicated by Innovation Communications.

Innovations are essential to ensure competitiveness over the long term. This involves the entire Siemens portfolio—from specific products to the solution business to other company offerings like service. Added value for customers is increasingly based on software and IT solutions. Therefore, the company is paying special attention to the field of digitalization: for example, with research and development in the area of software architecture and platforms. Innovation processes and an open culture of innovation are promoted using instruments like partner networks. The focus goes beyond new technology-driven growth fields and also includes innovative business models.

Ultimately, innovations play a key role in achieving Siemens' entrepreneurial mission of impressing with "innovations that make real what matters." Employees who communicate that message will need both a clear understanding and specific content. To that end, with the "objective of Innovation Communications," the Innovation Communications team has established an important foundation for a uniform understanding.

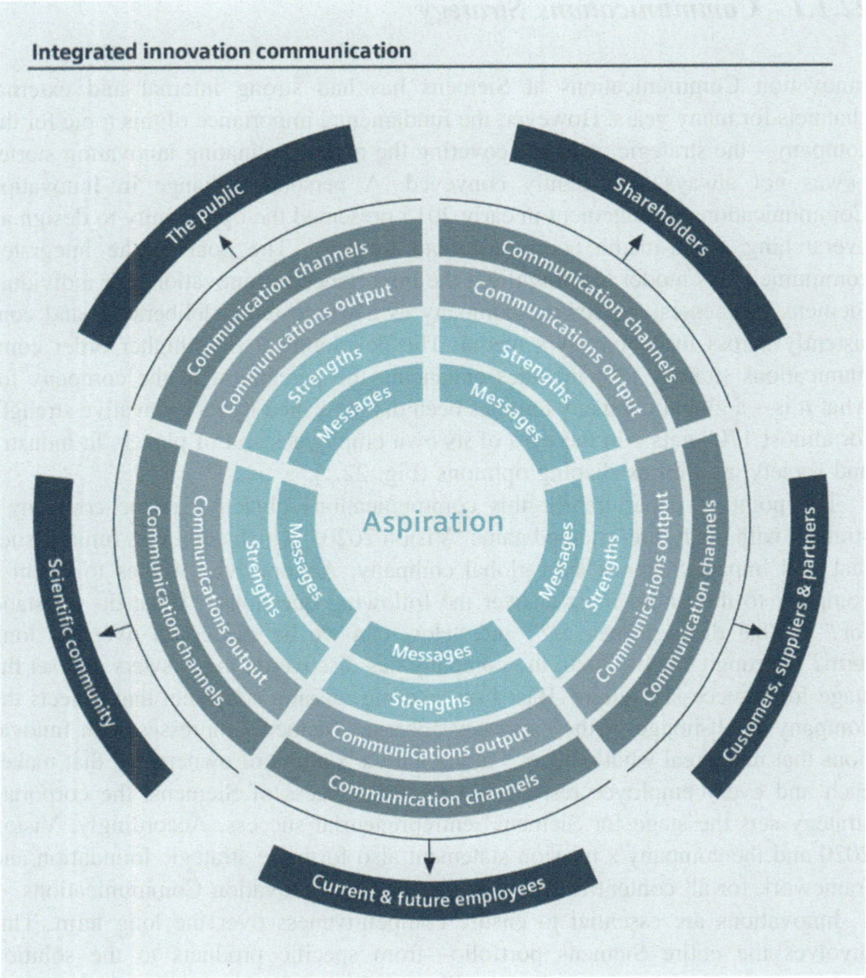

Fig. 22.1 Integrated innovation communication. *Source* Siemens

A clear understanding of innovation. Siemens defines innovation in accordance with the economist Joseph Schumpeter's meaning of the word: as the development and implementation of new technical, organizational, or business-based solutions—so innovation is holistic. Inventions are ideas, concept developments, and prototypes before introduction onto the market, whereas innovations represent their implementation or exploitation on the sales market. Innovation in the broadest sense refers to the successful process from idea to market maturity.

To allow everyone who talks and writes about innovations at Siemens to do so based on a uniform understanding and reliable information, the available facts about this subject—which are distributed across many areas of the company—have been combined, and unambiguous messages have been developed for communication.

22.1.2 Unambiguous Messages as the Backbone of Innovation Communications

The primary messages of the new strategy for Innovation Communications have been provided to all communicators, whether they work in internal communications (see Sect. 1.2) and orient their efforts to Siemens employees, or, like the Siemens press department, they are focused on outside target groups. These messages are not intended to be a straitjacket: Instead, they will form the backbone of all types of content that Siemens transmits to different groups, with form and language tailored to each group.

We make real what matters. For nearly 170 years, progressive ideas, new concepts, and successful business models have been the foundation of our success—innovations that stand out because they go beyond the initial idea, turning into compelling new products, solutions, and services in the market and setting the benchmark. They are what has made our company grow big and strong, and they help us shape a successful future.

Innovations generate profitable growth. Investments and accelerated innovation safeguard our existing business and are the real enablers of new growth that is profitable and generates value—for us and for our customers. That's why we invest in research and development. Our main focus here is on the forward-looking fields of electrification, automation, and digitalization.

Vision 2020 sets the course. Vision 2020 defines an entrepreneurial concept that will enable Siemens to consistently occupy attractive growth fields, sustainably strengthen our core business and outpace our competitors in efficiency and performance. Innovation and business excellence make a key contribution here, and are a clear priority within the live system of Vision 2020.

Business units and CT have an important role in the innovation process. As partners at eye level, our operating units and Corporate Technology (CT) make a key contribution here. Together, they create market-leading technologies and innovations on the basis of comprehensive trend analyses and scenarios of the future, shaping the present and future of Siemens and helping to tap our Company's potential with excellent business practices. By exchanging information with leading universities and non-university research institutes and cooperating extensively with start-ups, we make sure to tap important external sources of inspiration and consequently bolster Siemens' power of innovation. By giving legal protection to our innovations, we safeguard Siemens' intellectual property.

To supplement those messages, which resonate in the background and also influence the choice and placement of themes, communicators have access to up-to-date, reliable facts and audiovisual media and infographics. The corporate sources of that information are a set of slides with facts and figures on Corporate Technology and the public innovation path on the corporate website, which was redesigned in 2015: www.siemens.com/innovation (see Section 1.3.3).

In addition, Innovation Communications is now more personalized at the levels of both content and management. The Chief Technology Officer has been moved to center stage as an ambassador of innovation. An active proponent of innovation gives the topic authenticity, and shapes its attractiveness and our expertise and trustworthiness both within and outside Siemens. There are few people at the company who can convey Siemens' new technological developments as credibly and convincingly as Professor Siegfried Russwurm—to a broad public ranging from university professors to target groups less familiar with technology.

Innovation Communications sees itself as a mouthpiece for Corporate Technology, the CTO, and the research done throughout Siemens under his authority using various channels—internal and external—as described in the sections below.

22.2 Implementation of the Communications Strategy in Internal Communications

One thing was clear from the outset: The consolidation and clearer enhancement of Innovation Communications must involve all communications channels, both inside and outside the company. They must become even more strongly intermeshed—either by communicating themes across both pathways or by means of in-house reports on events reserved for a specific target group, in order to integrate all employees and obtain their buy-in. Communications support for in-house multi-departmental activities on the subject of innovation has proven to be very successful at showing the innovative strength of Siemens when employees serve as opinion-multipliers outside the company.

A good example of this is Quickstarter, the in-house competition for ideas. Introduced in 2015, the contest offers employees of Corporate Technology an opportunity to have their ideas and projects funded. The unusual feature is that company management or representatives of CT don't award the prizes. Instead, Siemens employees use an online tool to select the winners and allocate the total prize money of €500,000 to the individual entries. Employees submitted 111 ideas in four weeks in 2015.

Innovation Communications took much greater advantage of the opportunity offered by the 2016 contest. The cascade of communications ranged from a letter from CTO Russwurm to some 8000 employees of CT and a message on the intranet to articles in the outside innovation magazine *Pictures of the Future*. And that is just one example of how Innovation Communications at Siemens understands integration over all communications channels. More on that below.

22.2.1 The Intranet: SiemensWorld

Internal communication about innovation has also been reconfigured as part of the reorientation of Innovation Communications and the pooling of content. Innovation Communications is also part of reporting on the intranet, which is called "SiemensWorld." Each week, the innovation planners at Corporate Technology send three or four intranet messages (known as CT News) to Siemens' 350,000 employees all over the world. CT Magazine—an online company magazine and the largest in-house newsletter at Siemens AG, with 57,000 subscribers—combines those messages into larger thematic groups and sends them out to readers two times a month. Three or four feature articles inform readers about the important news in one thematic area. An additional factsheet provides the most important facts and figures on the theme.

Going beyond current research results in the fields of electrification, automation, and digitalization, new strategic orientations are also the subject of internal communications. One example of this is the launch of Innovation AG (working title) in summer of 2016, the next step toward opening up Siemens' innovation management. The unit will act as an independent advisor, promoter, and venture capital investor. It will take advantage of the overall worldwide strength of Siemens as a global company but will act like a start-up, free of policies, procedures, and hierarchical processes and structures.

CT News and CT Magazine are used for internal communications and are addressed to readers who are especially oriented to technology: heads of development, technical managers, decision-makers, and opinion leaders in Siemens' Divisions and Business Units. The facts speak for themselves: More than 140 articles were published on SiemensWorld and 15 magazines with links to 71 different articles were published between March and December 2015, reaching about 57,000 people. Corporate Technology supplies most of the articles posted on the Siemens global intranet. But as shown by the example of Quickstarter, those internal messages are often the basis for external innovation communications and are eagerly received by outside journalists and the scientific media.

22.2.2 Internal Social Networking: Siemens Social Network (SSN)

The Siemens Social Network (SSN) is an internal system popular with Siemens employees worldwide. Innovation Communications is now positioning itself in

SSN with content and specific activities. A good example of this is a live discussion between two members of the Siemens Managing Board, Labor Director Janina Kugel and Chief Technology Officer Siegfried Russwurm, and employees around the world in October 2015 during an event held for the Top 700 Siemens executives, known as the Siemens Business Conference (SBC). Before the conference, the Managing Board members called for a discussion on the subject of #ingenuity. During the conference and as an introduction to their joint talk on the innovative strength of every employee, the two top managers held a live discussion on stage using their laptops.

Prior to the launch of the new brand campaign on "Ingenuity," Board members Kugel and Russwurm invited selected groups to give them their feedback on the question "Are we ingenious enough? And why?" Five communities were specifically addressed and invited to participate (Fig. 22.2).

@CorporateTechnology, @GlobalHRCommunity, @Digitalization, @InnovationManagementSiemens, @CommunicationsandGovernmentAffairsCommunity.

This informal, interactive call to action revealed that the initiative hit a crucial point. Comments and posts from employees flooded in during the live event and throughout the subsequent hours. More than 400 were received in less than 24 h—all of them containing constructive content.

In addition to the very insightful and relevant content that Siemens employees posted, the initiative showed that SSN is a democratic tool that brings Board members eye-to-eye with employees, no matter where in the company's hierarchy they work. People from all over the globe gave their direct, unfiltered feedback. They spoke up!

One important detail that supports the credibility of this kind of initiative is that the initiator gives feedback—also on an eye-to-eye level. Board members need to take their employees' input seriously, and let them know that they listened—just as Kugel and Russwurm did. The debate continues, and it has carried over to a vital, democratic debate on #ingenuity (Fig. 22.3).

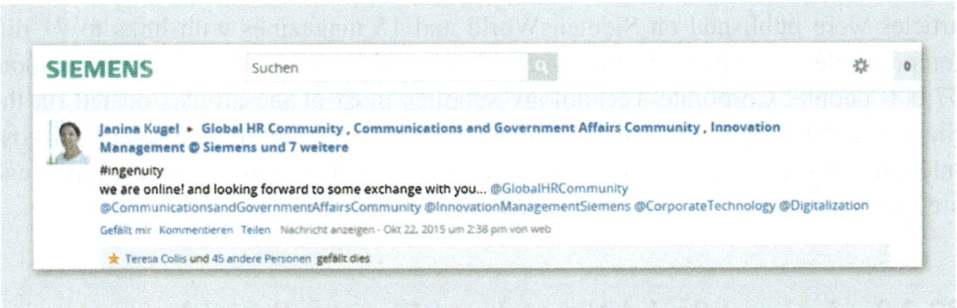

Fig. 22.2 Initiation of SSN-debate by board member Janina Kugel. *Source* Siemens

Fig. 22.3 Closing of SSN-debate by board member Siegfried Russwurm. *Source* Siemens

22.3 Implementing the Communications Strategy in External Communications

External communications on themes related to innovation also travel over multiple channels. It is closely intermeshed with internal communications and the Siemens press team. External channels were combined and their profile was enhanced during the restructuring of Innovation Communications. Unambiguous agreements among all participants have led to a clearer transmission of content and increased its range and topicality (see Sect. 1.4, Evaluation). The communications teams support each other with an integrated architecture and communicate content tailored to their target groups and their languages.

22.3.1 *Pictures of the Future Magazine*

The *Pictures of the Future* (PoF) online magazine is the leading corporate medium for innovation communications that addresses an external audience. The magazine is divided into sections like "Digitalization and Software" or "Research and Management" and has a dossier-based structure within those sections. There is a selection of articles on subjects such as smart cities, IT security, "From Big Data to Smart Data," innovations, and autonomous systems. PoF offers informed insight into research and development at Siemens along the entire value chain. It peeks into laboratories, and in general looks ahead to where the Siemens journey will lead. It offers a platform for the early recognition of technology trends, turning to both internal and external experts.

The content in PoF is also transmitted to readers in a biweekly newsletter that includes current issues in Corporate Technology. Where the choice of topics is concerned, PoF and CT News are closely connected to people working on the intranet, the press team, and the social media team. All channels place and publish themes in close coordination.

Fig. 22.4 Pictures of the future, landing page. *Source* Siemens

Target groups are the scientific community, international media, technology and science journalists, R&D managers, technical managers, decision-makers and opinion leaders in politics, public authorities, and associations. PoF Magazine counts an average of 110,000 page views and more than 62,000 visits per month and publishes a biweekly newsletter with about 28,000 subscribers (Fig. 22.4).

22.3.2 Events

Two high points during the first year of the new approach to Innovation Communications were two events in July and December 2015. Some 320 top managers and experts who are driving the subject of digitalization forward at

Siemens attended Digitalization Day in July at the invitation of CTO Russwurm. In keeping with its significance, Digitalization Day had its own press event, and the key topics and messages were mirrored in internal and external communications activities.

The Innovations at Siemens event was held on December 8, 2015, at the Deutsches Museum in Munich, where CEO Joe Kaeser and CTO Russwurm discussed the Company's innovation strategy. It was designed as a showcase for the latest innovative developments in the *key areas of digitalization, automation, and electrification*. The heads of three Siemens Divisions presented their latest innovation highlights and gave an overview of the company's current research projects. In addition, five start-ups from Silicon Valley and Europe presented their ideas and explained their working relationship with Siemens. The goal was to refresh and reinforce Siemens' reputation as a robust innovator—in the financial community, in the trade press, and among the general public.

Innovation Day was backed by an integrated 360° communications approach.

Content of the integrated communication approach:

- A press feature offered additional information, from press releases, fact sheets, and press photos to brief biographies of the speakers: www.siemens.com/press/innovation-at-siemens
- Targeted social media activities transmitted the company's messages to this realm of communications, including live broadcasting via Periscope.
- *Pictures of the Future*: "Innovation" dossier
- The dossier illuminates the high standing that innovations have at Siemens. Specifically, it contains more than 20 articles that explore the wide scope of Siemens' innovation philosophy and provide in-depth background information: www.siemen.com/pof-innovations
- "Innovations at Siemens" brochure:
- A supplementary printed brochure was prepared, demonstrating the innovative strength of Siemens by exploring specific innovation projects in the three key areas electrification, automation, and digitalization and by providing in-depth information about new innovation models as part of Open Innovation: www.siemens.com/innovation-brochure
- Internal reporting on the intranet and on the Siemens Social Network (SSN) using text, images, and video
- External live reporting via streaming
- "Inventor of the Year 2015" presentation: Communications activities involved a press feature containing media information, photos, and videos. At the same time, social media activities and a series in SiemensWorld were launched and inventors were highlighted on the new innovation path of the corporate website: www.siemens.com/innovation (Fig. 22.5).

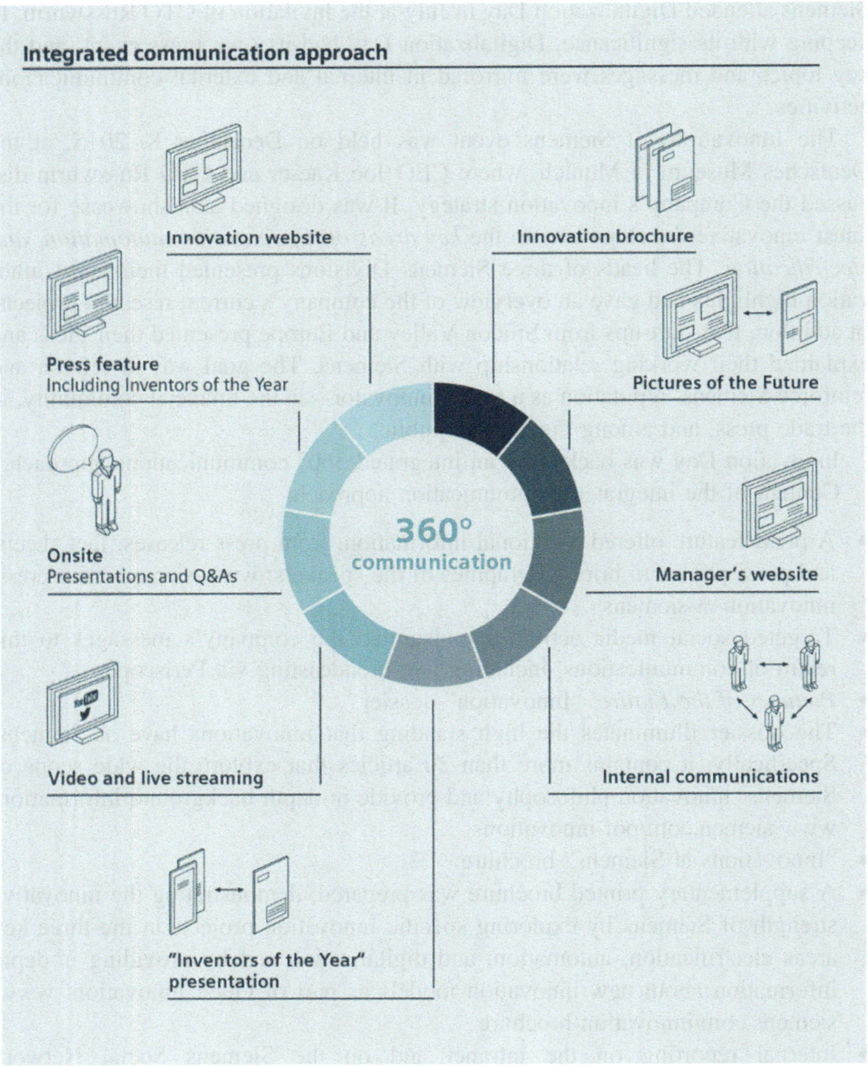

Fig. 22.5 Innovation at siemens: integrated communication approach. *Source* Siemens

22.3.3 *New Innovation Path on the Corporate Website*

The information architecture of the innovation path on the corporate website has been fundamentally revised. The goal was to offer users clearly structured access to the subject of innovation at different levels. The focus of the new site is on inventors, the minds behind Siemens' innovations. It is their ideas and concepts that are driving the company into the future. These employees are presented in video and text portraits. With the relaunch, the content pages have been attuned to the

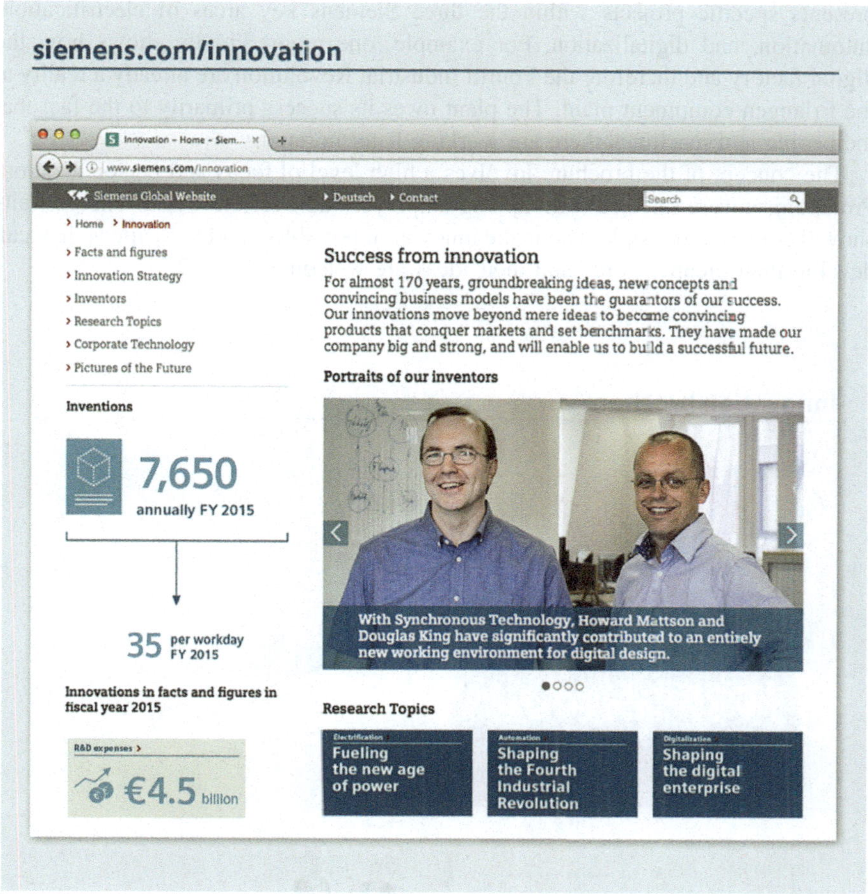

Fig. 22.6 Corporate website: innovation path. *Source* Siemens

company's mission and the strategic framework "Vision 2020." The innovation strategy is clearly explained. Current examples of videos, interactive 360° features, and reports provide insight into CT's research and into the operating units. These measures show CT's important role in the company, and the topic of recruiting is also addressed (Fig. 22.6).

22.3.4 Innovation Brochure

A print magazine was developed for the "Innovation at Siemens" press and analyst event, with content that transcends the specific occasion. It includes a double interview with the CEO of Siemens AG, Joe Kaeser, and CTO Russwurm and presents specific projects within the three Siemens key areas of electrification, automation, and digitalization. For example, one report vividly shows how the digital factory and therefore the Fourth Industrial Revolution are already a reality at the Erlangen equipment plant. The plant owes its success primarily to the fact that the people and machines there are working hand in hand.

The concept of the brochure involves a high level of focus and personalization. Protagonists were selected from among employees all over the world. The portraits show that these are people who make innovation possible—and at Siemens they can develop their creative force and their ideas are welcome (Fig. 22.7).

Fig. 22.7 Print brochure: innovation at Siemens. *Source* Siemens

The magazine represents the enhanced positioning of CT and offers examples from the operational units showing cooperation between CT and the Business Units, along with basic facts on the subject of innovation at Siemens. The current innovation model is explained in a separate area of emphasis: the increased support for Open Innovation. A journalistic feature discusses the actions Siemens took to "reinvent invention": competitions for ideas, collaborations with start-ups and the creation of the company's own start-ups, and internal platforms where employees all over the world can exchange views—not infrequently getting an answer to a question within a few minutes, or learning about a co-worker's smart idea that helps another employee's own work progress.

22.4 One Year of the New Siemens Innovation Communications: Evaluation

Barely one year after introduction of the new strategy for Siemens' Innovation Communications, the results are impressive—and not just with regard to quantity:

- More than 140 articles have been published internally at the SiemensWorld NewsCenter since March and December 2015.
- Between March and December 2015, 15 CT Magazines with links to 71 different articles were sent out to 52,000 people.
- PoF magazine counts an average of 110,000 page views and more than 60,000 visitors per month and publishes a biweekly newsletter with about 28,000 subscribers.
- Fourteen newsletters were sent out from April to December 2015. Each newsletter reached roughly 28,000 readers (Fig. 22.8).

The "Innovation at Siemens" event generated broad and very favorable media coverage: 336 articles have been published, and 90% of the coverage was positive or fairly positive. The close involvement of journalists and the effort to address agencies had a significantly positive effect on the reporting. The communication of messages proved effective. The key message that Siemens will strengthen its commitment to innovation was conveyed in 86% of event-related coverage. Quoted in 70% of the articles, Siemens representatives successfully pushed the event's visibility and message transmission. The event also highlighted Siemens' focus on topics beyond innovation: The plan to extend support for employees, enabling them to be more innovative (Figs. 22.9 and 22.10).

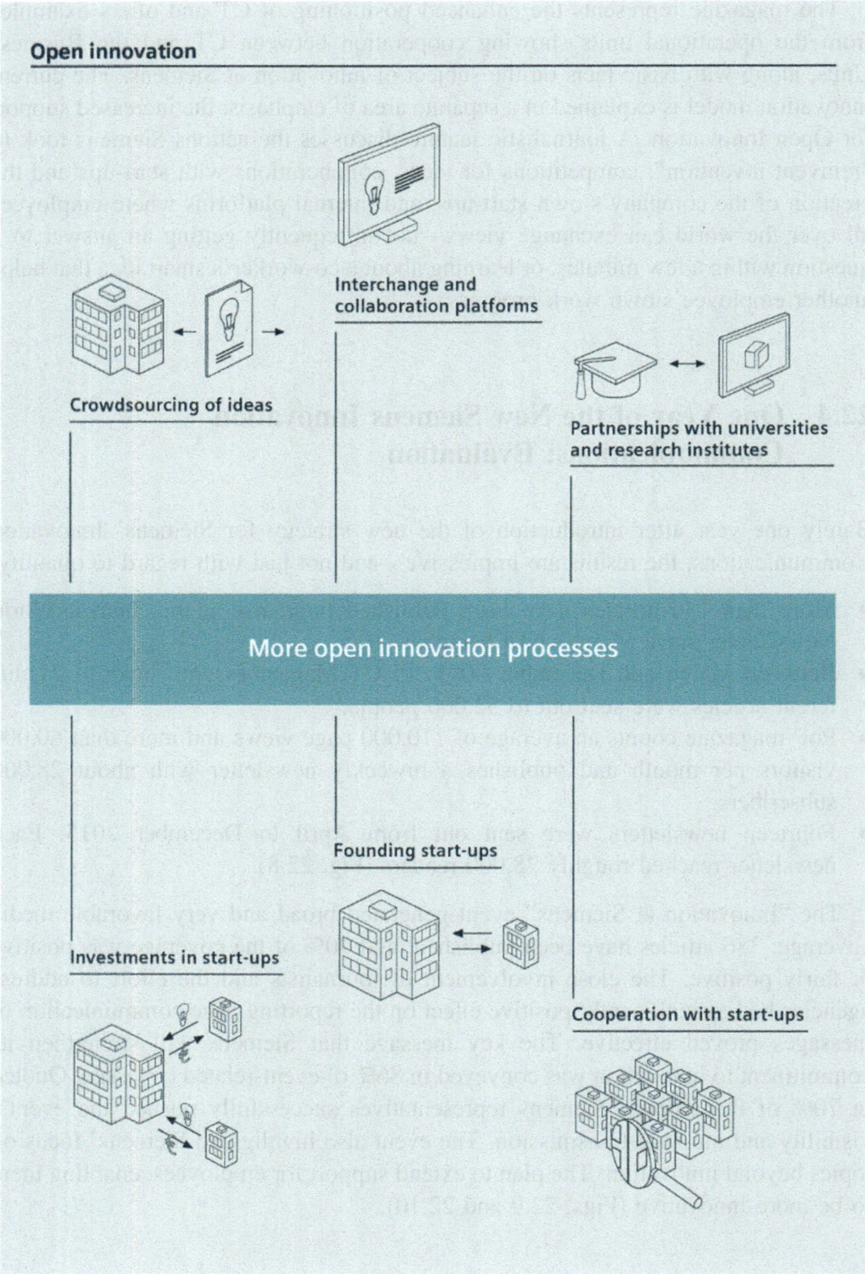

Fig. 22.8 Open innovation concept. *Source* Siemens

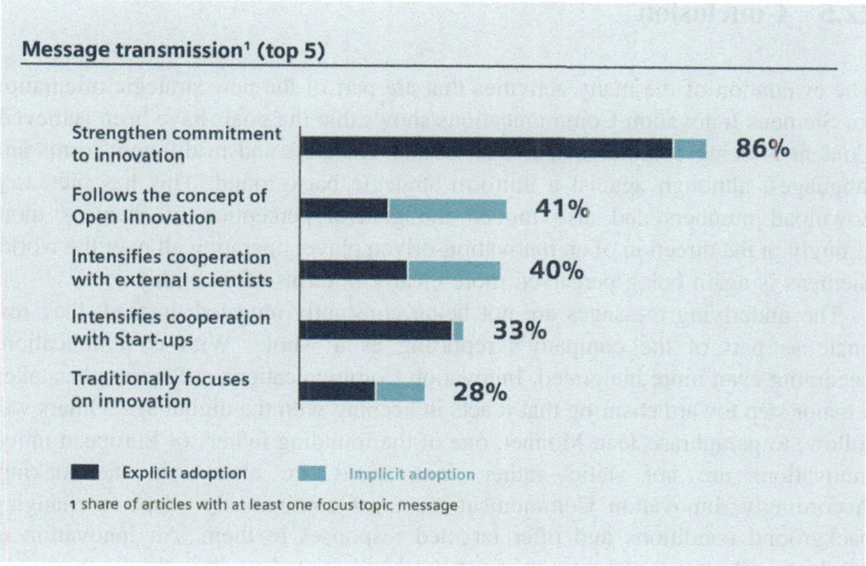

Fig. 22.9 "Innovation at Siemens" event 2015: press analysis. *Source* Siemens

Fig. 22.10 "Innovation at Siemens" event 2015: press review. *Source* Siemens

22.5 Conclusion

The evaluation of the many activities that are part of the new strategic orientation for Siemens Innovation Communications shows that the goals have been achieved. Content is being communicated over various channels and in different forms and languages, although against a uniform strategic background. This has increased download numbers and also moved the general perception of Siemens more strongly in the direction of an innovation-driven player operating all over the world. Siemens is again being perceived more clearly in terms of its agility.

The underlying messages are not being constantly repeated; instead, they resonate as part of the company's reporting as a whole. With communications becoming even more integrated, Innovation Communications at Siemens has taken a major step toward ensuring that it acts in keeping with the digital age. Others will follow: to paraphrase Jean Monnet, one of the founding fathers of European unity, innovations are not static—rather, innovations are always in the making. Accordingly, Innovation Communications must continuously adapt to changing background conditions and offer targeted responses to them. An innovation in speaking and writing about innovations must be initiated—within the company and also among the public.

Chapter 23
Co-innovation and Communication: The Case of SAP's Global Co-innovation Lab Network

Sabine Patsch and Ansgar Zerfass

Abstract This article describes a company case from the ICT industry and illustrates how collaborative innovation involving several actors is realized and why communication plays a crucial role within this process. Taking a social structural perspective on innovation and communication, it becomes clear that meaning and reality, which are the prerequisites of novelties, are constructed in communicative interactions between companies and their internal and external stakeholders. Consequently, communication cannot be considered as an instrument of innovation management that might be used or dropped in different innovation phases like a tool. The theoretical perspective also underlines that innovation-related actions are influenced by structures: rules and resources enable, modify, and limit innovation actions. The Global Co-Innovation Lab Network (COIL) of SAP, analyzed in the case study, can be seen as a corporate, communicative resource due to its role as an interface between internal and external stakeholders of the company. It enables all parties involved to define framework conditions of a shared co-innovation project and to execute it. Structurally, COIL connects stakeholders across the globe, such as certified or potential SAP partners, users or internals, such as existing and potential units. Therefore, collectively shared structures, like co-innovation projects, are created. They are considered as a prerequisite for future innovation and communicative actions. Based on previous expertise and experiences, COIL helps to specify shared structured for each project and thus enables successful co-innovations.

23.1 Introduction

Information and communication technologies (ICT) as cross-sectional technologies drive innovation and growth in many industries (Münchner Kreis et al. 2009, p. 192). It can be argued that they will even "play a more significant role as

S. Patsch (✉)
SAP AG, Walldorf, Germany

A. Zerfass
University of Leipzig, Leipzig, Germany

N. Pfeffermann and J. Gould (eds.), *Strategy and Communication for Innovation*,
DOI 10.1007/978-3-319-49542-2_23

software becomes a larger part of the company and the product" (Sarrazin and Sikes 2013, sidebar, para. 1). Therefore, companies have started to innovate their core business models based on ICT-driven opportunities (Kagermann et al. 2010). For example, ICT-related research and development (R&D) is of rising importance in industries like automotive, logistics, and health care (Dutta and Mia 2009, p. 102). In order to stimulate and realize ICT-based innovation across industries, different actors have to collaborate. Many companies have created structures and fostered actions to enable such collaboration processes with partners and customers in the past years. Concepts like 'open innovation' (Bogers 2012; Chesbrough 2003; Enkel et al. 2009), or 'networked innovation' (Swan and Scarborough 2005; Valkokari et al. 2009) explain how and why in-house R&D is complemented by innovation collaborations with outsiders. This chapter describes a company case study from the ICT industry and illustrates how collaborative innovation involving several actors is realized and why innovation communication (Zerfaß and Möslein 2009) plays a crucial role within this process. Taking a social structural perspective on innovation and communication, it becomes clear that meaning and reality, which are the prerequisites of novelties, are constructed in communicative interactions between companies and their internal and external stakeholders.

23.2 The Software Industry, SAP and Its Ecosystem

In general, the word software is used "to describe the digital instructions and operating information that are contained in *programs* serving to guide machines—especially computers—in implementing desired operations" (Lippoldt and Stryszowski 2009, p. 33). *Software products* are nonphysical and digital goods which can be reproduced at low cost, without any quality loss, and in any quantity desired. Another main aspect to describe software is its degree of standardization. Individual software and standard software can be defined as the two extreme cases, where individual software is tailored to the specific requirements of a user, and standard software is developed based on the lowest common denominator concerning the needs of potential users (Buxmann et al. 2013). The development of software can be described as a cumulative process because the transfer, reuse, and modification of code are possible (Lippoldt and Stryszowski 2009). *Software markets* are international, as software can be developed globally and distributed worldwide at low cost. Software markets include very few dominant players. In many cases, a standard or technology becomes prevalent, as the consumer's benefit of using a good—e.g., the same standard format in software systems—rises, when more consumers use the same good. The software market includes several kinds of actors, especially in the area of complex software solutions. Software companies, in a narrow sense, create software, whereas in a broad sense, they implement and operate software. Accordingly, there are several types of vendors competing on the market offering services such as implementation support, training or operating services (Buxmann et al. 2013). Collaborative approaches are used to foster software innovation, for example in partnerships between ICT firms, or between ICT companies and partners outside the sector (Lippoldt and Stryszowski 2009).

SAP AG is one of the main players in the software industry worldwide. In 2012, this global leader in enterprise software and software-related services had about 232,000 customers and more than 65,000 employees based in 130 countries in 2012. Table 23.1 summarizes SAP's products and service offerings in five market categories: Applications, Analytics, Cloud, Mobile, as well as, Database and Technology. Products across the five market categories are bundled into end-to-end solutions. These solutions are offered for all corporate functions, like finance, procurement or sales, and specifically for 25 industries such as automotive, banking, and health care (SAP 2013c). A main part of SAP's portfolio includes services ranging from the customized development of software solutions to support services, consultation on planning, implementing and optimizing business solutions, and educational services including IT training.

This short overview of SAP's portfolio indicates that the corporation collaborates with various partners. Besides partnerships with selling allies, SAP cooperates with service and implementation providers, as well as with development partners. Partners support the market adoption of SAP's solution portfolio by co-innovating on SAP platforms and embedding SAP technology, as well as reselling and/or implementing SAP software. SAP offers qualification and training programs for

Table 23.1 SAP's product and service portfolio

Market category	Description
Applications	Enterprise applications are the core competence of SAP, and SAP Business Suite as a business process platform. These include core software applications, like SAP ERP, which support critical business processes, such as finance and human capital management
Analytics	Analytics solutions enable users, e.g., to interact with business information and get answers to ad hoc questions without deeper knowledge concerning the underlying data sources
Cloud	Cloud applications and suites are provided as software as a service (SaaS) based on a subscription fee. SAP HANA Cloud is a platform as a service (PaaS) enabling SAP's customers, independent software vendors (ISVs), and partners to create software applications rapidly, e.g., for the needs of social and collaborative business networks
Mobile	Mobile solutions enable SAP's customers to deliver secure, real-time, business critical information to their employees, partners, and customers on mobile devices. SAP's mobile development platform also supports partners to develop their own applications for their employees and customers
Database and technology	The database and technology portfolio includes, e.g., the SAP NetWeaver® technology platform which enables the integration of SAP software with heterogeneous system environments, third-party solutions, and external business partners. In addition, the SAP HANA® platform, based on in-memory computing technology, processes huge amounts of data at a high speed

Source SAP (2013a) (adopted)

partners, as well as certifications for third-party offerings to underline the technical alignment with SAP solutions. This means that customers can benefit, for example, from pretested, certified partner offerings which extend the functionalities of SAP solutions or from accelerated integration projects (SAP 2013a, b).

Close relations with partners and the expansion of its partner ecosystem help SAP to increase its market coverage, improve its solution portfolio, and strengthen its innovation activities (SAP 2013a).

As a consequence, *collaborative innovation activities with stakeholder*s are supported by several initiatives at SAP. Examples are *SAP IdeaPlace*, a platform which invites stakeholders to submit, discuss, and rate ideas (SAP 2013d), as well as *SAP InnoJam,* a 30-h event to learn and discuss SAP technologies and finally apply them by creating a prototype (SAP 2013e). Another initiative is the *Global Co-Innovation Lab Network (COIL)* which has been established by SAP since 2007 to support co-innovation processes between SAP and external partners, customers, as well as other stakeholders (SAP 2013f). The COIL concept has been institutionalized for several years and has been adapted continuously. Consequently, a global network of Co-Innovation Labs has emerged, tackling a rising number of co-innovation projects with stakeholders. Therefore, the example of COIL offers an interesting insight in corporate co-innovation activities, its structures, and the role of communication, which is considered as a constituting element of innovation (Zerfaß 2009). To investigate the COIL concept in more detail, the theoretical concept of innovation communication in the era of open innovation is outlined first (Mast et al. 2005; Zerfass and Huck 2007; Zerfaß and Möslein 2009). Second, the best practice case of COIL is illustrated focusing on major aspects of innovation communication, which is understood as symbolic interactions between organizations and their internal and external stakeholders dealing with new products, services, and technologies (Zerfaß 2009).

23.3 Innovation and Communication: A Social Theoretical View

The term 'co-innovation' implies that at least two partners collaborate to foster innovation. According to Bogers (2012), *collaborative innovation* is a specific type of open innovation and combines knowledge inflows and outflows. The concept of open innovation is based on "the use of purposive inflows and outflows of knowledge to accelerate internal innovation, and expand the markets for external use of innovation, respectively" (Chesbrough et al. 2006, p. 1). Bogers refers also to the so called "coupled process" of open innovation understood as "co-creation with (mainly) complementary partners through alliances, cooperation, and joint ventures during which give and take are crucial for success" (Enkel et al. 2009, p. 313). Establishing a coupled process implies that the outside-in process to gain external knowledge is combined with the inside-out process to bring ideas to market, which

Table 23.2 Alternative conceptualizations of the core terms innovation and communication

Communication as…	Innovations as…
… ***transmission***: Companies transmit objective information via media channels to key stakeholders. This stimulus leads to the transfer of meaning and is intended to evoke desired reactions (e.g., knowledge, attitude change, behavior). *Communication* is a specific form of behavior that is mainly determined by systemic relations and psychological motives	… ***artifacts***: *Innovations* are novel products or processes that are marked as new by a company (or its leaders) and that are established on the market or within the organization. The innovative offer meets the demand of actors on the market and is used by them in different ways
… ***construction of reality***: Companies and stakeholders constitute social interactions by messaging and comprehension activities referring to each other, which link to shared symbolic structures (communicative schemes and competencies) and target understanding as a prerequisite for influencing each other. *Communication* is a form of interest-led action, where perceptions and orientations are shaped subjectively, but meaning and reality are socially constructed	… ***social constructs***: *Innovations* are technical, economic, or social novelties that imply a change of social practices and that are considered by the involved actors as new because they reach beyond ongoing adaption of practices. Shared meanings are a prerequisite for novelties; likewise, the meaning of technologies, innovations, markets or added value is only constituted in social interactions

Source Zerfaß (2009, p. 36)

drives the joint development and commercialization of innovation (Enkel et al. 2009). Coupled innovation processes entail interactions between at least two actors. Therefore, the role of communication in collaborative innovation processes is central and is considered in the following in more detail. At first, the two core terms 'innovation' and 'communication' are conceptualized based on different scientific perspectives. Subsequently, the theory of structuration (Giddens 1984) is outlined to reconstruct the role of communication within the innovation process.[1]

Innovation research uses mainly the conceptualization of communication as transmission and innovation as artifacts. More recent research underlines the social construction of technologies, innovations, and markets, but still uses the conceptualization of communication as *information transmission* and conveying of meaning. The same applies to scientific discussions on innovation communication, as they are mainly based on a traditional term of innovation. The different conceptualizations underline several aspects of corporate practice while complementing each other. The conceptualizations in the lower section of Table 23.2 are focused on the *creation and adaption of communication and innovation*, and the traditional concepts in the upper section of Table 23.2 underline the *management of communication or innovation processes* within the conditions of already prestructured social relations.

[1]This section is mainly based on an article previously published in German language (Zerfaß 2009).

Social theory (Joas and Knöbl 2004) enables us to understand the duality of acting within given structures and changing those structures at the same time. The *theory of structuration* by sociologist Anthony Giddens (1984) helps to overcome the limitations of both action theory and systems theory, which focus on either side of the process. Giddens' theoretical approach has been applied to various disciplines including corporate communications (Falkheimer 2007; Hahne 1998; Röttger 2005; Zerfaß 2010) and also sporadically in innovation research (Chanal 2004; Coopey et al. 2002; Duschek 2002).

Giddens (1984) points out that social coexistence is determined by both individual *actions* and societal *structures* (rules and resources). These structures enable and influence thinking, as well as acting, and are renewed and also partly adapted during each actualization.

Social interactions can only be successful if those involved can refer to a shared reservoir of rules and resources. This applies both to communication processes as well as to the creation and acceptance of innovations. *Resources* are, on the one hand, material aspects of the environment, a means of production and products (allocative resources), such as a method to create, convey, and understand messages or to build prototypes. On the other hand, resources can be competences (authoritative resources), such as the ability to communicate actively with certain stakeholders, to listen intensively in dialogs with customers and employees, to be able to cooperate, to solve engineering problems, and to create novel combinations of ends and means. *Rules* sanction social acting (legitimization) and constitute meaning (signification) (Giddens 1984).

Collective perceptions and interpretive schemes create cognitive structures that are necessary to enable social activities by various actors and to make them compatible with each other. Rules of communication are modes like articulation or symbolic schemes as well as cultural rooted modes of influence such as manipulations, instructions, or argumentative discourse (Zerfaß 2010, pp. 169–181). Rules of innovation are schematic visions of the novel, possible and valid combinations of materials, process operations, and applications, as well as cognitive associations of certain groups (e.g., farmers, consumers) linked with technologies and societal concepts (e.g., 'nature', 'ecology').

The following illustration shows the process of structuration (Fig. 23.1).

Structuration is a dynamic process. Collectively shared structures between involved actors are a prerequisite for innovation as well as for communication activities. At the same time, the actualization of rules and resources makes actions compatible and likely to be successful. Naturally, recurring actions and their prerequisites can be reflected. Parameters influencing the success of corporate communications or innovation activities can be identified and used in terms of implicit or explicit knowledge. As a consequence, strategies will be developed that help to change structures for the better.

Taking a social structural perspective, social change can be interpreted as a shift of rules and a change of resources over time. Accordingly, innovation management must focus on drivers that influence rules and resources. At this point, the fundamental role of communication becomes obvious: meaning and reality, which are the

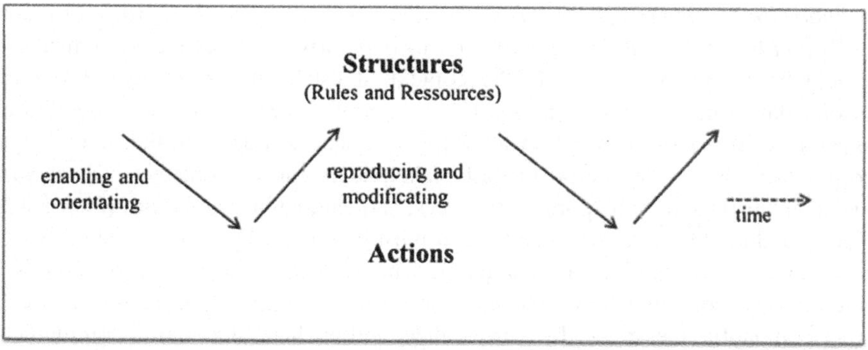

Fig. 23.1 The process of structuration. *Source* By the authors

prerequisites of novelties, are constructed in communicative interactions between companies and their internal and external stakeholders. Likewise, the meaning of technologies, innovations, and markets are created in social interactions between all actors involved. The role of communication within the innovation process can be reconstructed according to the following.

Communication is a constituting element in innovation management, as technical, economic, or social novelties always evolve in the context of interactions that are enabled and restricted by collective rules and resources. Cognitive schemes determine the meaning of new products, services, processes, and technologies, for whom they are beneficial and how they are evaluated. These cognitive structures can be formed, modified, and reproduced by communication processes.

Open innovation processes create social practices and spheres of communication that enlarge the potential for acceptable novelties and minimize the risks of closed reference systems. Traditional innovation concepts focusing on internal R&D departments decouple thought patterns of innovators from the world of potential users, multipliers, and critical stakeholders. In this case, communication departments are called in the final phase of the development process to translate visions of engineers and product managers into the cognitive frames of yet uninvolved stakeholders. This is often condemned to failure. If, by contrast, interfaces between the focal organization and its internal and external stakeholders are created in all phases of the innovation process, the probability of common reality constructions and shared values rises. Obviously, the success of such endeavors is never ensured. If the interests of different stakeholders meet, it is not always a win-win situation. However, early communication helps to recognize discrepancies sooner. Adjustments or cancellations of R&D activities are possible, and misallocations of resources are reduced. Accordingly, concepts of 'open innovation' (Chesbrough 2003; Möslein and Neyer 2009) obtain a new justification beyond economical purposive rationality. From a sociological and communicative perspective, there is much to be said for the inherent evidence of such concepts.

Innovation communication has to be situative, and it can be used both adaptively and by structuring in different contexts and phases. Strategic communications should be aware that it acts in the context of established rules and resources. Insofar, knowing the public opinion (Bentele et al. 2003), its structural conditions as well as drivers of change (Zerfaß 2007) is essential. Based on that knowledge, corporate messages can be positioned, stakeholders can be contacted via relevant media channels and with appropriate topics, and campaigns can be run (Fink 2009; Huck-Sandhu 2009). At the same time, it must be repeatedly reflected whether and how current structures and interpretation schemes can be destructed or modified. Accordingly, communication management must accomplish much more than creating compatible messages. To plan, realize, and evaluate innovation communication refers also—and with increasing frequency—to listening, establishing intelligent routines for monitoring opinions, as well as to identifying relevant publics and stakeholders. Feeding external opinions and interests into the organizational decision process ("inbound") is as important as conveying the corporate point of view to others ("outbound").

Thus, *innovation communication* can be defined (Zerfaß 2009, p. 42) as a strategic stimulation of communication processes with internal and external stakeholders to promote technological, economic or social novelties, (a) by creating, revising, or destructing socially shared patterns of meanings and communicative resources, and (b) by giving impulses for the development of novelties, and by promoting them professionally.

In summary, communication should not be considered as an instrument of innovation management that can be used or dropped in different innovation phases and situations like a tool. Communication has to be understood as symbolic interaction and reality construction. This perspective is also used in elaborated concepts of organizational communication, corporate communications, and public relations (PR) (Botan and Hazleton 2006; Zerfass 2008; Zerfaß 2010).

In the following section, the social theoretical perspective of innovation and communication is used to analyze a company case study. Central aspects of the innovation communication approach described above are outlined to demonstrate how corporations try to deal with the challenges of collaborative innovation by communicative means.

23.4 Innovation and Communication in SAP's Global Co-innovation Lab Network (COIL)

The social theoretical view on innovation as a social construct underlines that innovation-related actions are influenced by structures: rules and resources enable, modify, and limit innovation actions. One of SAP's key instruments to foster collaborative innovation between the company, customers, and partners can be described along this line.

The *Global Co-Innovation Lab Network (COIL)* consists of several teams and lab facilities spread globally. It is intended to be a platform of services enabling co-innovation projects between SAP product and field teams as well as other entities, like current or potential SAP partners and customers (Fig. 23.2).

The main aspects of the *working model* of SAP's Co-Innovation Lab Network are:

- COIL can be described as an *interface between internal and external stakeholders* enabling all involved parties to define framework conditions of a co-innovation project and to execute the project. Accordingly, COIL facilitates the extension of SAP's solution coverage as well as the acceleration of technology adoption and enablement by executing joint technical co-innovation projects and initiatives between SAP, its partners and selected customers. COIL focuses on how different parties can be connected successfully to a co-innovation topic within a dedicated project set up, whereas the topic ownership is not with COIL, but with SAP's respective product and R&D units.
- Throughout the years of its existence, COIL evolved to create structures and to enable actions in order to tackle the main challenges of collaborative innovation. Consequently, major *co-innovation enablement elements* have been identified to realize project-based co-innovation successfully, like *IP framework, IT infrastructure, subject matter expertise* in the co-innovation team, and *knowledge brokering* to determine experts out of the team, as well as *operations and project management* (Cruickshank 2010).

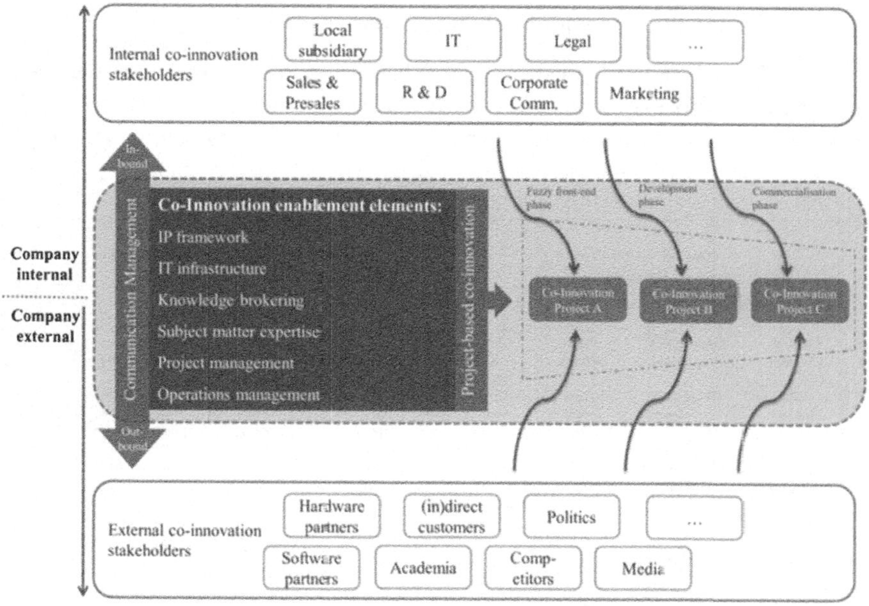

Fig. 23.2 Working model of SAP's Global Co-Innovation Lab Network. *Source* By the authors

- As COIL is intended as an interface platform between the SAP-internal and external world, *inbound and outbound communication* and *project-bound communication* need to be managed systematically.
- From SAP's point of view, co-innovation projects in COIL can be attributed to each *phase of the innovation process*. A co-innovation project can be an undertaking of SAP's 'fuzzy front-end phase', including a strong research focus, such as developing a proof of concept paper. A co-innovation project could also be assigned to SAP's development phase or the commercialization phase, such as providing a newly developed SAP technology to early testing and adopting partners and customers to foster a common innovation project. COIL projects are of mid-term length and typically last six months.
- The Co-Innovation Lab Network benefits from its global presence at currently nine locations *close to local and regional markets* in North and South America, Asia, as well as Middle and Eastern Europe. All nine locations include a project facility to work collaboratively, and a showroom to demonstrate ideas or prototypes. In four out of the nine locations, computing centers have been established. This set up ensures that local COIL employees with certain engineering competences, can act as intermediaries between local, external stakeholders and internal experts, e.g. from other locations of the Co-Innovation Lab Network or from product and sales teams. The onsite colleagues are able to liaise as they master the local language and know national as well as organizational cultures. These aspects help to ensure the necessary dialog between partners within successful co-innovation projects. Due to the local facilities, the respective managing director of the SAP subsidiary is a major stakeholder, fostering co-innovation projects with local and regional partners.

As indicated, COIL uses a project-based approach and has defined a co-innovation process containing seven individual steps (see Fig. 23.3). The elaboration of each process step depends on the individual project undertaking.

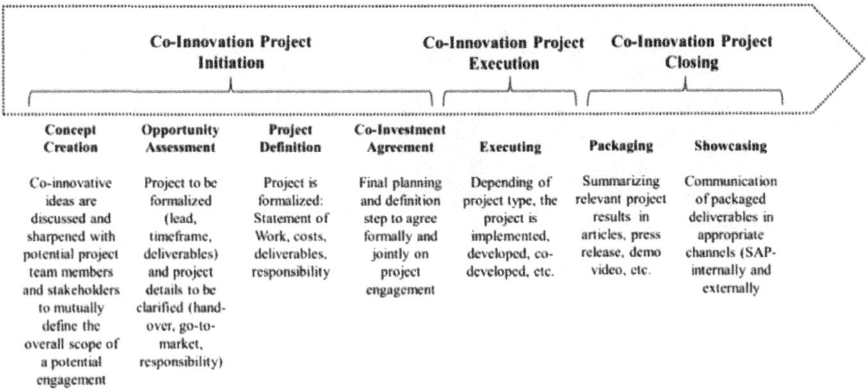

Fig. 23.3 Generic co-innovation process at SAP's Global COIL Network. *Source* SAP (2013f) (modified)

The process view of co-innovation projects at COIL underlines that *communication activities* play an important role:

- The *pre-project phase* and the *project initiation phase* are determined by monitoring internal and external project ideas and topics. Moreover, it is important to stimulate dialog between stakeholders, to evaluate ideas, and to attain appropriate project partners and promoters for a COIL project ensuring resources (e.g., budget, subject matter experts, infrastructure, portfolio integration, campaign integration, etc.). To prepare a COIL project, communication-related activities are discussed and planned with relevant internal and external stakeholders (e.g., the global or local marketing and communications teams).
- In the *project execution phase,* internal project communication between the COIL project partners and sponsors is focused. Interim or final project results are framed and communicated, for example, by writing a whitepaper or preparing a demo video.
- The *project closing phase* is characterized by showcasing results through the appropriate channels, for example, to foster the market launch of a (partner) solution or to prepare follow-up projects by stimulating dialog for further ideas.

Figure 23.4 summarizes the main communication-related activities associated with each project phase. In addition, the main *communication instruments* addressing company internal or external stakeholders are outlined.

Due to its working model and global set up, COIL uses *two main communicational pillars*:

- *Online Communication,* meaning all web-based communication activities, enables COIL to address its partners and customers independent of time and space on a *global scale*. The SAP Community Network (SCN)[2] is used especially to build a global co-innovation community. Dialogs within the SAP ecosystem are stimulated by sharing project ideas and project outcomes, for example with articles or podcasts. The partner program SAP PartnerEdge[3] includes a web-based platform and serves as another channel to attain new COIL projects and partners. To ensure internal project communication, online communication channels such as SAP Jam[4] or wikis as well as online conferencing tools are used to exchange knowledge or document project results and challenges. Within the SAP organization, one main communication instrument available on the intranet are the *internal communities* which are especially used

[2]SAP Community Network (http://scn.sap.com) is the social network for both internal and external SAP professionals, such as software users, developers, or consultants, with more than two million members globally.

[3]SAP PartnerEdge (www.sappartneredge.com) is SAP's partner program, offering business enablement resources and benefits to support implementing, selling, marketing, developing, and delivering SAP products.

[4]SAP Jam (www.sap.com/jam) is an enterprise social network solution.

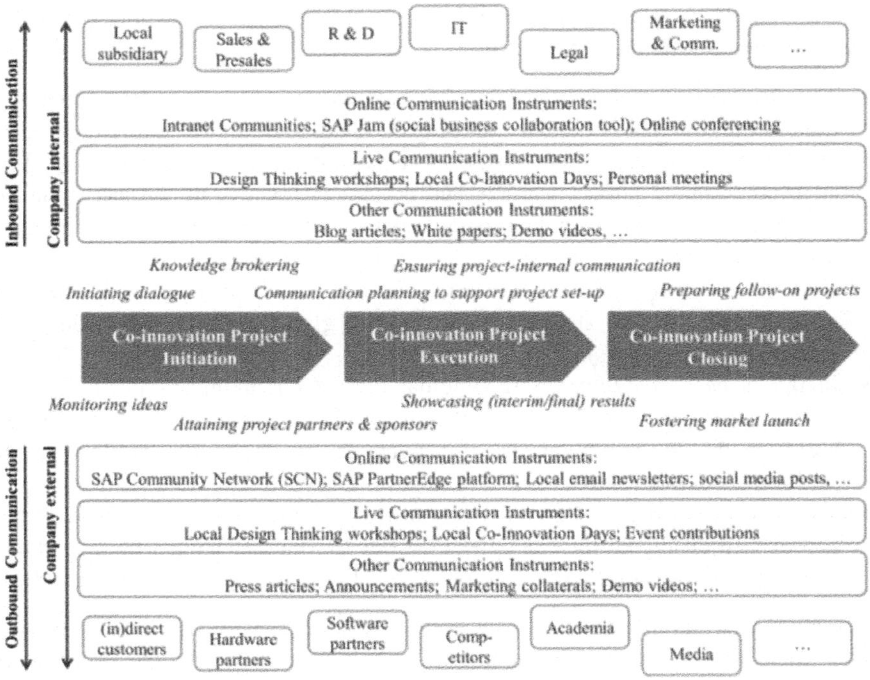

Fig. 23.4 Sample communication activities and instruments during the co-innovation process involving various co-innovation stakeholders. *Source* By the authors

to determine subject matter experts for COIL projects or share results, e.g., by posting blog articles, white papers, or demo videos. Online Communication is also used on a *local or regional* level, e.g., by creating local COIL newsletters or using other communication channels offered by subsidiaries or regional representatives.

- *Live Communication*, understood as communication activities in copresence, is used to strengthen COIL's activities on a *local and regional level.* Strategic decisions to establish a new co-innovation lab as part of the global Co-Innovation Lab Network are always made to benefit from the proximity to respective customers and partners in local or regional markets. Besides project facilities that execute co-innovation projects with externals, each lab location has a showroom to demonstrate interim and final results to stakeholders or to discuss new project ideas. Personal meetings, local events, and workshops are hosted, and design thinking techniques are increasingly applied as an iterative innovation approach to support teams during co-innovation processes (Plattner et al. 2011). On a global scale, live communication is especially used at industry events, e.g,. SAP's global annual events SAPPHIRE NOW[5] and SAP TechEd.[6] In most cases, co-innovation prototypes and demos are shown in strong alignment with the respective marketing and communications teams at SAP and other project stakeholders, enabling potential customers and future partners to experience co-innovation first hand.

A manufacturing project (Odlozinski 2013) can be used to illustrate a typical co-innovation project at COIL. After discussing rough project ideas with stakeholders, COIL connected with internal development and product teams at SAP with a manufacturing hardware firm and an industry standards organization to develop a co-innovation project. Besides knowledge brokering to identify appropriate project members and sponsors, COIL provided the IT infrastructure and project management enabling the co-innovation partners to use existing manufacturing solutions from SAP and its integration capabilities. Besides ensuring shared project structures, COIL fostered a joint, transportable demo board to showcase aspects of a real-world manufacturing process. In addition, communication activities with internal marketing and communications teams were executed, such as press activities, collaterals, videos, and SCN posts, as well as contributions to industry fairs and SAP sales events. SAP-internal stakeholders from sales and development especially valued the prototype showcasing how potential customers from the manufacturing industry can implement SAP solutions combined with the partners' extensions. The partners contributing to the project valued the co-innovation outcome, as a novel, market-ready solution that was created by bringing together the partners' capabilities.

In the light of the social theoretical view described above, COIL can be considered as an interface with specific expertise to facilitate co-innovation projects. Furthermore, COIL enables access to existing resources, and promotes the creation of new resources, as well as fosters the understanding and change of rules necessary to realize collaborative innovation. Figure 23.5 illustrates the structures and actions in SAP's Co-Innovation Lab Network to enable co-innovation between internal and external stakeholders.

COIL is able to give project partners access to resources like *IT infrastructure* based on the latest engineering and system landscapes provided by SAP and key partners, as well as to an *IP framework* adaptable to specific project requirements. COIL enables stakeholders to access existing resources and to create new ones because it identifies and mobilizes internal and external *subject matter experts* as well as *potential project sponsors or promoters*, such as product and field teams or

[5]SAPPHIRE NOW (www.sapphirenow.com) is a customer-facing event where SAP generally has announced major product changes and strategic direction.

[6]SAP TechEd (www.sapteched.com) is an annual conference hosted by SAP, aimed at the company's ecosystem of consultants and software development partners. It is a technical education conference for IT architects, administrators, and developers.

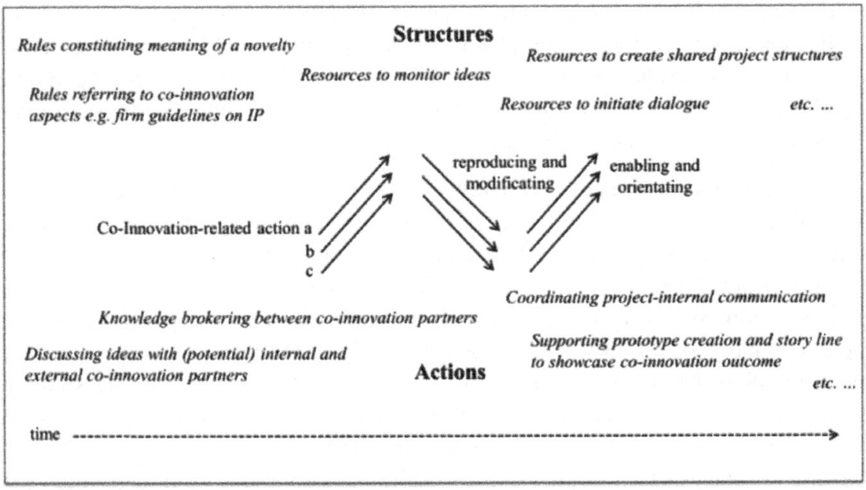

Fig. 23.5 Examples of structures and actions in SAP‘s Co-Innovation Lab Network enabling co-innovation. *Source* By the authors

marketing and communications teams driving COIL project topics within their areas of responsibility. Furthermore, COIL's ability to prepare and support projects with *operations and project management* activities can be considered an important resource. Based on COIL's experiences from former undertakings, the establishment of a new co-innovation project is constantly streamlined. Therefore, involved actors can focus on the core process of innovation.

Project-based co-innovation facilitated by COIL supports joint thought patterns that bind involved actors to each other. Common constructions of reality and shared values among internal or external partners, potential users, multipliers, or critics become more probable. Acting as an interface between the internal and external world of the corporation, COIL is a vehicle for inbound and outbound communication. On the one hand, COIL enables SAP to scout the external environment. The network benefits from its local labs functioning as points of contact especially for local stakeholders interested in co-innovating with SAP. Bringing in external project ideas and facilitating project planning and execution involves managing *inbound communication* by addressing suitable internal project sponsors, promoters and potential team members. On the other hand, COIL fosters *outbound communication* by stimulating dialogs with (potential) co-innovation partners on topics where SAP searches for allies or by supporting market launch activities. Communication processes help to influence cognitive schemes that determine what novelties mean.

23.5 Conclusion

Collaborative innovation between several actors is a complex undertaking and offers challenges in both theory and practice. Taking a social theoretical view on innovation and communication, the case of SAP's Global Co-Innovation Lab Network (COIL) indicates that structures are created to enable co-innovation actions in organizational practice. A global player like SAP uses several instruments to foster innovation, and opens up organizational boundaries to gain access to internal and external knowledge necessary for innovation (Chesbrough 2003). This case study shows that an appropriate mix of internal and external infrastructure, domain expertise, and processes is needed to capitalize on novelties.

Furthermore, the case underlines the theoretical argument that communication is a constituting element of innovation (Zerfaß 2009): The Co-Innovation Lab Network can be conceptualized as a corporate communicative resource due to its mission to act as an interface between internal and external stakeholders, enabling all involved parties to define structural frameworks for a shared co-innovation project and to execute it. As innovations are created in social interactions, collaborative innovation is based on communication processes. Structurally, COIL as network of currently nine co-innovation labs connects stakeholders across the globe, such as certified or potential SAP partners, users or internals, such as existing and potential business units. Therefore, collectively shared structures are created and are considered as a prerequisite for innovation and communicative actions to take place. Co-innovation projects are a form of collectively shared structures, and the expertise and experience of the global Co-Innovation Lab Network helps to specify shared structures per project to enable successful co-innovation actions. Accordingly, actions within co-innovation projects facilitated by COIL refer to existing structures, e.g., meaning of novelties, internal topic responsibilities, or IP guidelines, and influence the modification of existing structures.

In conclusion, different perspectives in theory and practice should be taken into account when investigating collaborative innovation processes. The practice of corporate innovation management and communication may benefit from these considerations by deriving best practices, including other industries, to implement and adjust appropriate structures that successfully enable collaborative innovation actions. From a theoretical point of view, the concept of collaborative innovation and the role of communication in innovations should be researched further, both theoretically and empirically, to enlarge the body of knowledge (Bogers 2012; Ernst and Zerfaß 2009; Lee et al. 2012; Pfeffermann 2011). An interdisciplinary view combining communication science, social theory, and management theory helps to reflect on the communicative dimension of actions and structures constituting today's organizations and their collaborative efforts to innovate.

Bibliography

Bentele, G., Brosius, B., Jarren, O. (eds.): Öffentliche Kommunikation. Westdeutscher Verlag, Wiesbaden (2003)

Bogers, M.: Knowledge sharing in open innovation: an overview of theoretical perspectives on collaborative innovation. In: de Pablos Heredero, C., López, D. (eds.) Open Innovation in Firms and Public Administrations: Technologies for Value Creation, pp. 1–14. IGI Global, Hershey (2012)

Botan, C., Hazleton, C. (eds.): Public Relations Theory II. Lawrence Erlbaum, Mahwah (2006)

Buxmann, P., Diefenbach, H., Hess, T.: The Software Industry. Economic Principles, Strategies, Perspectives. Springer, Heidelberg (2013)

Chanal, V.: Innovation management and organizational learning: a discursive approach. Eur. J. Innov. Manag. **7**, 56–64 (2004)

Chesbrough, H.W.: Open innovation. The new imperative for creating and profiting from technology. Harvard Business School Press, Boston (2003)

Chesbrough, H.W., West, J., Vanhaverbeke, W. (eds.): Open innovation: researching a new paradigm. Oxford University Press, Oxford (2006)

Coopey, K., Keegan, O., Emler, N.: Managers' innovations and the structuration of organizations. J. Manag. Stud. **35**, 263–284 (2002)

Cruickshank, D.: Open innovation and co-innovation—enablement platform processes to align innovation to the right business model. Retrieved from http://scn.sap.com/docs/DOC-16932 (2010)

Duschek, S.: Innovation in Netzwerken. Renten - Relationen – Regeln. Deutscher Universitäts-Verlag, Wiesbaden (2002)

Dutta, S., Mia, I.: The global information technology report 2008–2009. Mobility in a networked world. Retrieved from http://www3.weforum.org/docs/WEF_GITR_Report_2009.pdf (2009)

Enkel, E., Gassmann, O., Chesbrough, H.W.: Open R&D and open innovation: exploring the phenomenon. R&D Manag. **39**(4), 311–316 (2009)

Ernst, N., Zerfaß, A.: Kommunikation und Innovation in deutschen Unternehmen. In: Zerfaß, A., Möslein, K.M. (eds.) Kommunikation als Erfolgsfaktor im Innovationsmanagement. Strategien im Zeitalter der Open Innovation, pp. 57–81. Gabler, Wiesbaden (2009)

Falkheimer, J.: Anthony Giddens and public relations: a third way perspective. Pub. Relat. Rev. **33**, 287–293 (2007)

Fink, S.: Strategische Kommunikation für Technologie und Innovationen. Konzeption und Umsetzung. In: Zerfaß, A., Möslein, K.M. (eds.) Kommunikation als Erfolgsfaktor im Innovationsmanagement. Strategien im Zeitalter der Open Innovation, pp. 207–223. Gabler, Wiesbaden (2009)

Giddens, A.: The constitution of society. Outline of the theory of structuration. University of California Press, Berkeley (1984)

Hahne, A.: Kommunikation in der Organisation. Westdeutscher Verlag, Opladen/Wiesbaden (1998)

Huck-Sandhu, S.: Innovationskommunikation in den Arenen der Medien. Campaigning, framing und storytelling. In: Zerfaß, A., Möslein, K.M. (eds.) Kommunikation als Erfolgsfaktor im Innovationsmanagement. Strategien im Zeitalter der Open Innovation, pp. 195–208. Gabler, Wiesbaden (2009)

Joas, H., Knöbl, W.: Sozialtheorie. Suhrkamp, Frankfurt am Main (2004)

Kagermann, H., Osterle, H., Jordan, J.M.: IT-Driven Business Models: Global Case Studies in Transformation. Wiley, Hoboken (2010)

Lee, S.M., Olson, D.L., Trimi, S.: Co-innovation: convergenomics, collaboration, and co-creation for organizational values. Manag. Decis. **50**(5), 817–831 (2012)

Lippoldt, D., Stryszowski, P.: Innovation in the Software Sector. OECD Publishing, Paris (2009)

Mast, C., Zerfaß, A. (eds.): Neue Ideen erfolgreich durchsetzen. Das Handbuch der Innovationskommunikation. Frankfurter Allgemeine Buch, Frankfurt am Main (2005)

Mast, C., Huck, S., Zerfass, A.: Innovation communication—outline of the concept and empirical findings from Germany. Innov. Journalism **2**(7), 1–14 (2005)
Möslein, K.M., Neyer, A.K.: Open innovation. Grundlagen, Herausforderungen. Spannungsfelder. In: Zerfaß, A., Möslein, K.M. (eds.) Kommunikation als Erfolgsfaktor im Innovationsmanagement. Strategien im Zeitalter der Open Innovation, pp. 85–103. Gabler, Wiesbaden (2009)
Münchner Kreis, e.V., European Center for Information and Communication Technologies (EICT), Deutsche Telekom, TNS Infratest (eds.): Prospects and opportunities of information and communication (ICT) and media. International Delphi study 2030. Retrieved from http://www.zukunft-ikt.de/wp-content/uploads/2_Prospects_and_Opportunities_of_ICT_and_Media.pdf (2009)
Odlozinski, H.-J.: Manufacturing COIL. Retrieved from http://scn.sap.com/docs/DOC-23561 (2013)
Pfeffermann, N.: Innovation communication as a cross-functional dynamic capability: strategies for organizations and networks. In: Hülsmann, M., Pfeffermann, N. (eds.) Strategies and communications for innovations, pp. 257–289. Springer, Berlin (2011)
Plattner, H., Meinel, C., Leifer, L. (eds.): Design Thinking. Understand, Improve, Apply. Springer, Berlin (2011)
Röttger, U.: Kommunikationsmanagement in der Dualität von Struktur. Medienwissenschaft Schweiz/Science des mass média suisse **15**(1–2), 12–19 (2005)
Sarrazin, H., Sikes, J.: Competing in a digital world: four lessons from the software industry. Retrieved from http://www.mckinsey.com/insights/business_technology/competing_in_a_digital_world_four_lessons_from_the_software_industry (2013)
SAP: 2012 Annual Report. Retrieved from http://www.sapintegratedreport.com/2012/fileadmin/user_upload/2012_SAPintegratedreport/downloadcenter/13_03_27/SAP_AR2012_en.pdf (2013a)
SAP: Benefits of SAP certified integration. Retrieved from http://scn.sap.com/docs/DOC-25281 (2013b)
SAP: Explore SAP solutions. Retrieved from http://www54.sap.com/solution.html (2013c)
SAP: SAP idea place. Retrieved from https://ideas.sap.com/ (2013d)
SAP: SAP InnoJam. Retrieved from http://scn.sap.com/community/events/innojam (2013e)
SAP: Introducing the Global SAP Co-Innovation Lab Network. Retrieved from http://scn.sap.com/docs/DOC-8660 (2013f)
Swan, J., Scarborough, H.: The politics of networked innovation. Hum. Relat. **58**(7), 913–943 (2005)
Valkokari, K., Paasi, J., Luoma, T., Lee, N.: Beyond open innovation. The concept of networked innovation. Paper presented at 2nd ISPIM innovation symposium: stimulating recovery—the role of innovation management, New York, 6–9 Dec 2009
Zerfaß, A.: Von der Einkanal-Kommunikation zum Dialog – wenn Empfänger zu Akteuren werden. In: Ellerbeck, T., Siebenhaar, K. (eds.) Vernetzte Welt –Veränderungen der Kommunikation durch Neue Medien und Mobilfunk, pp. 31–48. B&S Siebenhaar, Berlin (2007)
Zerfaß, A.: Kommunikation als konstitutives Element im Innovationsmanagement. Soziologische und kommunikationswissenschaftliche Grundlagen der Open Innovation. In: Zerfaß, A., Möslein, K.M. (eds.) Kommunikation als Erfolgsfaktor im Innovationsmanagement. Strategien im Zeitalter der Open Innovation, pp. 23–55. Gabler, Wiesbaden (2009)
Zerfaß, A.: Unternehmensführung und Öffentlichkeitsarbeit. Grundlegung einer Theorie der Unternehmenskommunikation und Public Relations, 3rd edn. VS Verlag für Sozialwissenschaften, Wiesbaden (2010)
Zerfaß, A., Möslein, K.M. (eds.): Kommunikation als Erfolgsfaktor im Innovationsmanagement. Strategien im Zeitalter der Open Innovation. Gabler, Wiesbaden (2009)

Zerfass, A.: Corporate communication revisited: integrating business strategy and strategic communication. In: Zerfass, A., Van Ruler, B., Srira-mesh, K. (eds.) Public Relations Research. European and International Perspectives and Innovations, pp. 65–96. VS Verlag für Sozialwissenschaften, Wiesbaden (2008)

Zerfass, A., Huck, S.: Innovation, communication, and leadership: new developments in strategic communication. Int. J. Strateg. Commun. **1**(2), 107–122 (2007)

Chapter 24
Empowering Members of a Brand Community to Gain Consumer Insights and Create New Products: The Case of the Vorwerk Thermomix Research Community

Madeleine Kröper, Volker Bilgram and Ramona Wehlig

Abstract Online Research Communities (ORCs) have ushered marketing research into a new era and are one of the fastest growing segments in the industry. ORCs offer a selected number of consumers a closed online environment in which they interact and co-create with the company not only in "one-off" projects but over an extended period of time. The approach distinguishes itself by the flexibility and diversity of market research methodologies that can be applied. In our case study, we describe the set up, live phase, and post-processing of the Vorwerk Thermomix Research Community. Over the duration of 3 weeks, Thermomix invited customers, who are also registered users of the brand's official online receipe community, to participate in a moderated ORC. There were two major goals. First, Thermomix wanted to understand consumers' "cooking journeys," i.e., learn how they decide what to cook, plan their cooking, buy groceries, etc. Second, Thermomix aimed to co-create with its customers and develop new features and functionalities for the Thermomix recipe community which serves as a central meeting place for cooking enthusiasts and brand fans. In the study, we show how the ORC was designed to answer our research questions and what methodological benefits the method yields. Furthermore, we investigate the user types that contributed to the platform most and provide insights into how to boost activity levels on the platform. In today's social media context, ORCs are a valuable tool for brands to connect with their customers and engage them in value creating activities. This chapter sheds light on this new research approach and unveils the potential of brand community members for insight generation and co-innovation.

M. Kröper · V. Bilgram (✉)
HYVE, Munich, Germany

R. Wehlig
Vorwerk, Wuppertal, Germany

N. Pfeffermann and J. Gould (eds.), *Strategy and Communication for Innovation*,
DOI 10.1007/978-3-319-49542-2_24

24.1 Introduction

Today, to many consumers it feels natural to actively produce content and publish it online (so called 'user-generated content'). This change in media behavior is even more impressive considering that only a few years back, the majority of people mainly consumed (i.e., read/watched) content online. When Wikipedia was launched—widely unnoticed at that time—users notified the webmasters that there was a security issue on the platform because they could actually change the content of the website themselves. What appears to be alienated and ridiculous in times of daily Facebook updates reflected a very common understanding of roles between consumers and producers—in the media business as well as in the new product development process in any other industry. With the rise of social media, a vast majority of consumers have gradually adopted the role of producers and gained experiences. For instance, they stage themselves on Facebook, write restaurant and hotel reviews, share updates about their lives on Twitter and comment on newspaper articles. This fundamental change in consumer behavior induced a radical shift of the role of consumers from purely consuming toward prosuming (produce + consume) actors on the market.

Confronted with this new breed of consumers, many companies are at a loss. The widespread and often purposeless striving for a large number of Facebook fans as a key performance indicator in social media embodies the current ambivalent attitude toward social media. Alerted by the growing impact of social media, companies precipitately implement social media strategies, but fall short of tapping the full potential. Just as in the old days, they often only see a crowd of willing recipients of communication messages and use social media channels as yet another way of distributing their own content. The game has changed, but they still play with the same old methods and tools from the pre-prosumer era.

24.2 Tools to Empower Consumers

More and more innovative companies, however, manage to overcome the learned definition and separation of consumer and producer roles and reciprocate appropriately (Bernoff and Li 2008; Fuchs et al. 2010). Consumers have proved to be valuable partners in new product development. They help companies to learn about future consumer needs (Bartl et al. 2012) and to innovate more effectively and efficiently, i.e., create desirable products in a shorter time (Bilgram et al. 2013). A number of different interaction modes have evolved which serve different purposes.

In the ideation process, for example, Innovation Contests are conducted to turn to a large crowd of creative or knowledgeable people with a specific innovation task or problem (Hutter et al. 2011). These contests are based on two core principles: Crowdsourcing and the concept of tournaments. The first principle refers to the

Table 24.1 Comparison of empowerment tools. *Source* By the author

Empowerment tool	Research community 	Innovation contest 
Method	• Asynchronous online research platform • Collaborative sessions and discussions	• Online crowdsourcing of an innovation task in a 'call for ideas' • Selection and rewarding the winners of the competition
Phase in value creation	• Definition of innovation fields • Selection and refinement of ideas	• Ideation of new product ideas
Involved individuals	• Users of a defined target group or existent customers	• Creative and skillful users
Outcome	• Consumer insights and needs • Evaluated and enriched ideas	• Hundreds of (visualized) innovative ideas • Underlying needs
User compensation	• Equal rewards for all active participants	• 'The winner(s) take(s) it all' principle of tournaments

outsourcing of creative tasks traditionally performed by the company itself to a large crowd of individuals outside the company by means of an open call. In order to address highly innovative users or even lead users, companies turn to established specialist communities such as Innocentive or Nine Sigma or call for ideas in relevant online communities and other social media sites. The second principle describes the reward system and the temporary nature of Innovation Contests. Only the best ideas submitted within the contest phase are rewarded or in other words 'the winner(s) take(s) it all' (Morgan and Wang 2010). Rewards can range from monetary incentives to brand-related prizes, internships at the company or the invitation to a special event (Füller 2010). This form of crowdsourcing is particularly suited to efficiently collect ideas or solutions to solve complex and novel problems (Boudreau and Lakhani 2013).

In contrast, so called Research Communities (Bernoff and Li 2008) are used to gain consumer insights and feedback on new product concepts (see Table 24.1). For this task, companies usually do not address the most innovative people, but the average customer who buys their products.

Research Communities have a collaborative nature encouraging a free-flowing discussion between participants and the company. An agenda for each day comprising various questions and tasks helps to guide the collaboration. Consumers not only interact with the company in a one-to-one question and answer type of research but enter into a many-to-many collaboration mode. Accordingly, participants of this collaborative research platform are also rewarded differently. As opposed to the tournament principle, every participant of a Research Community receives a small incentive as a 'thank you' and a compensation for her efforts.

Online Research Communities have disrupted the market research industry and have become a significant mosaic of many companies' empowerment strategies. The fundamental change that Research Communities bring about is that they establish a co-creation environment enabling a meaningful dialog and interaction mode. They go beyond short encounters between brands and consumers and create a different experience for consumers than 'one-off' research initiatives such as surveys or focus groups. The continuity of the platform-based approach, the versatile research modules and the exchange with the company as well as with other consumers creates an inspiring atmosphere of being part of something meaningful—something that has an impact and adds value for oneself and the company.

24.3 The Vorwerk Thermomix Research Community

The Thermomix is a multifunctional food processor which has up to 12 functions, enabling consumers not only to mix ingredients but also to cook, knead, mix, beat, chop, grind, purée, weigh, and roast them. For years, the company Vorwerk Thermomix has been integrating users in innovation processes in order to better understand problems, needs, and wishes as well as to co-create new products and services. Among other things, the open business model involving thousands of enthusiastic Thermomix users who evangelize and sell the product may account for the company's consumer-centricity. Thus, the Thermomix business model is exclusively based on word of mouth referrals. The Thermomix cooking appliance cannot be bought in official stores or via online channels but is exclusively distributed by sales representatives.

The 'fuel' of the Thermomix usage are the special recipes which are specifically tailored to the Thermomix appliance. These recipes enable users to get the most out of the appliance and fulfill the promise of 'success guarantee.' Over the years, an enormous variety of recipes has been generated both by the company as well as by users. In the Vorwerk Thermomix brand community these recipes can be explored, shared, discussed and downloaded. More than 600,000 users worldwide are registered users of the brand community. In the frame of a global initiative aiming to improve the user experience of the recipe communities in particular search, understand user needs and preferences, Vorwerk Thermomix leveraged an Online Research Community to involve its customers.

In particular, Vorwerk Thermomix pursued the following objectives: First, understanding and analyzing users' cooking journey was a key goal. The cooking journey comprises several steps: Planning, (online) recipe search, decision making, cooking, evaluating the recipe, and finally managing and organizing the recipes. The main focus of the research was on the search phase within the cooking journey. The goal was to analyze user habits and identify different search types among users.

Second, the ORC aimed at collaboratively developing the new search for the community so that users find recipes they are looking for quickly and easily and collecting feedback on the interface, and attributes.

The case study is structured in five parts dealing with the people who participated in the ORC, the agenda that was conceived for the interaction with the customers and the platform which was designed to serve as a research environment. Subsequently, we describe the outcome of the ORC and share some lessons learned.

People

A key challenge at the beginning of a co-creation project is to identify and recruit suitable participants to collaborate with. Here, the Vorwerk Thermomix community 'Rezeptwelt' in Germany served as a base for the recruitment process. Out of a sample of 107,686 registered members (October 2012), 800 people were preselected as potential participants. Among other things, having been active in the Rezeptwelt community in the past 6 months was one of the preselection criteria to filter. The goal was to ensure quality of participants' feedback and to identify the most suitable participants for the ORC.

The selected customers received an invitation via direct message to fill in a prescreening questionnaire. In this message, the project's goals and requirements (active participation in the forum over 3 weeks) were briefly described. The prescreening questionnaire assessed a number of demographic variables (age, gender and Thermomix model), the activity level, cooking behavior, and the motivation to participate.

From a total of 113 Rezeptwelt members who completed the prescreening questionnaire and expressed their desire to participate in the project, 69 particularly motivated people were selected and invited to the Research Community. The majority of participants were female (94 vs. 6% male), with ages ranging between 21 and 70 years, while most of the participants were between 31 and 50 years old. Two-third of the participants visit the Rezeptwelt community and use their Thermomix at least three times a week and have been members of the Rezeptwelt community for more than 24 months.

Participants of the ORC were also analyzed regarding their activity in Vorwerk's continuous brand community Rezeptwelt (i.e., the community they were recruited from). Activities in the Rezeptwelt comprise the up- and download of recipes, comments and evaluations of recipes, and exchange of messages with other community members or friends. As shown in Fig. 24.1, 12% of the participating consumers are rather passive users in the Rezeptwelt and therefore called "Marginal activity users." An even smaller number of participants belong to the segment of "Heavy upstreamers." The main activity of this group relies on the active creation of content which they contribute to the community. This includes the number of published recipes, given comments, and recipe evaluations. In contrast to these relatively small groups, 38% of the ORC members belong to the group of "Heavy downstreamer"—a user segment best characterized by its downstream activities in terms of saving recipes. The largest user segment—with 40% of the participants—represents the group of "Extreme users." This group shows considerably higher

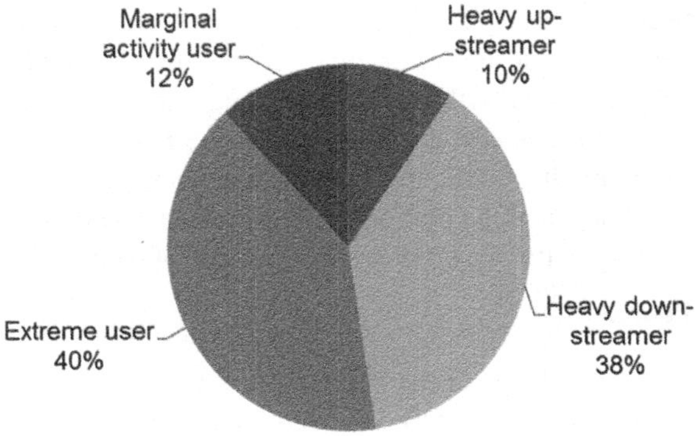

Fig. 24.1 User types participating in the ORC

activity levels in all previously mentioned activity dimensions. For instance, the average "Extreme user" uploaded approximately ten times as many recipes as the average user.

In the course of the online research phase of 3 weeks, 47 participants very actively contributed to the discussion with at least 15 posts in total. Due to the participants' strong identification with and passion for the brand, a nonmonetary incentive closely related to the Thermomix world was chosen to compensate them for their efforts. Each participant could choose a Thermomix cookbook which is highly coveted by Thermomix fans.

Agenda

Following a discussion guideline, which captured all relevant research topics, the Research Community moderator posted a daily task to the forum, inviting participants to share their thoughts and ideas. Typically, the task comprised one main question with several sub-questions, to ensure comprehensibility and to encourage people to elaborate on their thoughts in greater detail. Besides posting the daily task each morning, the moderator followed the ongoing discussions and intervened only in cases of misunderstandings or questions. In doing so, the moderator could adapt the structure of the discussion reacting to directions the conversation took or to aspects which emerged during the discussion. Due to the asynchronous and unobtrusive nature of this interaction, people could individually decide when to work on the tasks. At the same time, the approach fostered deliberate consumer responses on a cognitive level.

Within the first research week, the research focused on the 'status quo'. People were asked to talk about their general cooking behavior, recipe search and management and to evaluate the current Rezeptwelt. In the second week, people were introduced to the future online recipe platform ('Rezeptwelt 2.0'). Besides an overall evaluation of the design and structure, participants were invited to test and evaluate new search features and functionalities. In the third and final week, people's creative anticipation potential was addressed by showing and discussing different future features and services with the aid of visualizations and mock-ups. In addition to these smaller daily tasks, on Friday participants were given a detailed weekend task on which they could work until Monday morning. One weekend task for example included the visit of other, competing online recipe platforms in order to compare special recipe search features. In a diary, participants were asked to capture their experiences revolving around their daily cooking and search behavior. In contrast to the general open discussions—where participants could see and comment on each other's posts—only the moderator and the individual participant had access to the personal diary.

Platform

A simple and user-friendly interface ensured easy access and guided participation over the three-week research phase. Displaying not only the overall project description but also the task of the day and an overview of all three weeks, the start page served as a hub for all activities within the ORC. Participants could directly access the task of the day prominently displayed on this page. By clicking on each research week, five different threads for each weekday appeared (see Fig. 24.2). In order to allow for simple and clear tasks, access to the daily tasks was only possible for the present (and past) research tasks, but not for future ones. Each thread was composed of background information, task description and a comment field where participants could write their answers and reactions to the task. The postings could be enriched by embedding videos, pictures or emoticons. Besides referring to the daily question, participants were encouraged to comment on others' posts or 'like' other participants' comments.

Outcome

The data produced during the research phase was impressive due to its particularly rich and insightful qualitative content, and also regarding its sheer quantity. Within

Fig. 24.2 Vorwerk online research community: start page and weekly overview

the set time frame of the three-week online phase, 1540 posts were written including 350 diary entries. Hence, consumers' contributions amount to a total number of 191,346 words which equal approximately 350 written A4 pages. Each day, at least 52 comments were posted with a peak of 91 comments on two particular days: On one of these days, the beta version of the future Vorwerk Rezeptwelt was introduced for the first time. Of course, this was an exciting day and the customers were eager to explore the new platform and share their first impression and feedback. The second peak could be observed on the final day of the online research phase, where participants were asked to express their general and future wishes for the Rezeptwelt and Thermomix, and to evaluate their overall experience within the Research Community.

Each participant posted 24 comments to the forum per day on average including one outlier with 74 posts (see Fig. 24.3 for the average division of participants' activity levels). In line with the amount of content generated during the live phase, the average participant visited the forum 36 times (total of 2232 visits for all participants), and spent 1076 min on the platform. Calculated in man days, this corresponds to 2.2 man days per participant, or 139 man days as a total for all members.

Discussions were summarized daily and qualitatively analyzed. Feedback on the future Vorwerk community was collected to eventually refine features and functions. A special feature of this co-creation project was the collaboration of the HYVE research team with the Vorwerk programmers and creators. They closely worked together during the whole research project and were able to transfer user feedback immediately into the new community platform. When the new German community website was launched after finalizing the research project, several functionalities already were adapted based on the research results and users' feedback. As an example, the rank of search fields was modified, showing the most important search features on the top of the list. Also small pieces of advice were taken seriously. For instance, the search button's color was changed from a green color to a light and signaling orange as people complained about the rather unremarkable search button.

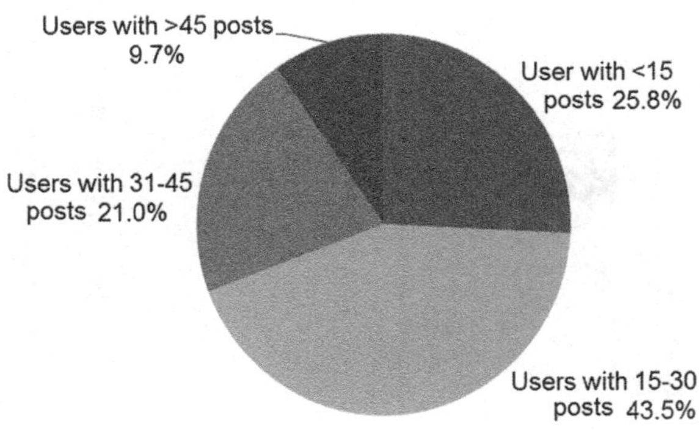

Fig. 24.3 Participants' activity levels in the ORC

Furthermore, quantitative data from the prescreening questionnaire was linked to the comments of every participant. Through this, different user search types surfaced: One example is the 'confused user', who needs a simple structure or extra guidance and explanations. Identifying and understanding user types differing in terms of the cooking journey and search behavior in particular is a key strategic lever to drive activity on the platform and usage of the Thermomix. By offering unique experiences tailored to the individual user visiting the recipe community, the relevance of recipes can be increased and inspire consumers to use their Thermomix more often.

Although the ORC required a high level of engagement from participants over a longer period of time, almost all invitees participated actively up to the final day. Hence, they were highly involved and showed motivation to be part of the Research Community, which they also expressed in their final feedback. The research community was perceived as a great opportunity to learn and discuss not only with other people involved, but also with Vorwerk Thermomix employees. People felt honored to be part of a selected and exclusive group of people which could influence the future development of the Vorwerk community. While they were looking forward to receiving their reward—a Vorwerk cookbook—people stated that this was not the main driver for participation. Instead, they were glad to contribute to the improvement of the new community website and expressed their excitement about the launch of it as can be seen in the following original quotes.

> Thank you for choosing me and providing the opportunity to discuss and share ideas. Time flies - I really enjoyed every second of the three weeks. Discussing and exchanging – or just reading – others' members' opinions was really nice (love). Consider me ever interested in future research projects. Best wishes for success to the team and good luck with the implementation of our propositions and wishes.[1]So this is the final day – what a pity! I really enjoyed being part of this forum (any time again). I am looking forward to the new Rezeptwelt. Besides my cookbooks and recipe folders, it will remain the place where I can relax in the evening.[2]Joining this discussion forum and keeping the diary, I was intensively dealing with the recipe search. Thus, I've learned to use various unknown or not frequently used functions and tricks. It has been fun to test these things out![3]

---

[1]Original quote in German: "Zum Abschluss möchte ich mich dafür bedanken, dass auch ich ausgewählt wurde und hier mitschreiben und mittüfteln durfte! Es waren recht schnelle drei Wochen, die mir wirklich sehr gut gefallen haben. Auch der Austausch mit anderen Mitgliedern bzw. das Lesen anderer Meinungen gefiel mir gut! (Love) Ich würde jederzeit gerne wieder an einer solchen Befragung teilnehmen! Nun wünsche ich dem Team viel Erfolg und Spaß bei der Umsetzung unserer nicht wenigen Vorschläge und Wünsche…".

[2]Original quote in German: "der letzte Tag -schade-, denn es hat mir viel Freude bereitet an diesem Forum teilnehmen zu dürfen.(Immer wieder gerne). Ich freue mich schon auf die neue RW. Sie wird neben meinen Kochbüchern und meinen Rezepteordnern, der Ort bleiben, in dem ich mich abends entspannen kann."

[3]Original quote in German: "Durch die Teilnahme am Diskussionsforum und dem Tagebuch habe ich mich doch intensiver mit den Inhalten der Rezeptsuche beschäftigt. Ich habe Funktionen und Tricks kennengelernt, die ich vorher noch gar nicht kannte oder nicht aktiv genutzt habe. Das Probieren hat Spass gemacht."

The example quotations also indicate how customers' passion for the Vorwerk Thermomix brand and their brand knowledge was positively influenced. Empowering consumers therefore not only helped to gain insights and create new community features but also gave a boost to the consumer-brand relationship.

24.4 Conclusion and Outlook

The Vorwerk Research Community illustrates the enormous benefits of involving customers both in innovation research and new product development processes. Online Research Communities enable the collaboration with customers over a specific time—several days, weeks or even years. Correspondingly, this intense collaboration requires the engagement of a highly affected community like brand enthusiasts. Bringing together this strongly committed group of people helps to establish in-depth dialog, in which people not only perform a particular task assigned to them, but also exchange, share and develop ideas.

Whereas the key factors of successful ORCs are its members, some additional factors should be considered to tap the methods' full potential. First, the moderation of the forum should be unobtrusive and avoid influencing, but rather provide consumers with all relevant information to conduct the tasks. Second, the website's interface and structure should be clear and uncluttered, to be intuitive and suitable environments for people with little social media knowledge. Additionally, an appealing layout encourages people to visit the forum frequently—daily, or even more often. If desired, the design of the ORC can be customized to the specific corporate design of an already existing brand community to create a consistent brand experience. This may give people the feeling of familiarity and comfort. Finally, the tasks should be diverting and fun to work on. This can be achieved, for example, through visually rich input and pictures.

In times of online social networks and organized brand communities, companies can and should tap this potential and collaborate with consumers in order to gain consumer insights and translate them into solutions for new product development. Establishing Online Research Communities as virtual places for a deep and possibly ongoing exchange and discussion between a company and consumers will help to develop user-centered products and services, establish strong relationships and retain them long-term.

Outlook

In the past two decades, the Internet has significantly transformed the market research industry in terms of efficiency. New methods have been conceived to harness the easy access to consumers and electronic communication channels via the Internet. Online panels with millions of users willing and ready to fill out questionnaires have reduced the duration of field phases and considerably accelerated feedback loops within innovation processes. Although the Internet is known to be an enabler of consumer empowerment, it has not changed the role of consumers in market research to a larger extent. Rather, the introduction of online panels has commoditized the participation of consumers. Answers can be bought in hundreds and thousands and participating consumers are considered mere providers of data.

Online Research Communities offer a new opportunity to continuously engage with consumers for market research purposes and build meaningful relationships. Given the chance to play a more intensive role within the innovation process, consumers feel they work on an important task, are taken seriously and have an impact on the future of a company. Instead of outsourcing the access to consumers to specialists only, companies started to build and maintain relationships to consumers not only for sales reasons but also for research purposes. Here, the Internet and social media platforms, in particular, are a promising source of interested consumers. Also branded communities, such as Vorwerk Thermomix's recipe and cooking community, can be used to invite consumers to participate in innovation endeavors. These consumers are usually more involved than panel users, they know the products and services very well and often feel honored to work together with the brand. They have decided to follow the brand in social media and indicated that they would like to enter a relationship.

In many cases, companies are overwhelmed with the crowd of consumers willing to interact with them on various social media websites. Instead of looking for the really urgent questions and tasks they could tackle together with these users, they start involving them in shallow 'small talk' conversations. Although consumers often appreciate any type of interaction with a brand, a meaningful dialog has several positive effects. For one, motivated consumers or brand fans may provide valuable information and ideas which can help the company propel their innovation efforts. Secondly, maintaining the relationship with these consumers is a core task of marketing anyway, thus, including some innovation tasks can provide relevant discussion topics and spur the conversation. In order to capitalize on this effect, companies need to intertwine innovation and marketing capabilities and disciplines to holistically utilize consumer empowerment. Third, empowering consumers has also been found to have an impact on consumers' product demand (Fuchs et al. 2010). Therefore, involving consumers in meaningful tasks and inviting them to co-create value can increase sales, deepen the consumer-brand relationship and strengthen the brand's image. Fourthly, building one's own community of users may also yield cost cutting effects, especially when research communities are centrally orchestrated to serve multiple departments within a company.

Consumer empowerment strategies are an imperative for companies nowadays. By staging authentic co-creation experiences centering on meaningful task rather than superficial encounters between brands and consumers, empowerment may yield multiple positive effects for insight generation, innovation management and customer relationship management.

Bibliography

Bartl, M., Füller, J., Mühlbacher, H., Ernst, H.: A manager's perspective on virtual customer integration in new product development. J. Prod. Innov. Manage. **29**(6), 1031–1046 (2012)

Bernoff, J., Li, C.: Harnessing the power of the oh-so-social web. MIT Sloan Manage. Rev. **49**(3), 36–42 (2008)

Bilgram, V., Füller, J., Bartl, M., Biel, S., Miertsch, H.: Eine Allianz gegen Flecken, pp. 62–68. Harvard Business Manager, März (2013)

Boudreau, K.J., Lakhani, K.R.: Using the crowd as an innovation partner. Harvard Bus. Rev. 61–69 (2013)

Fuchs, C., Prandelli, E., Schreier, M.: The psychological effects of empowerment strategies on consumers' product demand. J. Mark. **74**, 65–79 (2010)

Füller, J.: Refining virtual co-creation from a consumer perspective. Calif. Manag. Rev. **52**(2), 98–122 (2010)

Hutter, K., Hautz, J., Füller, J., Mueller, J., Matzler, K.: Communitition: the tension between competition and collaboration in community-based design contests. Creativity Innov. Manage. **20**(1), 3–21 (2011)

Morgan, J., Wang, R.: Tournaments for ideas. Calif. Manag. Rev. **52**(2), 77–97 (2010)

Chapter 25
The Integration of Art and Design, Creativity and Professionalization

Rolf Sachsse

Abstract This chapter describes the importance and integration of innovation in today's art and design education, leading to radical changes in the curricula of the universities of fine arts. Modularization replaces not only the old system of master classes but opens the individual choices of subjects to a system of projects. Adding to this, the installation of nonconsecutive curricula integrating science and art as well as co-operative curricula between classical universities and art institutions offer new horizons in both creative as traditional industries and sciences. Another step is the installation of In- and At-House-Institutes acting as professional links between education, management and politics. All of these elements that have been developed within the past decade are described on the example of the Saar University of Fine Arts (HBKsaar) which was the first Fine Art University in Germany to be fully modularized. Creativity has, at last, gained high importance as a key issue of industrial and economic development, and the professional education in this field will acknowledge the creative entrepreneurship in both arts and design.

25.1 Introduction: Art and Design Education as a Lesson in Entrepreneurship

»Marry a teacher« was the common response given by German academy professors when asked how to survive as an artist. For two centuries, since the installation of art academies in the late eighteenth century, the academic professor had to run a class at the academy and introduce his students into the art world where they would either pursue a career or disappear into obscurity. Until the late twentieth century, this art world was a market visible at a glance, with fixed rules, a good dozen of dealers and a group of roughly two hundred collectors—worldwide.[1] This changed

[1]De Marchi and Van Miegroet (2006).

R. Sachsse (✉)
Hochschule der Bildenden Künste Saar, Keplerstr.3-5, 66117 Saarbrücken, Germany
e-mail: r.sachsse@hbksaar.de

N. Pfeffermann and J. Gould (eds.), *Strategy and Communication for Innovation*,
DOI 10.1007/978-3-319-49542-2_25

radically after World War II. When the art market opened into full globalization with the basic rules of a stock market but until recently there was nearly no reaction by the state-funded art education which caused for a twofold result: Private institutions grew into the educational business like mushrooms after rain, and the artist himself became the key model of the absolutely freelancing entrepreneur. Even though Renaissance and Baroque artists like Albrecht Duerer, Rembrandt Harmenszoon van Rijn, and Peter Paul Rubens had run manufacture-like studios with great success for at least one decade each, they did not fulfil the role model of the mad but quiet, deadly addicted artist that romantic visions had focussed on.[2]

Today, artists have crashed all boundaries of the common presuppositions in their profession: They work in whatever field global circumstances will allow. Artists have to be non-specialized experts in any area of aesthetic practice, and they have to survive under economic pressure on both their person as on their radical propositions from which they have to start when becoming an artist, designer, musician, performer, etc. One of the smallest institutions of higher learning in Germany, the Saar University of Fine Arts has successfully expanded its Bauhaus-derived integrative approach to education and research to engage with these issues.

25.2 Against Specialization

The fields of Art, Design, and Media education have merged both from the epistemic background in their theories and practices as from the outcome in today's society: Artists are working as architects, designers, and create stage settings; designers have lost the emphasis of their work in anonymosity but have become stars like artists and create installation in large museums or exhibition spaces. Aesthetic innovations have neither acknowledged nor respected the constraints of disciplinarity, since the twentieth century modernisms even less than before.[3] Thus, it does not make much sense to differentiate between these fields of interest, especially not with young people at the start of their careers. There are basic terms of aesthetic questions to be taught, there are ethics of media and the arts to be taught, and there are individual fields of interest slowly emerging from a personal development within an educational process. Each education that promises to generate exactly the job scale proposed has to be closed down immediately: It is out of its time.

Rapid developments in technology and engineering have created a bulk of professional perspectives than cannot last longer than the existence of the ground formation—too short for one life. This is also true for the arts, design and especially for the media that change their perspectives in short periods. It is, e.g., not really

[2]Deresiewicz (2015).

[3]Lippard (1973).

necessary to teach display typography when you know that there will be no displays within the next twenty years. But there are basics in the arts that have not changed since the caves of Lascaux, 22.000 years ago: Aesthetic innovation must surprise as well as survive short periods of fashion and interest; then it can be named and canonized as Art (with a capital A).

25.2.1 Projects Instead of Classes or Curricula

Teaching Art is impossible. This banality was known to the first authors of curricula in art schools. Art means a full and free creativity without restrictions in any field except for the lives of other human beings. Thus, academies first of all have the task to let their students develop individually, to free their creativity and help them define the areas suitable to their personal vision of life and work. Design is a different field with similar ethic and aesthetic implications: Nobody can ask who designed the first wheel, chair, bed, box, or any other item of everyday life. At best, design disappears into its social use—and designers have to be happy being well payed for instigating these processes. Media is, by definition, a field in between; its production is to be consumed quickly but with a high recycling quote which might lead into the consideration of being art but not necessarily so when at production. All of these fields merge in what we call art education, and they should not be specified too much at the moment of individual formation.

Four of the five areas of study at the HBKsaar have more or less the same, modularized curriculum fully according to the Bologna process[4]; the fifth area—teacher training in art education—has to follow specific legal constructions. The four areas—fine arts, communication design, product design, media arts and design—consist of a common core of aesthetic concepts and practices, established in the first term, and then of a number of projects run within each term and beyond.[5] Projects can be commissioned by industries and administrations, competitions in arts, film, and design, or they can start from scratch by some kind of self-definition by either the students or the teachers. A large part of these projects consist of co-operations with other universities and there are a number of Master programs such as, e.g., Media Informatics run by the HBKsaar and the department of computer science at Saarland University hosting the German Institute of Artificial Intelligence. As the Bachelor program is scheduled for eight terms, students can change their areas of study within the first seven terms according to their own development. A number of nonconsecutive Master programs fulfil needs in theoretical approaches to the subjects of the studies such as, e.g., a Master in Net

[4]http://www.ehea.info/ [2016–04–14].

[5]http://www.hbksaar.de/fileadmin/hbk/download/pdf/Modulare_Studiengaenge/gemeinsames_modulhandbuch2012.pdf [2016–04–14].

Culture or a Master in Public Design with a strong emphasis on the social relevance and interventionist registers of fine arts.

25.2.2 *Education and Innovation*

At the first glance, education and innovation are contradictions in adjectives: Either you teach something by conditioning and training or you produce innovations by the Schumpeterian trias of invention, adaptation and diffusion.[6] In cultural theory, the process has been described by Roland Barthes with the distinction of »studium« and »punctum«, the latter being a shock of sudden recognition without prior warning, thus not being able to be taught.[7] A University of Fine Arts like the HBKsaar has to meander through these oppositions and serve a form of fuzzy logic with open ends, mirroring Umberto Eco's semiotic definition of an art work as a set of signs open to any new interpretation at any new confrontation.[8] Basically, an educational institution in this field has to create a situation that enables shocks to come, that offers reflections on any subject in human existence, and, of course, serves the needs of students with technical and practical workshops as well as specific trainings. Any new technology both in media and materialization has to be integrated into daily practices as the offerings of old and often forgotten crafts returning the students' minds into the origins of their professions.

These basic principles may sound trivial but have a number of consequences within the curricula formed from projects: When a project is introduced it has to be discussed from its ethic grounds up to aesthetic values, it may have to be turned over into its opposite or re-thought on a totally different level of perception. Pottery and ceramics, e.g., may be some of the oldest crafts of mankind but they are completely re-defined on the base of 3D printing with nanoceramic materials—these do not only enable totally different forms from those handed down by tradition but re-constitute the use of materials from both ecologic and economic assumptions. The HBKsaar runs a *Center of Digital Production* in conjunction with according institutes of the Saarland University where all forms of digital preparation to material production can be imagined and tested either on models or prototypes. Rapid prototyping as a method has consequently changed the old processes of sketch, study and modelling within all forms of industrial design and is offered not only as a part of education but as a services to the local industries as well.

A similar perspective can be found in Fine Arts as well when important issues are raised in politics and economics: At the Voelklingen branch of the HBKsaar, a preliminary living shelter for refugees was erected by the class in sculpture (Georg Winter) in co-operation with the architectural department of the Saar University of

[6]Mahdjoubi (1999).
[7]Barthes (1981).
[8]Eco (1989).

Applied Science (HTW Saar), completely made from recycled materials in wood, glass, and metal. As a test, this shelter houses one of the students himself being a migrant from Brasilia. This project was accompanied by a number of performances in the city of Voelklingen as well as by a theoretical symposium on terror and racism in conjunction with the Institute of Philosophy of the University of Bonn. In all of these cases—which are samples of typical projects at the HBKsaar—the innovation arises from a field of ideas and reflections prepared by both the teachers and the students. They show that innovation cannot be trained but be prepared by education.

25.3 Postgraduate Professionalisation

When academies of fine arts were founded, their aim was the aesthetic service to an emperor, king, or noble who financed the venue like a state theater, opera house, or museum.[9] From the beginning of industrialization, the professionalization of artists developed in two directions: With the establishment of copyright acts the work of designers became an important part of economics,[10] and the artists were subject of a market with dealers and collectors. Acting in these fields was supposed to be a form of free entrepreneurship, and roughly ten percent of the graduates were able to make a living from what they had learned. Thus, academies were accused of producing nothing but taxi drivers or bored housewifes with a hand for furnituring beautiful homes. Thing had to change, and they have changed.

At the HBKsaar, all students of the sixth to the eighth term are strongly advised (of course, everything is voluntary in an institution like this…) to take part in a three-term professionalisation program. The program is prepared by courses in rhetorics and by a wide ranging schedule in ethics of art, design and media. The curriculum starts with the introduction of all possible partners in the approached market, from gallerists to curators to marketing experts to creative directors in agencies and industry. The second term is concerned with legal aspects of the everyday life in business, from copyright questions over insurance and tax regulations to business contracts and their obligations. The HBKsaar seats a professorship in the philosophy of law and aesthetics which fulfils all the needs of this curriculum. The third term consists of project type tests of individual cases: What is needed to produce and finish a documentary movie? How is the conception and the realization of an advertising campaign classified? Where are the frontiers of viral marketing, where the borders of mockumentary? These areas of personal advice take a lot of energy and capacities from the board of teachers but they proved to be fruitful since the professionalization program was installed in 2010.

[9]Ryan (2008).

[10]Gronert (1989).

For those who have finished their studies, the move into free entrepreneurship can be fast—as is with all those who had the most brilliant ideas of their lives during study times—or comparatively slow but nonetheless bearing fruit by time. The students can take part in postgraduate programs named Deepening studies or Master classes, and when their interests reach move on to levels of theory and reflection, they can start to prepare a thesis for a PhD in philosophy or art, design, and media history or theory. For those who used the practical postgraduate programs for pursuing their career within their fields of interest, the universities of art in the Saar country (besides the HBKsaar, it is the University of Music HfM Saar), there was an institute founded hitherto unknown in Germany: K8.

25.3.1 K8 Institute for Strategic Aesthetics

The K8 Institute for Strategic Aesthetics was founded in October 2014 as a non-profit company. K8 was initiated by both universities of art and music in the Saar country, with a small amount of public funding (provided by the Ministry of Economics) to cover the costs of setting up the company as a legal entity combined with project-based funding for the co-organization of conferences and workshops. According to its managing directors Julia Hartnik and Soenke Zehle, K8 operates as transfer agency facilitating the internationalization of existing arts-and-research projects and seeks to involve new actors in multidisciplinary co-development processes. In accordance with the state's long-term strategies for creative industry development and innovation, K8 also offers a variety of professionalization, scenario development and project management services. These activities draw on the design-driven approaches of open innovation and speculative design to pursue a core theme—the collaborative exploration of co-development strategies in the digital society that are based on a comprehension of user (citizen, worker) agency broader than that of the "critical user" targeted, for example, by common media literacy programs. Locally, current activities include a variety of transfer projects for the two arts academies (trade fair representations, summer schools, international exhibition projects) as well as a workshop series with the state's innovation agency on themes related to the policy and technology development framework "Industry 4.0", and the organization of practice-based workshops with local secondary schools around the framing concept of a "School of Things" developed by the academy's Center for Digital Production to facilitate the experimental exploration of processes of digitization and informatization. Already involved in European and international networks organized around multidisciplinary research and development, K8 is currently establishing a partner network in the so-called "Greater Region" around the Saarland, with a particular emphasis on digital culture organizations in France and Luxemburg.

25.3.2 Theories of Art, Design, Media and Theoretical Innovation

»Design Thinking« has become a recent hype in problem solving areas of economics and commerce.[11] Applying methods used in the preparation of design projects to other fields of concern does not necessarily form a new epistemology: Both in economic and social sciences as in information technology have thinkers like Otto Neurath or Emanuel Goldberg introduced non-verbal forms of developing ideas by design as practiced at the same time in art schools like the Bauhaus or at Stuttgart, Krefeld, and Hamburg.[12] At the same time, the former design practice of enhancing given models by a »good form« with even moral aspects of educating society as had been the rule of the »Deutscher Werkbund« (German Work Federation) when industrial production and form giving found their way to each other, had ceased to exist as the base of understanding both industry as business.[13] This has lead to a strong desire to understand design as a matter of shaping society, economics and post-heavy-industry's production. A number of design processes such as rapid prototyping or the use of game engines in developing architectural, urban, or transportation solutions have called for a reconsideration of all the elements involved, from the structure of the application of computer sciences to the socio-psychological implications of labour in the times of »Industry 4.0«.

The PhD program at the HBKsaar reflects, at least in parts, the role of these design evolutions and revolutions. Besides classical monographs on important artists and subjects like fashion, besides the implementation of classical methods like semiotics to new fields of interest, there are a number of theses prepared that contemplate immediate results of design thinking or its criticism to the understanding of innovation. There is a thesis on the juxtaposition of anonymous versus star design and its impact on society; there are studies on »disruptive innovation as design theory« and on the language of Facebook marketing as well as on the optimization of self-marketing under net conditions. Other studies work on surveys of marketing instruments and their use of communication design in certain businesses or on the impact of retro-designs on the automotive industries; some studies aim at epistemic understandings of digital imagery and psychoanalytical relations between dreaming and drawing. All of these theses are products of cooperations between several and distinctive sciences, be it computer science, economics or fields of engineering. To strengthen the impact of these debates on the quality of innovation within the fields of art, design and media, the theses have to be strictly philosophical dissertations under the conditions of the good practice of German sciences as formulated by the Deutsche Forschungs-Gemeinschaft (DFG). At first glance, one might consider a huge difference between the freedom of arts and the

[11]Brown and Katz (2009).

[12]Hartmann (2015) and Buckland (2006).

[13]Schwartz (1996).

discipline of scientific work but in these cases they simply form the two sides of the same medal.

25.3.3 *Places and Spaces for Innovation and Entrepreneurship*

As there should be an open brief for all methods of education within the arts, design, and media, another question rises to recognition: the question of space. This is not only the design of offices and workspaces; first of all there is a lot of emphasis on the metaphorical space in allowing students to become free in their ideas and disciplined in the realization of them. As art today is more but making objects to be exhibited and sold, there are different needs for workshops, tools and workspaces than the average studio of the past two centuries. The same goes for designers, and media people can only work in large teams including a multitude of professions and their needs. But creativity—especially when it will be expanded to any form of entrepreneurship—asks for places to start of and for spaces to extinct from. They have to be kept at hand in any community that wants to incorporate creativity as a definitive part of its economy and social environment. Providing places and spaces therefore is an important in art, design and media education—without context there will never be a ground for creative innovations. Places include areas where unexpected meetings can happen, without any organization before. In the 1960s and 1970s, the British »Artist Placement Group« (APG) had prepared contracts for artists and the industry or civil services where—after a feasibility study of 6 months —artists joined companies and administrations for one or two years with a >carte blanche< and no obligation to produce anything that could be considered a work of art.[14] In the roughly one dozen cases when contracts and placements happened they changed the contractors substantially—and art, too.

With APG's founder and theoreticist John Latham, the relation of creativity, time and space can be described as an eventstructure, an extinction of time in space and vice versa.[15] And, as he had borrowed his terms from theoretical physics, these extinctions cannot be described by a program—they are simply unforeseeable. This is exactly what all education in art, design and media has to face: The progression of the creative ground goes in all directions which includes success but extremely high risks as well. Preparing places and setting up spaces in which things happen in time has the indefiniteness of all thermodynamic processes but is the only ground for creativity to go, and there is no innovation without invention—this is the end of any prepared, scheduled, modularized or curricular education.

---

[14] Berry and van Mourik Broekman (2003) and Sachsse (1991).

[15] Hudek and Velios (2010).

25.4 Conclusion

Once again: Art cannot be taught. Providing studio corners with occasional suggestions in style and techniques are not enough instigation for a personal future in creativity and innovation—art schools have to offer a certain amount of professional foundation for a successful life in arts, design or media. The subjects of art and design have changed so much that the old class system cannot cope with that—imitation is no longer the right way to invention, especially not in matters of aesthetics. The ethics of art, design and media clearly define today's values: Besides the art market, besides the film and entertainment industry, there are a lot more fields to plough for creative minds. The result is a twofold situation: By the introduction of professional skills within the politically shaped and given economics the curricula in art, design and media shape themselves into the educational normativity as financed by the state, but by doing so they help the students and graduates to find their role in this given society. On the other hand, economics, industry and civil services can learn a lot from what art schools do: They prepare the ground from which creative innovations can arise, and by proposing open briefs for them in their form of education, society can help them set up a class of really creative young human beings that are able to offer our society innovative solutions to the many problems we currently have on earth.

Acknowledgement The author wishes to acknowledge the help of Soenke Zehle PhD with the preparation of this article.

Bibliography

Barthes, R.: Camera Lucida. New York (1981)
Berry, J., van Mourik Broekman, P.: Countdown to zero, count up to now. Mute. Cult. Polit. After Net **9**(25), 80–91 (2003)
Brown, T., Katz, B.: Change by design. How design thinking can transform organizations and inspire innovation. New York (2009)
Buckland, M.: Emanuel Goldberg and His Knowledge Machine, Information, Invention, and Political Forces. Westport CT London (2006)
De Marchi, N., Van Miegroet, H.J.: The History of Art Markets. In: Handbook of the Economics of Art and Culture, vol 1, pp. 69–122 (2006)
Deresiewicz, W.: The Death of the Artist and the Birth of the Creative Entrepreneur. In: The Atlantic Retrieved at http://www.theatlantic.com/magazine/archive/2015/01/the-death-of-the-artist-and-the-birth-of-the-creative-entrepreneur/383497/ 14 Apr 2015 (2015)
Eco, U.: The Open Work, Cambridge, pp. 44–82 (1989)
Gronert, S.: The best patterns at the cheapest rate, Studien zum englischen Design im 19. Jahrhundert, Diss.phil. Cologne (man.) (1989)
Hartmann, F. (ed.): Sachbild und Gesellschaftstechnik. Otto Neurath, Bauhaus-Universität Weimar, Forschung Visuelle Kultur, Band 3, Hamburg (2015)
Hudek, A., Velios, A. (ed.): The Portable John Latham, London (2010)
Lippard, L.: Six Years: The dematerialization of the art object. London (1973)

Mahdjoubi, D.: Schumpeterian Economics and the Trilogy of Invention—Innovation—Diffusion Retrieved at https://www.ischool.utexas.edu/~darius/17-Schumpeter-innovation.pdf 16 Apr 2016 (1999)

Ryan, C.: The history of academies. In: Astle, J. Ryan, C. (eds.), Academies and the Future of State Education, pp. 1–13, London (2008)

Sachsse, R.: From 0 − 1 to 0 + 1. The artist placement group. In: von Maurm, K., Elliott, D (eds.), Exh.cat. John Latham, Art After Physics. Stuttgart Rejkjavik Oxford (1991)

Schwartz, F.J.: The Werkbund: Design Theory and Mass Culture Before the First World War. New Haven CT (1996)

The manufacturer's authorised representative in the EU is Springer Nature Customer Service Centre GmbH, Europaplatz 3, 69115 Heidelberg, Germany. If you have any concerns regarding our products, please contact ProductSafety@springernature.com

Printed and bound by CPI Group (UK) Ltd, Croydon, CR0 4YY
30/09/2026
02225655-0013